TINY WORLDS

of the Appalachian Mountains

ROSALIE HAIZLETT

TINY WORLDS

of the Appalachian Mountains

AN ARTIST'S JOURNEY

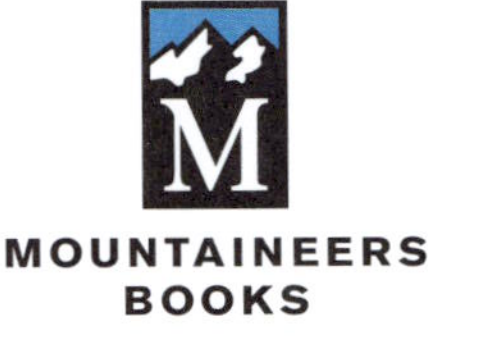

MOUNTAINEERS
BOOKS

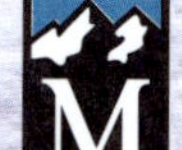

MOUNTAINEERS BOOKS is dedicated to the exploration, preservation, and enjoyment of outdoor and wilderness areas.

1001 SW Klickitat Way, Suite 201, Seattle, WA 98134
800-553-4453, www.mountaineersbooks.org

Mountaineers Books and its colophon are registered trademarks of The Mountaineers organization.

Printed in China

27 26 25 24   2 3 4 5 6

Design and layout: Jen Grable and Rosalie Haizlett
All illustrations and maps by the author
Photo, page 237, bottom, by Leah Stankus

Library of Congress Cataloging-in-Publication Data is available at https://lccn.loc.gov/2024005744

Mountaineers Books titles may be purchased for corporate, educational, or other promotional sales, and our authors are available for a wide range of events. For information on special discounts or booking an author, contact our customer service at 800-553-4453 or mbooks@mountaineersbooks.org.

Printed on FSC®-certified materials

ISBN (hardcover): 978-1-68051-635-7

An independent nonprofit publisher since 1960

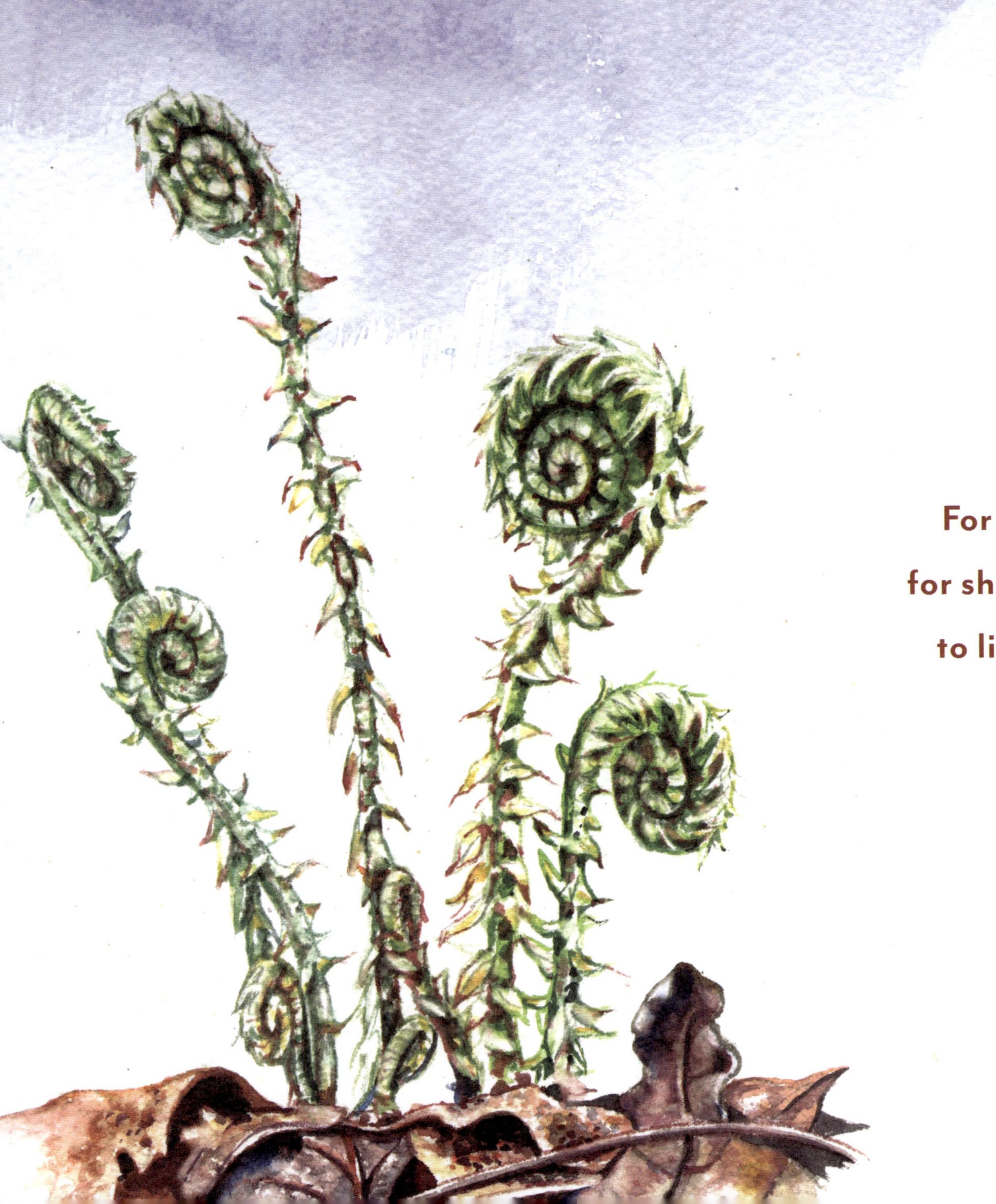

For my parents,

for showing me how

to live creatively

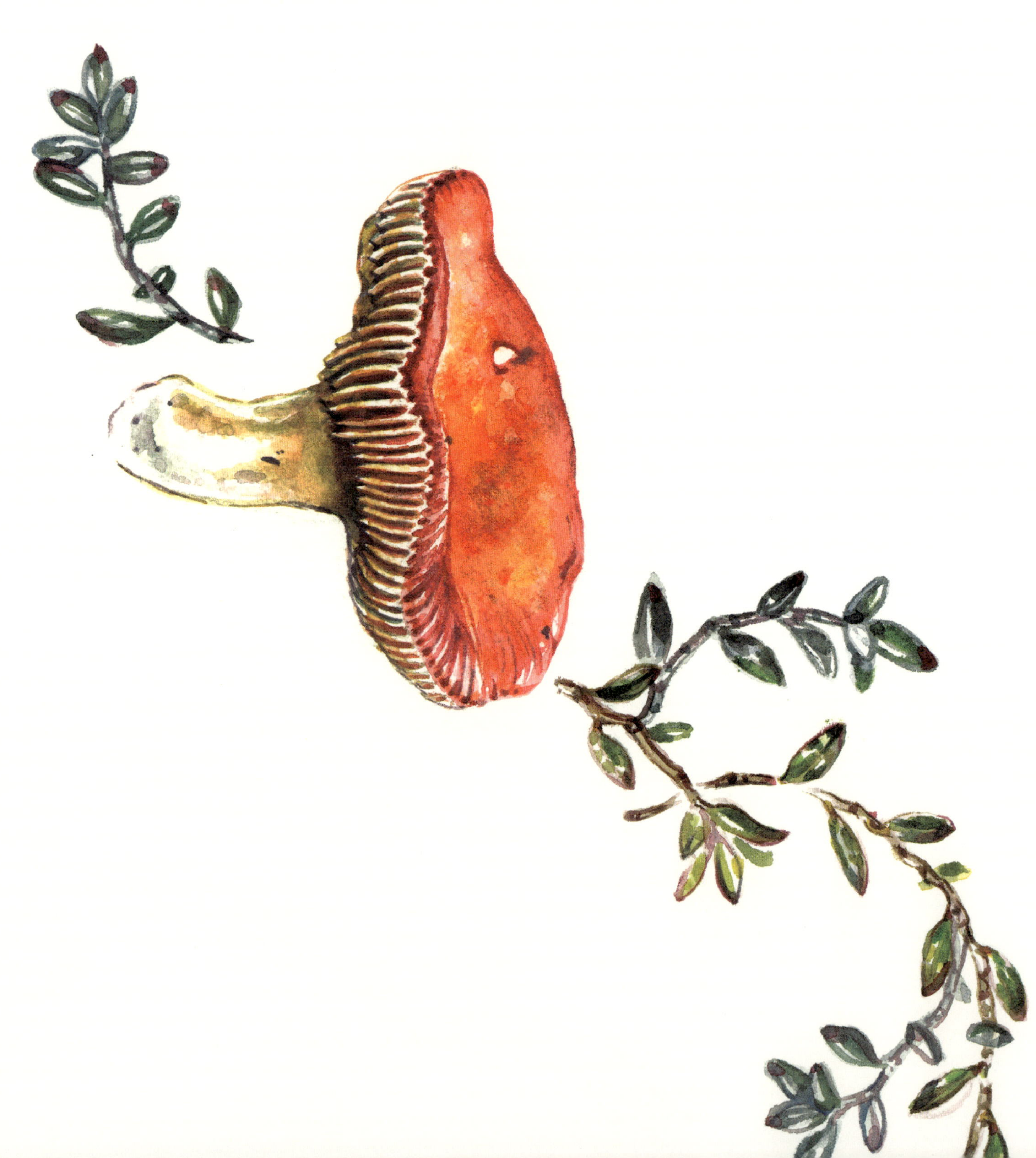

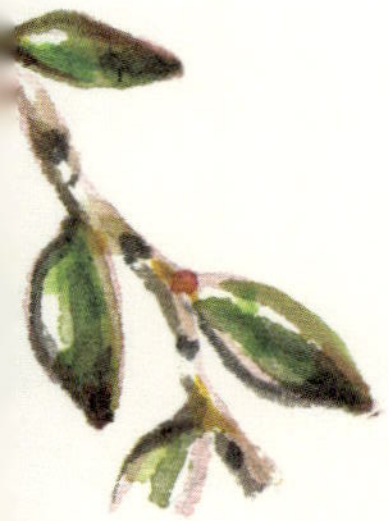

CONTENTS

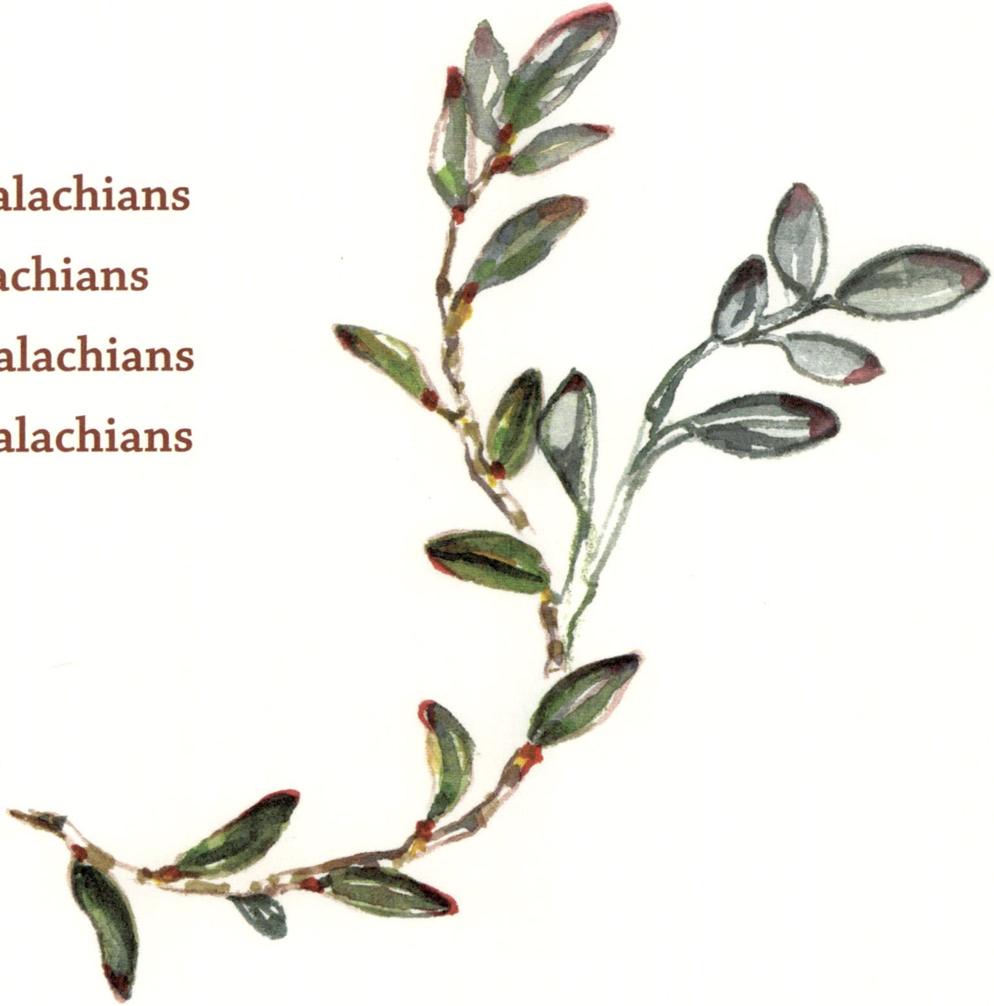

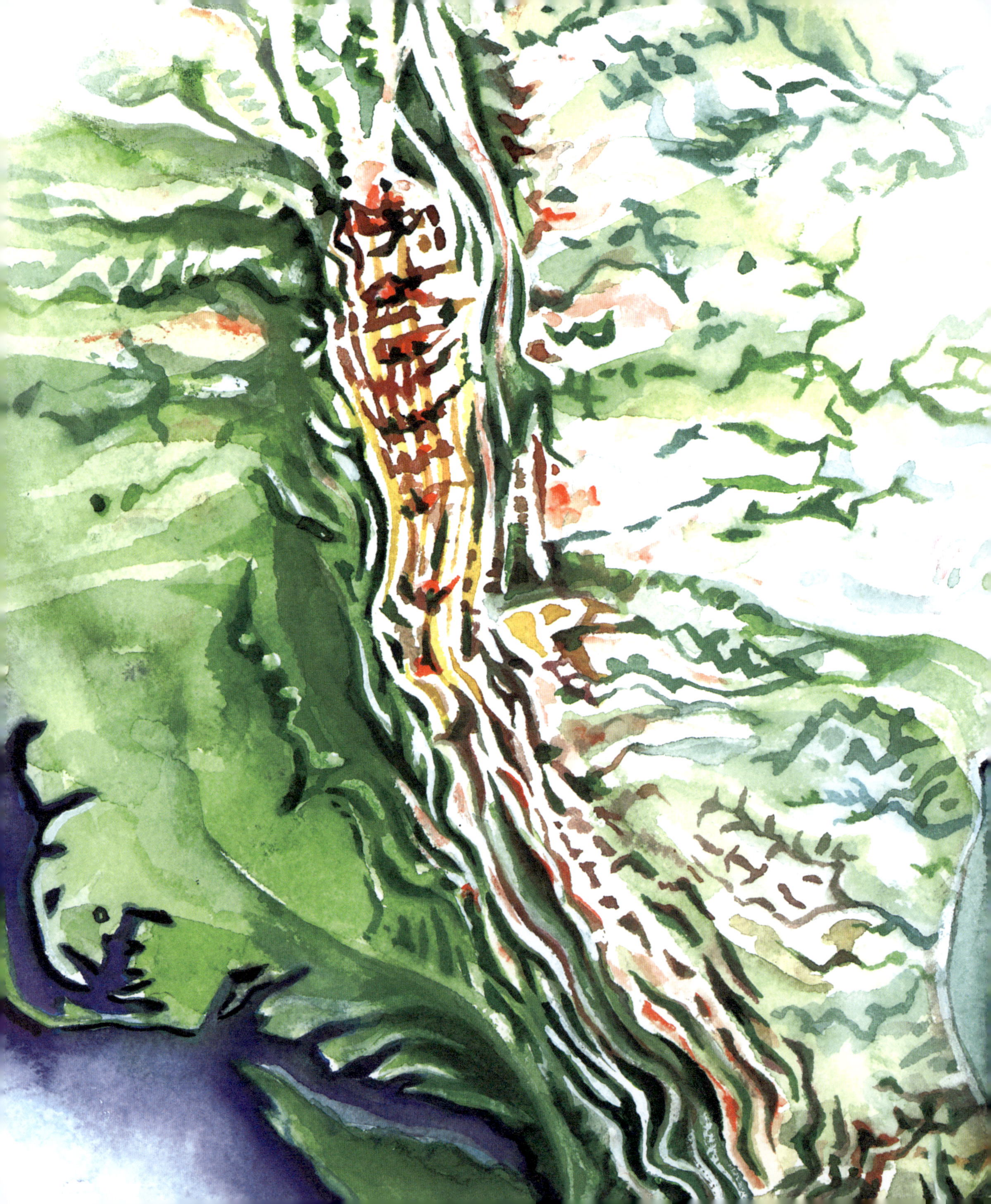

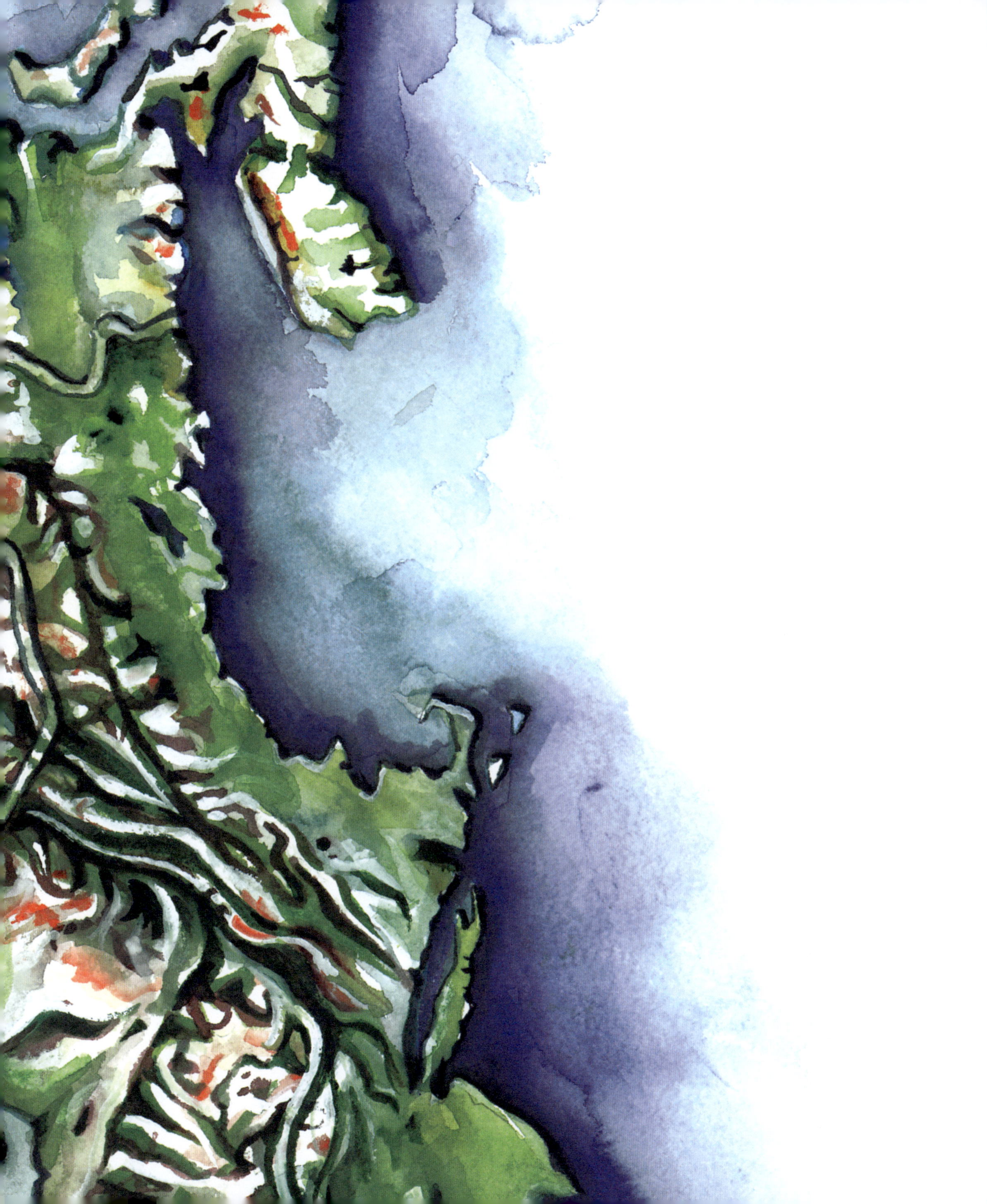

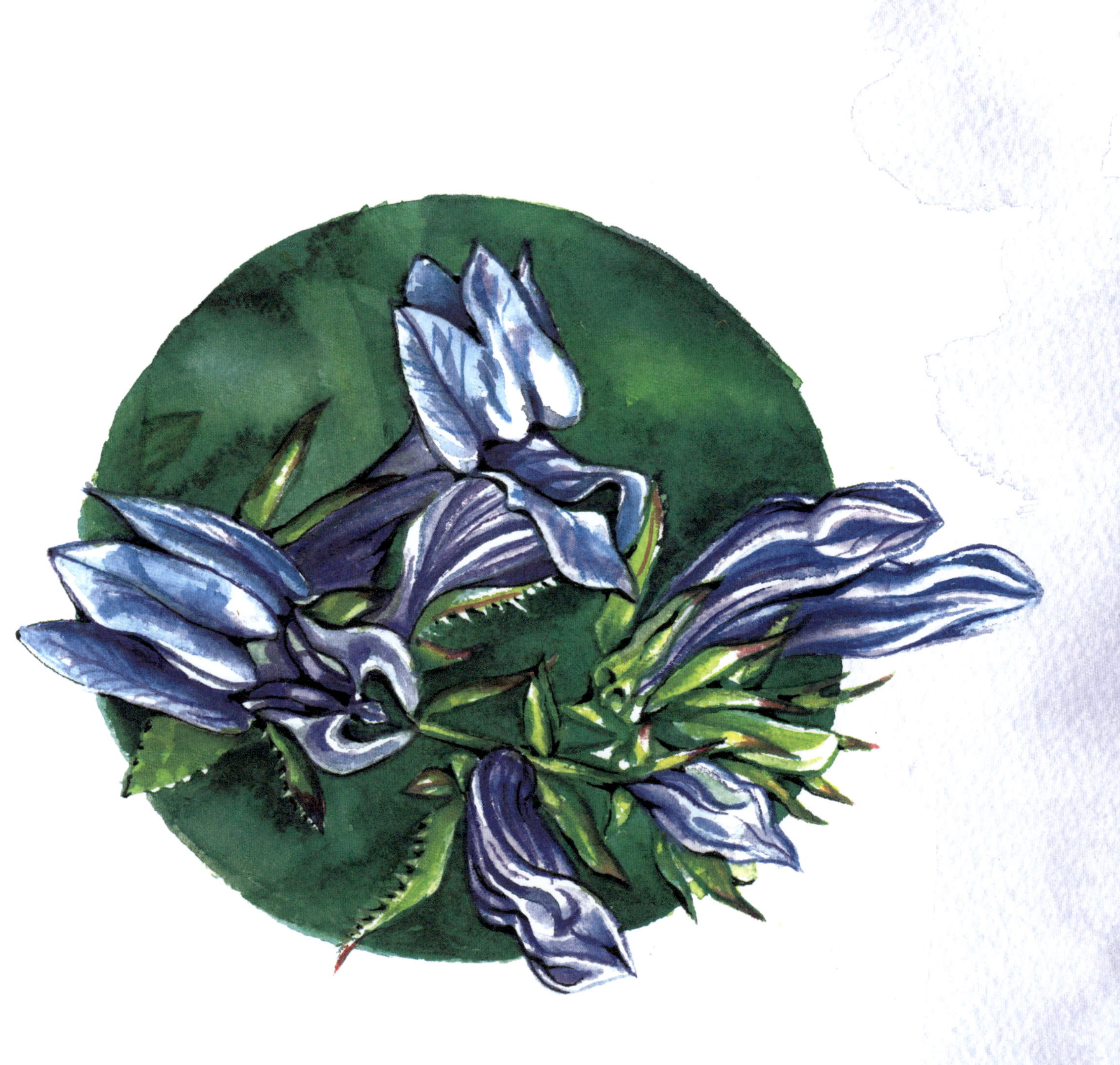

AUTHOR'S NOTE

My educational and professional background is in illustration, not ecology. So while I took great efforts to ensure the accuracy of my art and words by having scientists review my work, consulting trustworthy sources, and studying my reference photos closely while I painted, it is possible a stray detail escaped my attention. I take personal responsibility for any errors. Likewise, my map illustrations are whimsical, hand-sketched interpretations of our beloved parks and forests and should not be used as a guide to exploring these locations.

I personally observed—and took my own reference photos of—every subject depicted in this book during my six-month trip through the Appalachian Mountains. The only exceptions are in the Public Lands Spotlight pages, where I incorporated a few plants and animals that I didn't see but that I know are important members of the featured area's ecosystem. Since all of my selected subjects are very small, I chose to illustrate a majority of them at a larger scale to showcase their unique textures and markings.

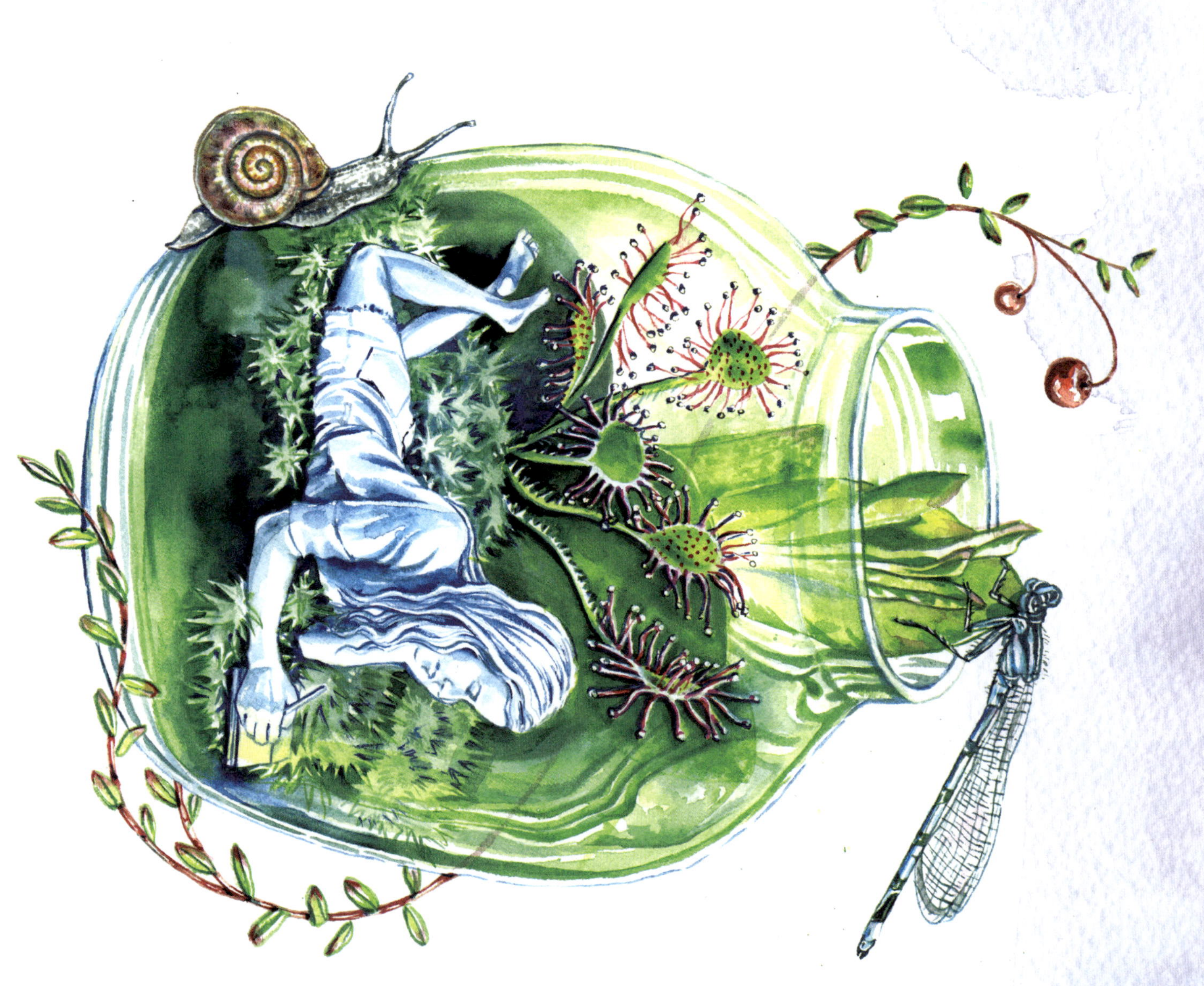

INTRODUCTION

A CLOSER LOOK

My fascination with nature's smallest wonders blossomed in my late teenage years. I had been experiencing chronic migraines, and a neurologist suggested that I seek relief by taking gentle walks outside anytime I felt a headache coming on. Although I thought that I was simply following the doctor's orders, these unhurried walks gave me a gift that changed my life: I learned to observe my natural surroundings with a deeper awareness of and appreciation for even the littlest of details. I realized that the tiny, often-overlooked worlds of insects, songbirds, wildflowers, fungi, and amphibians (and so much more!) are just as interesting and awe-inspiring as the larger, more charismatic flora and fauna that usually get the spotlight.

Before this, I had approached nature with greater emphasis on the grandiose—a powerful waterfall churning over a craggy mountainside, a mighty bald eagle sweeping above a river, a sunset so vibrant that it stopped me in my tracks. And while these kinds of experiences are worth seeking out, they aren't often present in our everyday lives. As a result, nature observation becomes something we reserve for designated settings and times, such as an annual vacation to a national park or a weekend hike. The rest of our lives is carried out separately from the plants and animals that we share our home regions with, and sometimes we don't even realize they're there at all.

From a distance, a butterfly is just a butterfly, like a million I've seen before. But when I tiptoe closer to get a better look, I'm rewarded with a vibrant, metallic color scheme that makes me want to grab my paints; I marvel at the delicate,

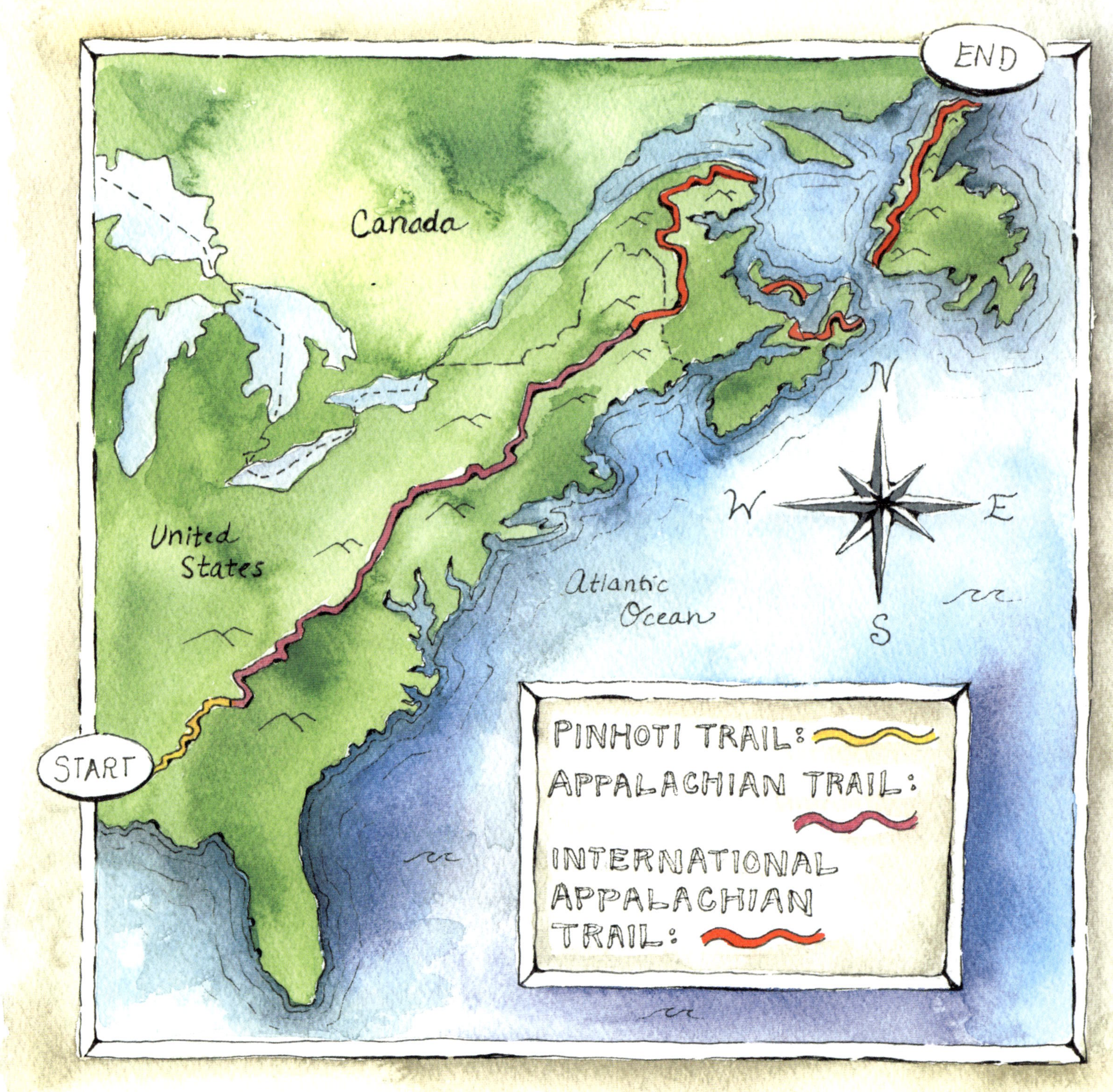

END
Canada
United States
Atlantic Ocean
START
N
W
E
S
PINHOTI TRAIL:
APPALACHIAN TRAIL:
INTERNATIONAL APPALACHIAN TRAIL:

hairlike scales that provide a soft texture to the wings, each one boasting a whimsical pattern that would make any textile designer envious. After a few minutes, I have picked up on other information, too, such as which plants it likes to nectar on and how it camouflages itself when threatened.

When we see the world through this lens, an ordinary stroll through the neighborhood becomes an opportunity to look for beauty in the familiar. A bench along a river trail encourages a pause to listen to the melody of birdsongs or to be still and see what critters might emerge. The lacelike veins of a fallen leaf invite us to trace new lines.

A WILD IDEA TAKES ROOT

I am an artist, and my paintbrush and an observant eye are my two most important tools. They help me understand and interpret my surroundings. Over the years, I've focused most of my work on the hidden wonders of the natural world—the intricate textures of plants and fungi or the endangered species that don't get much attention. Artist residencies and hiking trips to parks and forests all over the United States have exposed me to new landscapes and inspirations, many of which have found their way into my paintings.

But despite my far-flung travels, a familiar topography continued to beckon—the Appalachian Mountain range that I've always called home.

Spanning roughly 2,000 miles from Alabama to Newfoundland, the Appalachian Mountains are some of the oldest in the world. Their formation began between 500 and 300 million years ago with the slow-motion collision of the North American

and African continental plates, which forced the thick crust of both plates to crumple and fold upward. To put this hard-to-fathom timeline into context, the Appalachians existed before the presence of mammals, dinosaurs, and even trees. They are millions of years older than the Rocky Mountains and the Alps.

In their infancy, these peaks rose to lofty elevations, comparable to those of the Himalaya, but then they gradually eroded down to the relatively humble heights that we see today. Now, the Appalachian Mountains' highest peak is North Carolina's Mount Mitchell at just 6,684 feet. In the meantime, volcanoes erupted, earthquakes contorted the earth into new shapes, and ice sheets carved deep ravines as they thawed. Can you imagine watching a time-lapse video of these dramatic shifts? This dynamic geologic history resulted in an array of specialized habitats for plants and animals, from forested coves and craggy boulder fields to deep gorges and boreal forests.

But in addition to housing unique natural environments that enable myriad flora and fauna to thrive, these storied mountains contain vast reserves of fossil fuels—coal, oil, and natural gas—that have contributed greatly to the United States' economic and industrial development over the last 300 years. Mining, drilling, the logging of valuable timber, and urban expansion have transformed the Appalachian landscape, with widespread deforestation and watershed contamination representing two of the starkest by-products of this pursuit of financial growth.

In many places, the seas of trees that once blanketed the mountains have been reduced to isolated islands of green. This habitat fragmentation has contributed to dramatic population declines in birds, salamanders, crayfish, freshwater mussels,

The Appalachian Trail

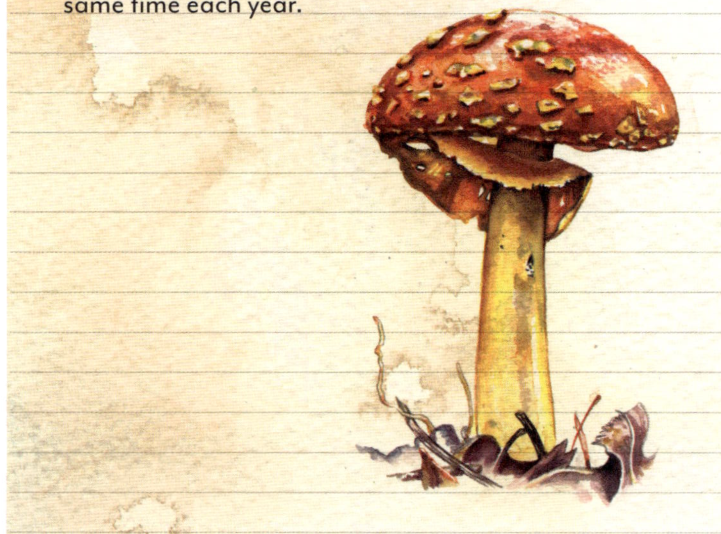

The Appalachian Trail was the ambitious brainchild of forester and planner Benton MacKaye, who proposed it in 1921 as the connective trunk of a larger network of camps—escapes for city-dwellers from the urban grind. Since its completion in 1937, the AT has become the most famous long-distance hiking trail in the world. The foot-path zigzags through twenty-two traditional territories of Indigenous peoples, a good reminder to tread lightly and respectfully. It also passes by dozens of small towns, providing locals with an eccentric spectacle as waves of dirty, adventure-seeking hikers pass through around the same time each year.

and insects, pushing some to the status of federally endangered. And as the climate warms, new challenges enter the picture. Northern species will be squeezed out of many areas of the Appalachians, and invasive species will expand into new territories at increasing rates. Many plants and animals will be vulnerable to destructive fires and floods.

Yet despite it all, the Appalachians are still considered one of the most biodiverse areas of the United States. But for how much longer? While the public lands that dot the range create precious pockets of refuge, almost three-quarters of the wider region is currently unprotected. Its plant and animal inhabitants tell a story of resilience and hope, one that needs to be amplified to encourage further safeguarding.

So one spring day, after drinking a big cup of coffee and taking my morning walk, I came up with an ambitious project: What if I set out on a journey to slowly explore the entire length of the mountain range, using my illustration skills to document the area's tiny worlds—those creatures, plants, and fungi that often go unobserved? I envisioned an illustrated time capsule of sorts, a visual celebration of the life that can be found in the Appalachians, right here and now. At first, the idea seemed completely daunting—too expensive, overly time-consuming, and logistically impossible—but gradually the details came into focus, and in April 2022 my husband, Ben, and I set out from our West Virginia home on a meandering six-month road trip along the spine of the Eastern Seaboard.

There are three primary long-distance hiking trails that connect the full Appalachian Mountain range. From south to north, the first is the Pinhoti Trail, which begins in northern Alabama and traces 335 miles to where it ends near the start of the Appalachian Trail (AT) in Georgia. The AT then stretches roughly 2,190 miles north to its terminus in central Maine. Finally, the International Appalachian Trail picks up where the AT leaves off and continues 1,580 miles through the Canadian Appalachians, all the way up to and along the western coast of Newfoundland. While we didn't thru-hike any of these trails, we used them as a guide to plan our travels and section-hiked many segments. We also explored countless nearby parks, forests, and wildlife refuges that the trails don't pass through.

EVERY DAY IS A TREASURE HUNT

Our Appalachian Mountains adventure inspired more than just the hundreds of hand-painted illustrations and journal entries that make up this book. I also discovered that every day can be a treasure hunt. Some days, you'll encounter an unfamiliar plant or animal that will generate the nervous excitement of meeting a new friend. Other days, you'll come across familiar subjects that require you to look more closely, observe a little longer, or read about them in order to learn something new. But if you apply a dose of creativity to your adventures, there will always be at least one new detail to delight in.

Viewing every day as a treasure hunt can also be a powerful way to balance out the harder aspects of life. It is far too easy in this modern age to numb, distract, and avoid; instead, by paying closer attention, by going outside with the intention of noticing something new, it is possible to sit with feelings of grief or discomfort. The focus enables you to contemplate these raw emotions, even as your gaze is turned outward.

Healing, hope, and inspiration are all available in the natural world. It's not escapism—it is simply waking up to the wonder all around.

As you read this book, I encourage you to move slowly through the pages, as if on your own treasure hunt. Allow yourself time to admire the littlest details of this ancient, beautifully biodiverse mountain range. Between the pages where I zoom in on specific animals and plants, you'll find short essays in which I trace the thoughts that swirled through my mind as I walked the trails—musings, often rooted in the natural world, that branch outward into insights about family, friendship, fear, partnership, adventure, and more.

But this book is not just a documentation of my own adventures and observations. At the end of each section, I share a nature journal page from my sketchbook and three prompts to encourage your own up-close nature exploration. Scan them

Helpful Tools for Nature Journaling

- A pencil, pen, or any colorful art supplies that you enjoy using or have on hand; colored pencils are especially convenient for on-the-go sketching.
- A notebook or sketchbook
- Binoculars
- A loupe or magnifying glass/hand lens
- A camera or phone for taking reference photos

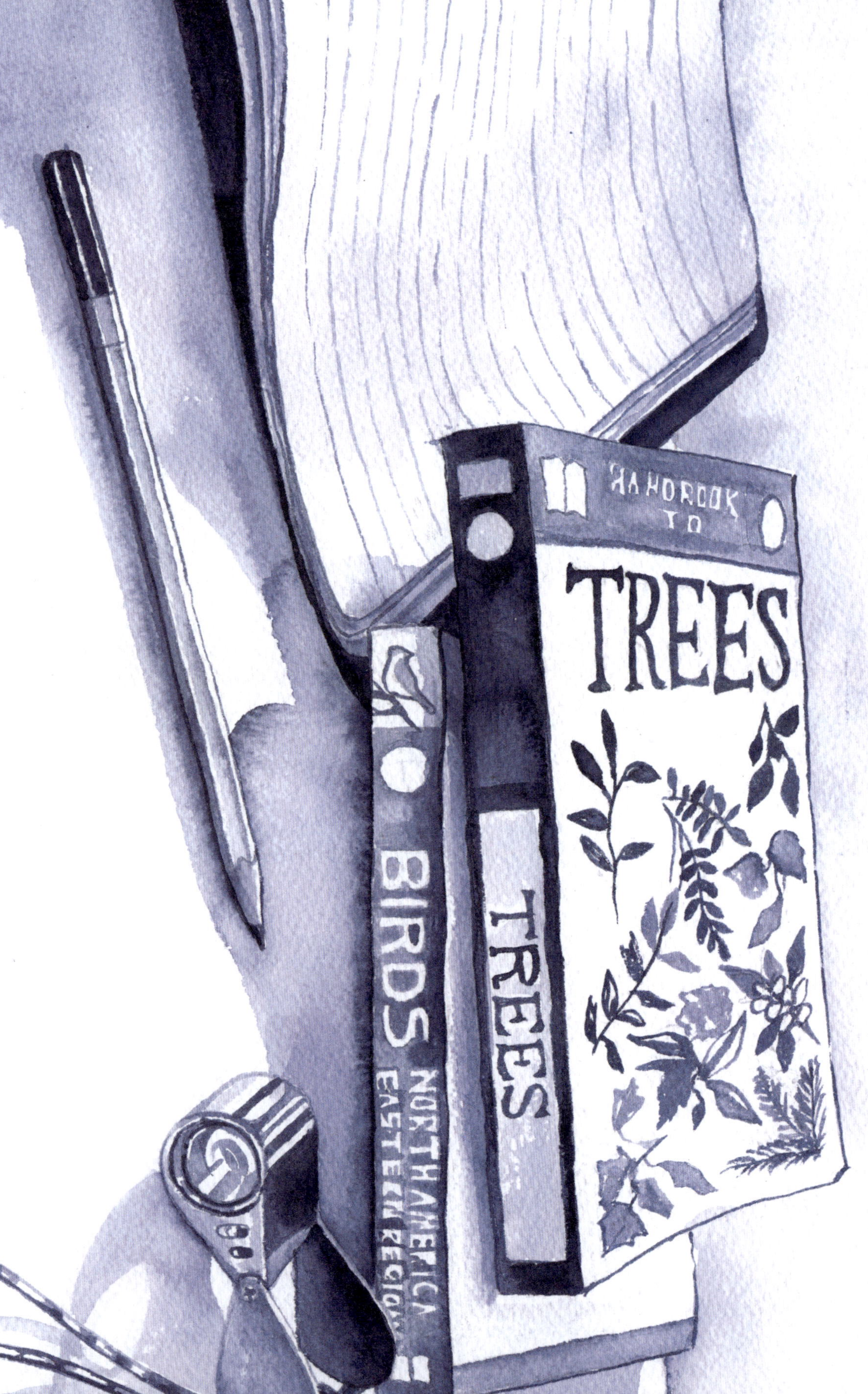
HANDBOOK TO
TREES
BIRDS NORTH AMERICA EASTERN REGION
TREES

and choose one that inspires you, or come up with your own idea for your nature journal.

Nature journaling is a great way to reconnect with our senses and become more in tune with our environment. Using the simplest, most analog tools—a pencil, a notebook, binoculars, and a hand lens—we can slow our pace and learn to move through the world in a more conscious way.

It's definitely not about creating perfected works of art, like the nature illustrations found throughout this book, many of which took me about a week to paint. Instead, nature journaling involves simply recording what we see while we're outside, whether that's through written notes, quick sketches, or an impromptu poem—no artistic or writing experience required.

At the end of the book, you'll find directions for ways to share your nature journal pages with me and your fellow readers, if you so desire. In this way, we can celebrate our tiny worlds together, delighting in one another's discoveries.

Thank you for joining me on this journey through the mountains. Who knows what we will find!

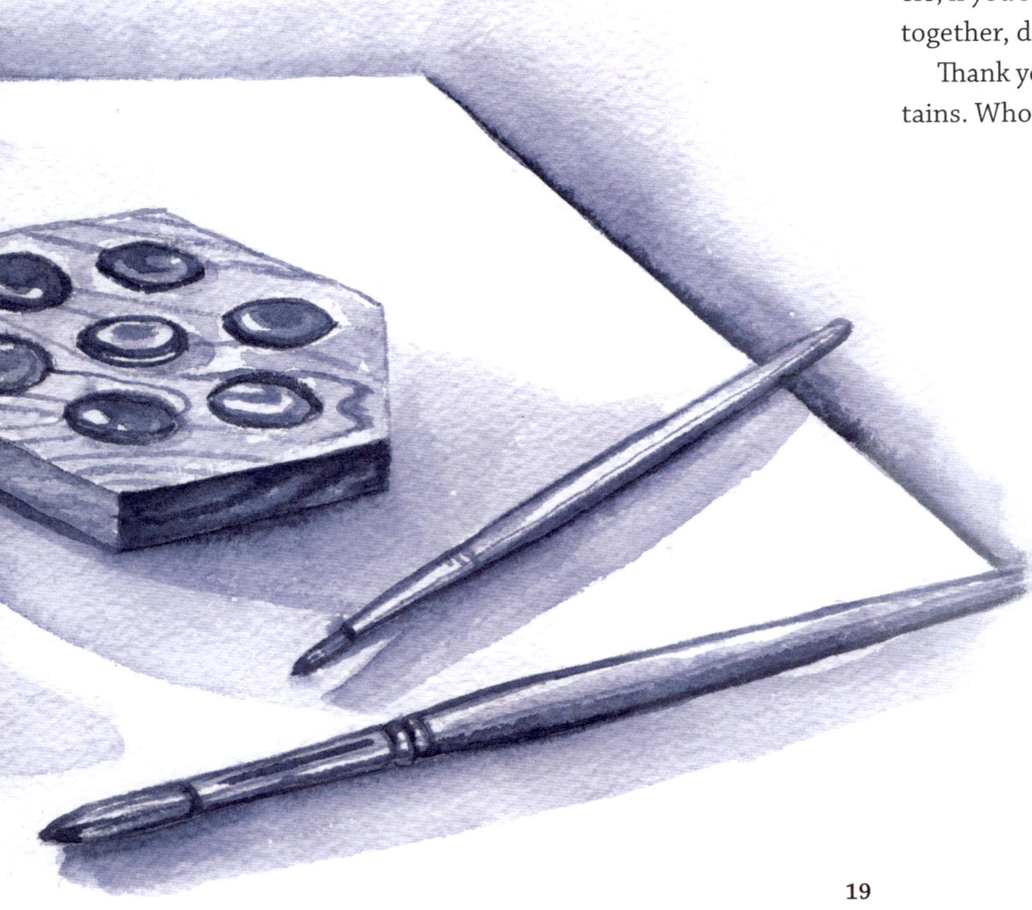

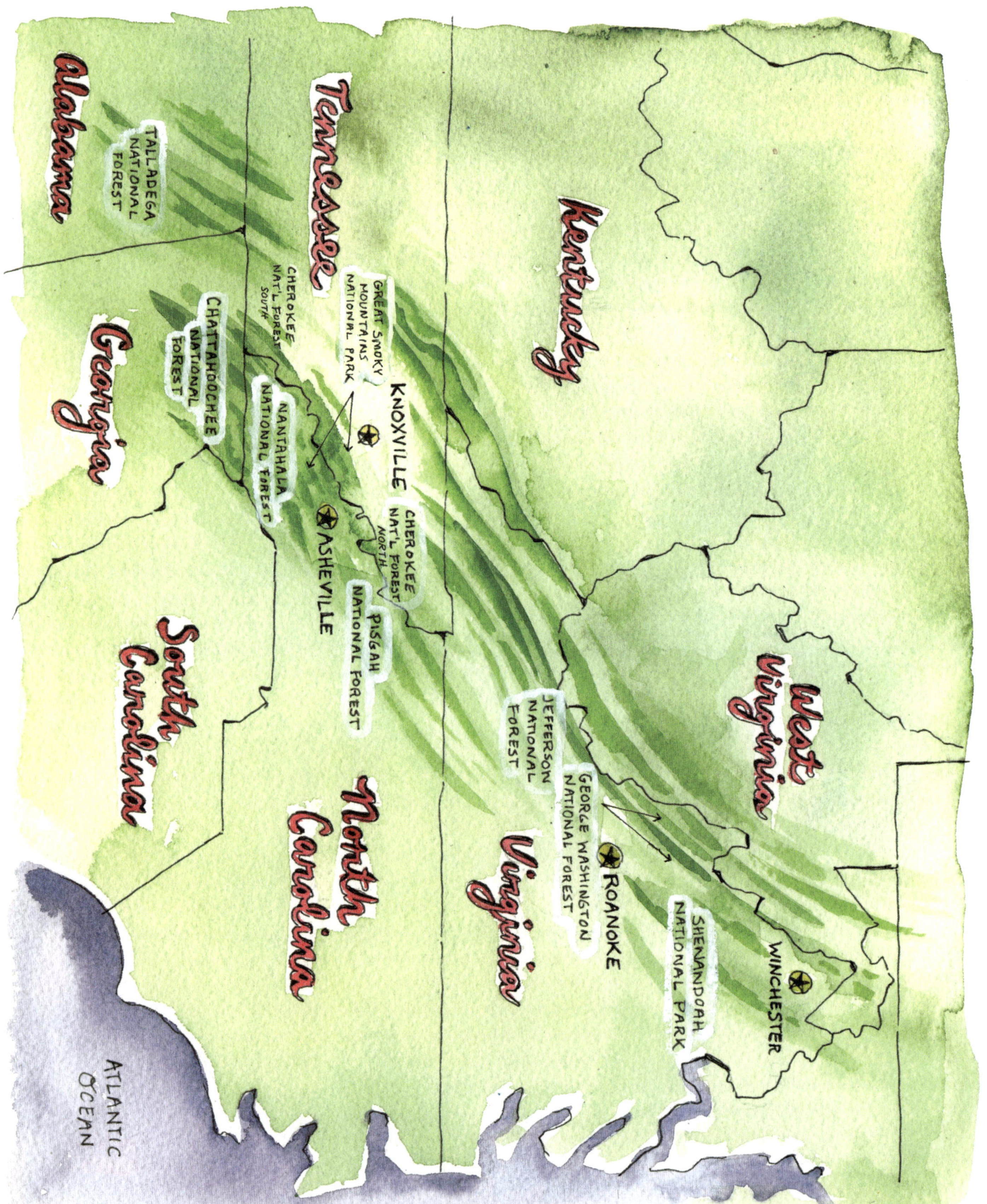

Alabama
TALLADEGA NATIONAL FOREST
Tennessee
CHEROKEE NAT'L FOREST SOUTH
CHATTAHOOCHEE NATIONAL FOREST
Georgia
NANTAHALA NATIONAL FOREST
GREAT SMOKY MOUNTAINS NATIONAL PARK
KNOXVILLE
Kentucky
CHEROKEE NAT'L FOREST NORTH
ASHEVILLE
PISGAH NATIONAL FOREST
South Carolina
North Carolina
Virginia
JEFFERSON NATIONAL FOREST
GEORGE WASHINGTON NATIONAL FOREST
ROANOKE
West Virginia
SHENANDOAH NATIONAL PARK
WINCHESTER
ATLANTIC OCEAN

Southern Appalachians

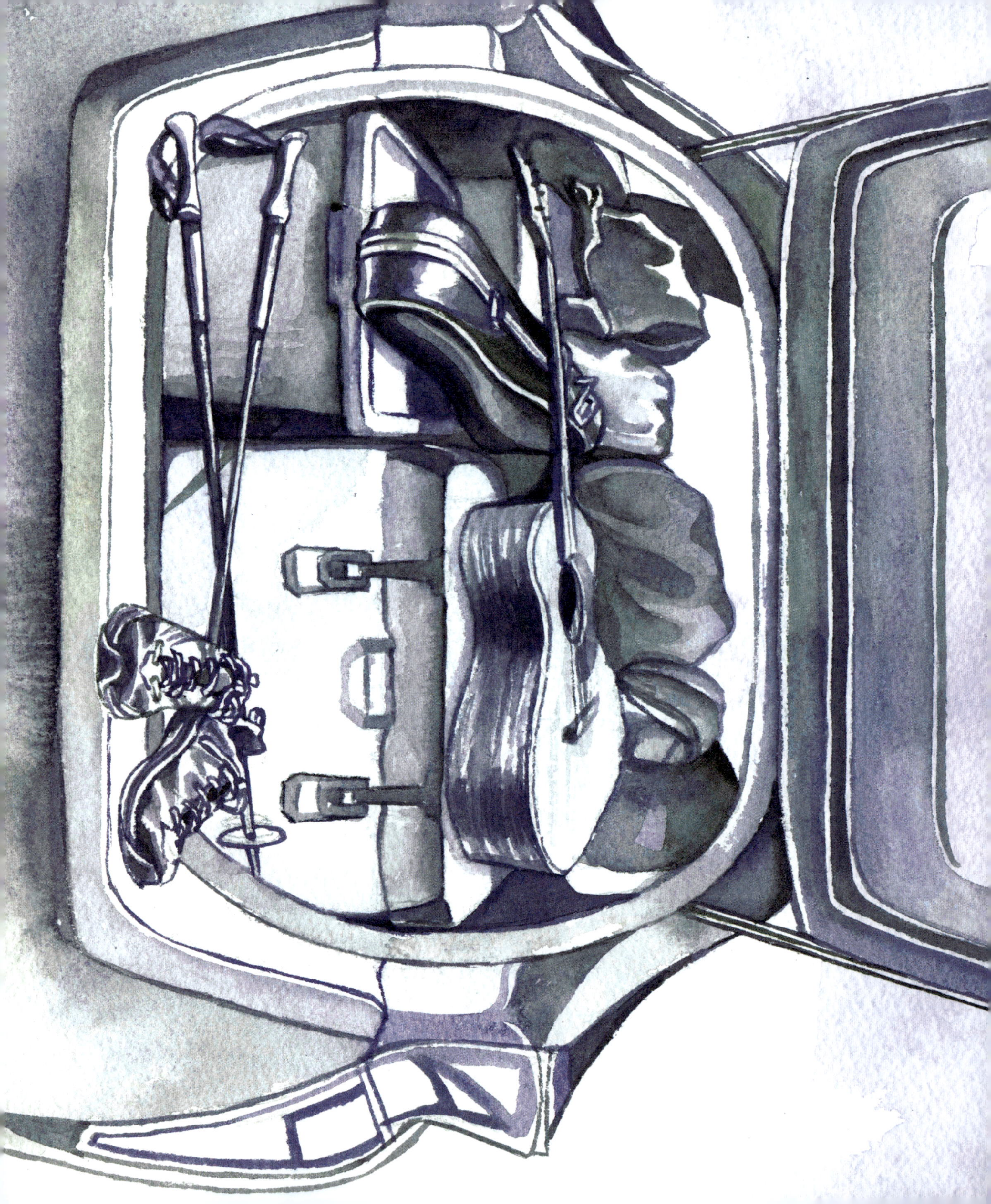

THE ADVENTURE BEGINS

APRIL 1

Ben and I wedge everything we need for the next six months into our Prius and head out, "Sweet Home Alabama" blasting from the speakers.

I recall a comment a friend made when I told her about our plan to explore the Appalachian Mountains for six months. "Don't you want to go somewhere a little different?" she'd asked. "I'm surprised you don't want to use this time to go abroad or something." I understood. If we were going to make financial sacrifices to travel for an extended period of time, why not make it the most exciting experience possible? But I have an inkling that traveling regionally, with my attention turned to the plants and wildlife, will be packed with excitement and discovery. I feel like I know this landscape, but do I really? It's my home, but when I break down my surroundings into the individual members of the forests and streams, the wetlands and mountaintops, how much do I actually know about my nonhuman neighbors?

As nighttime nears, we pull into Alabama's Cheaha State Park and get to work setting up camp. During our trip, we'll do a mix of car camping, backpacking, and staying in short-term rentals. The sun slips behind the still-naked trees and I feel a wave of gratitude that I will get to see the full awakening of spring in the southern Appalachians. I'm ready to learn the story of these mountains, one observation at a time.

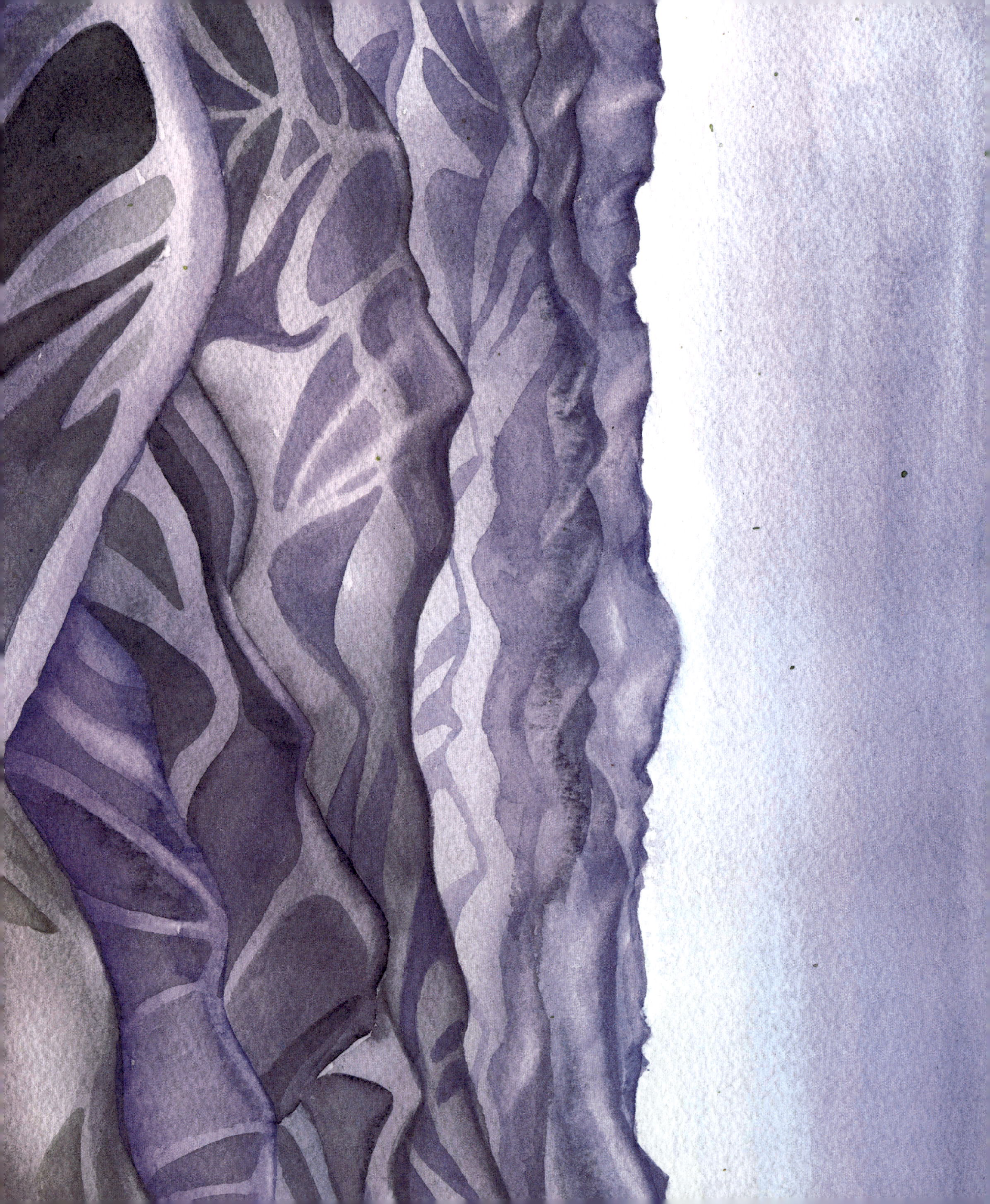

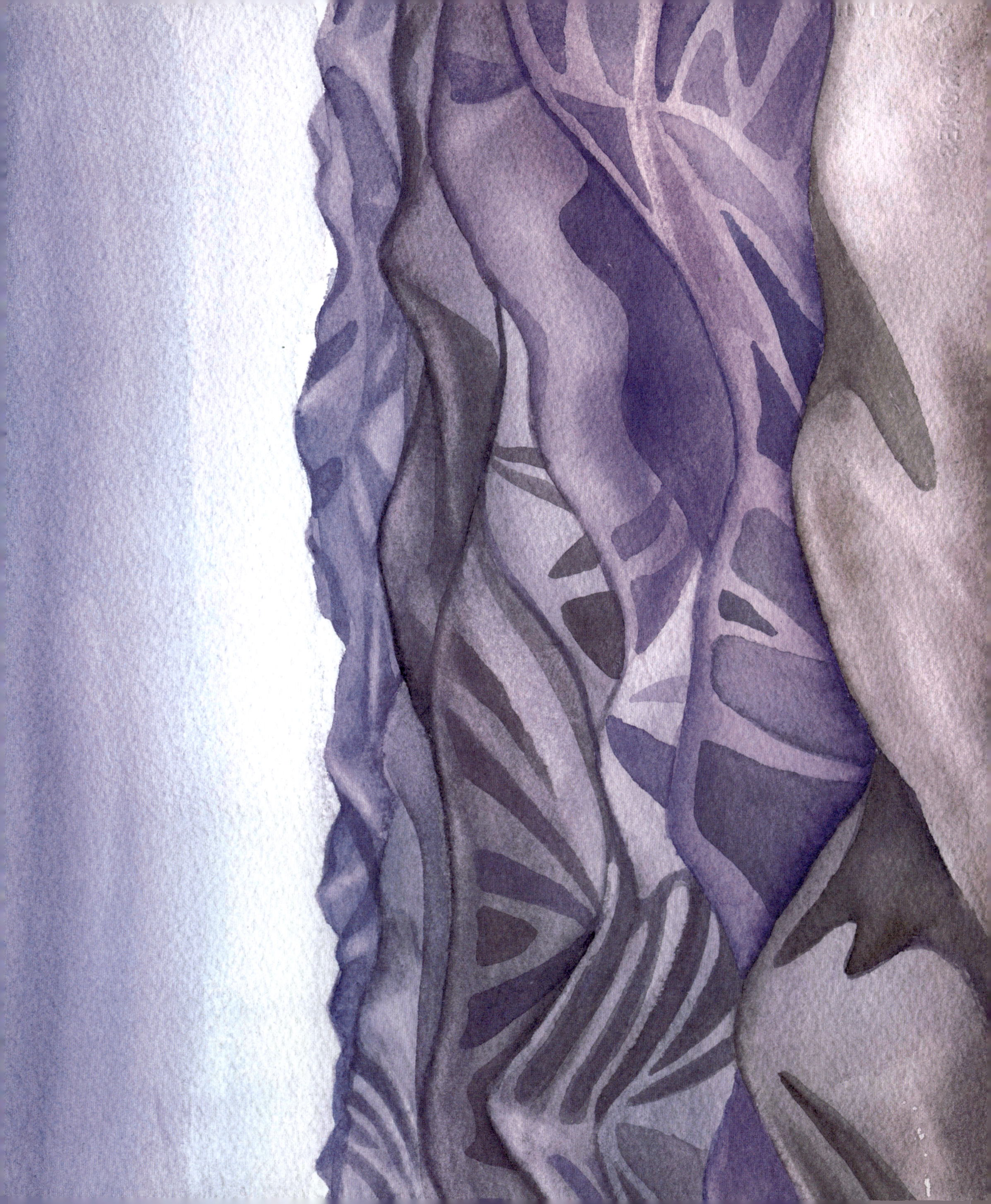

Quick-Change Artist

I stop in my tracks as a tiny critter scampers across my path and up a nearby tree. As smoothly and stealthily as possible, I reach into my backpack pocket to grab my phone. Then I take one large, slow-motion step toward the tree. It's a lizard, small and dark brown with a thick white stripe running down its back. My thighs burn as I hold a squat for many minutes, but I'm elated that this cooperative lizard is allowing me a full-blown photo shoot.

Suddenly, it lunges from the branch into a pile of leaves and grabs an unsuspecting orange beetle in its mouth, ripping into it in a brutal, dinosaur-like fashion. In a moment of mid-meal self-awareness, the lizard seems to notice that I'm watching and drags what's left of the beetle under a leaf to polish it off in privacy. I leave it alone for a few minutes until my curiosity gets the best of me and I remove the leaf so I can continue my observations.

The lizard is now a stunning shade of lime green! In utter disbelief, I whisper to Ben to come take a look. "It's *green* now," I tell him. "The lizard was *brown* and now it's *green*." There's no response. I turn my head in his direction and realize that I've been watching the lizard for so long that Ben has officially fallen asleep on a bed of leaves beside the trail, using his backpack as a pillow.

I later learn that green anoles are native to the subtropical southeastern United States, and they change color based on their temperature and mood, as well as to send social signals. Perhaps this lizard suddenly turned green because catching her juicy beetle snack had filled her with happiness.

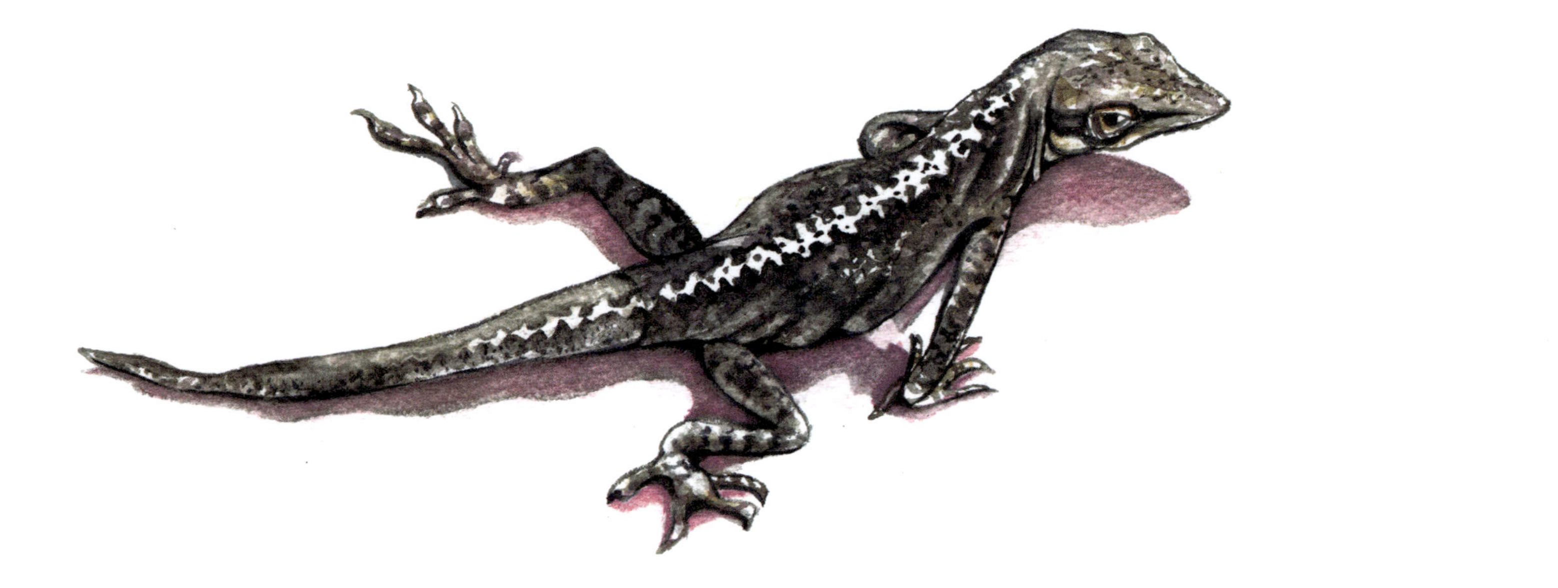

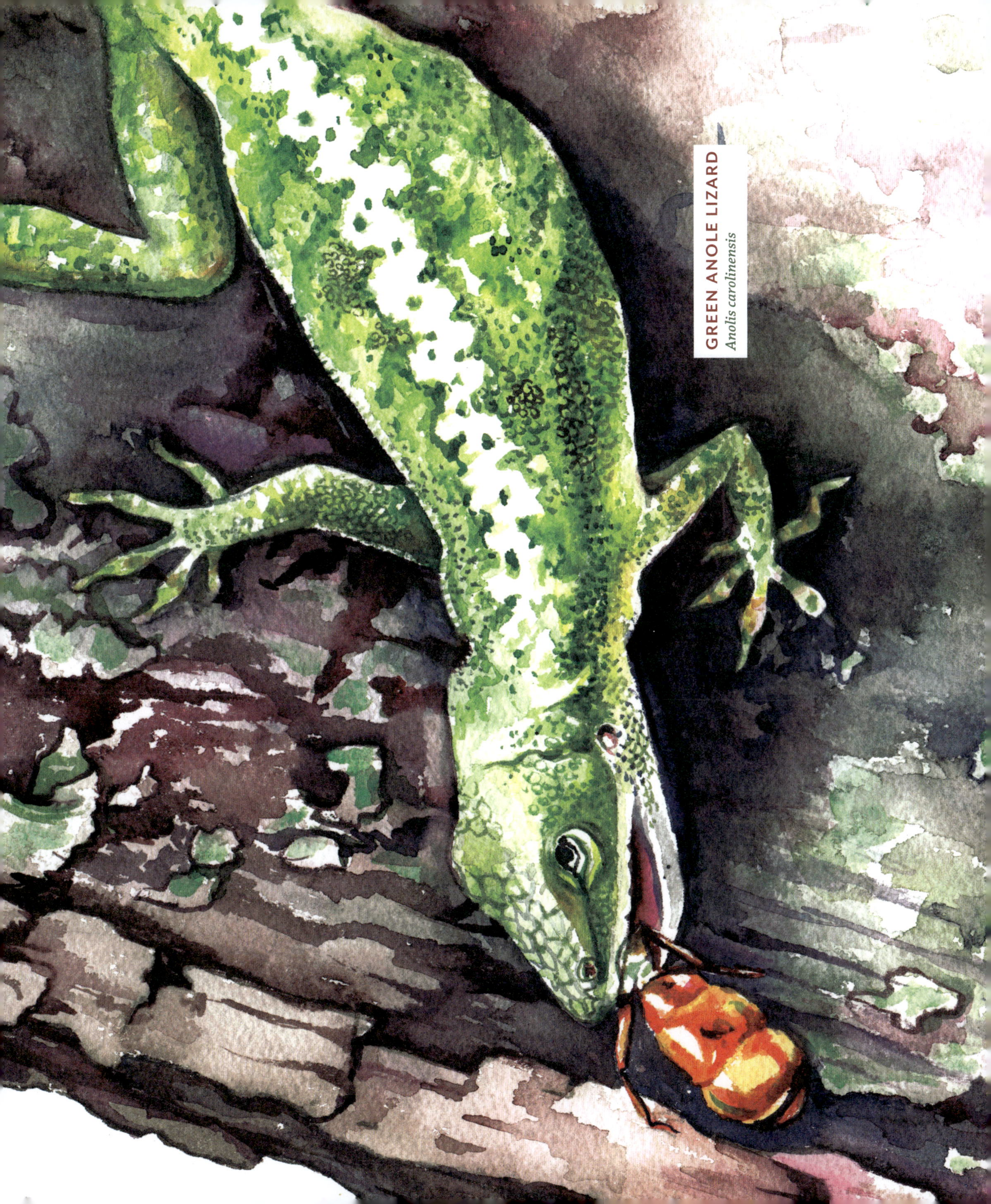
GREEN ANOLE LIZARD
Anolis carolinensis

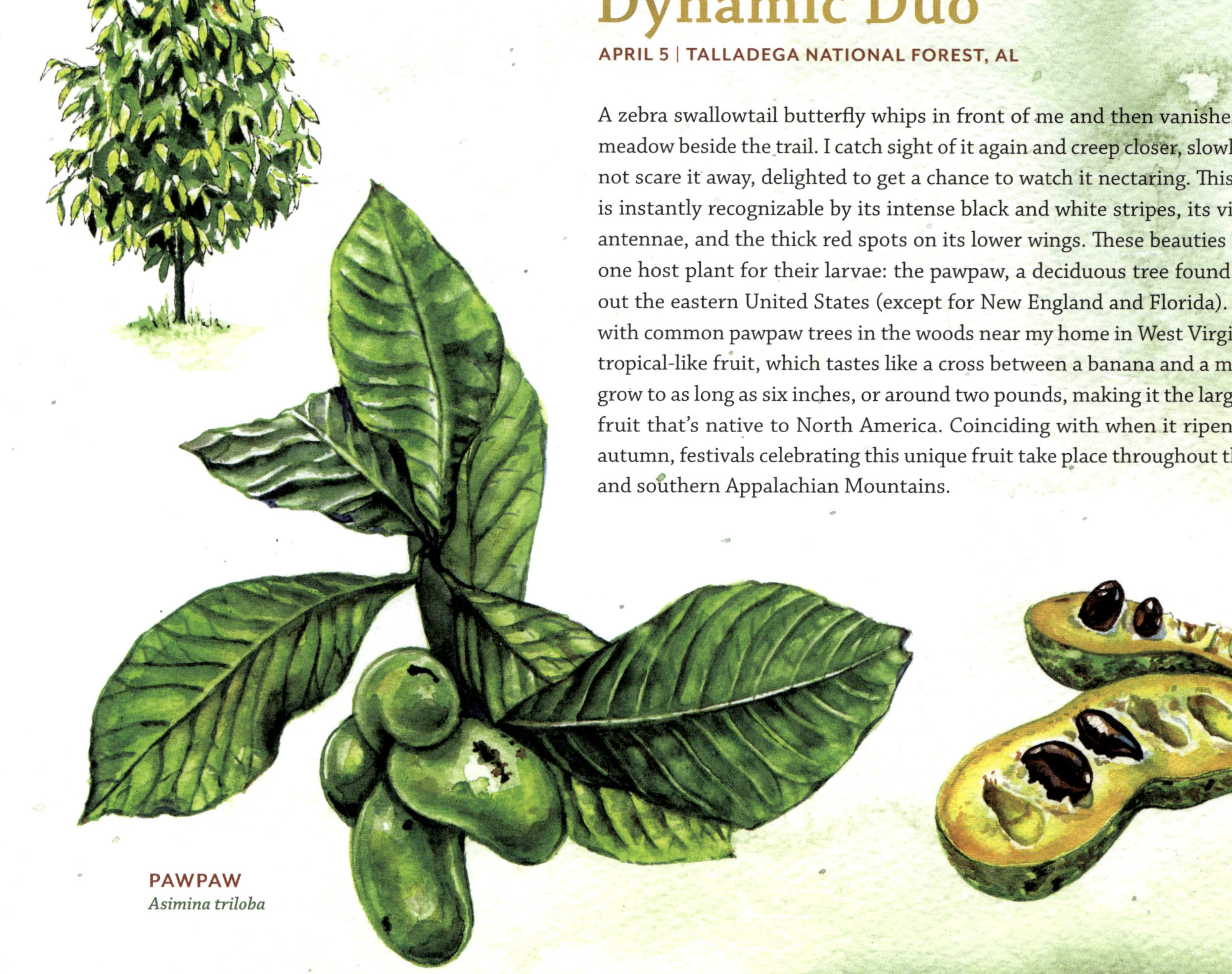

Dynamic Duo

APRIL 5 | TALLADEGA NATIONAL FOREST, AL

A zebra swallowtail butterfly whips in front of me and then vanishes into the meadow beside the trail. I catch sight of it again and creep closer, slowly, so as to not scare it away, delighted to get a chance to watch it nectaring. This butterfly is instantly recognizable by its intense black and white stripes, its vibrant red antennae, and the thick red spots on its lower wings. These beauties have only one host plant for their larvae: the pawpaw, a deciduous tree found throughout the eastern United States (except for New England and Florida). I grew up with common pawpaw trees in the woods near my home in West Virginia. Their tropical-like fruit, which tastes like a cross between a banana and a mango, can grow to as long as six inches, or around two pounds, making it the largest edible fruit that's native to North America. Coinciding with when it ripens in early autumn, festivals celebrating this unique fruit take place throughout the central and southern Appalachian Mountains.

PAWPAW
Asimina triloba

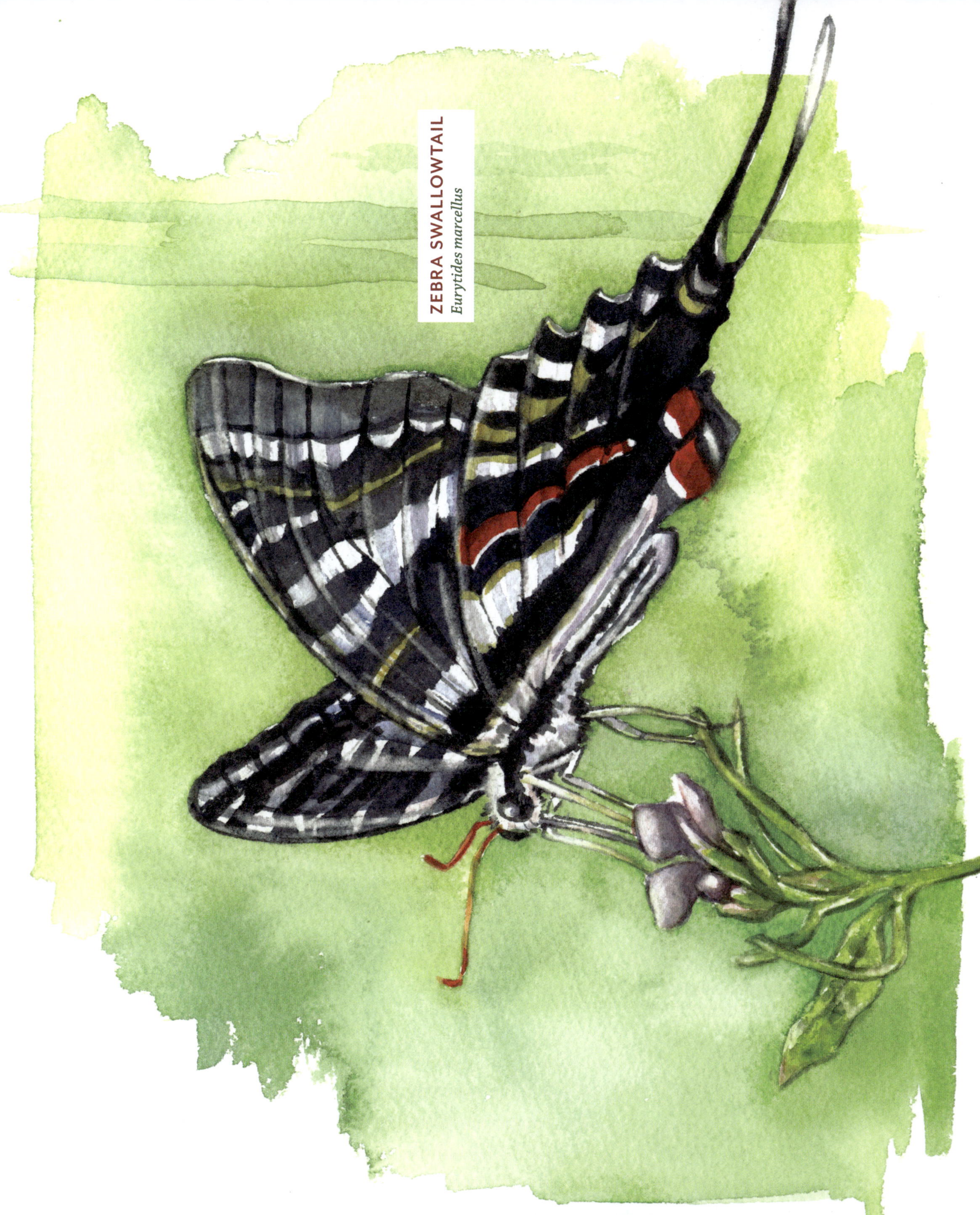

ZEBRA SWALLOWTAIL
Eurytides marcellus

Wildflower Weirdo

With a backpack full of snacks and adrenaline pulsing through my veins, I soak up the magic of hiking in a place I've never been before. Looking down, I pause on the strange structures of bear corn, a plant I've seen poking up through the soil on many occasions but have never really taken the time to investigate. This cone-shaped wildflower is a parasite that feeds on the roots of oak trees. Because it doesn't photosynthesize, it lacks the green color that chlorophyll produces. Most of the plant is concealed underground, but the corncob-like flower structures show themselves each spring, growing four to eight inches tall. Bears love to munch on them, but only a few pollinators are attracted to these indistinct flowers. However, bear corn has its own solution to make seeds for the next generation: its male and female reproductive structures are located near one another and are able to self-pollinate.

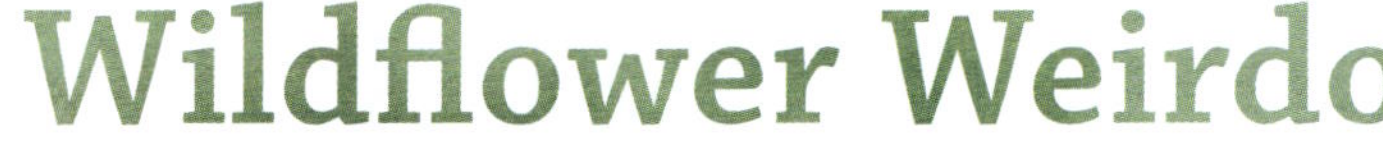

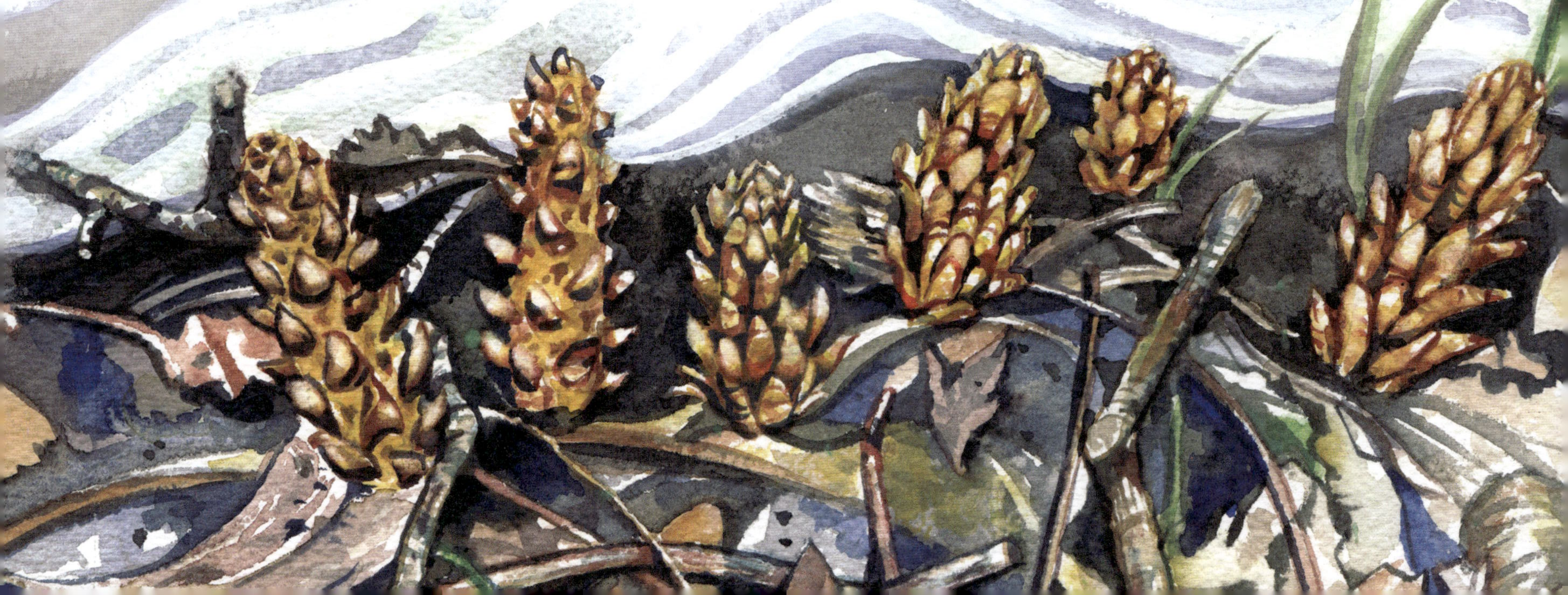

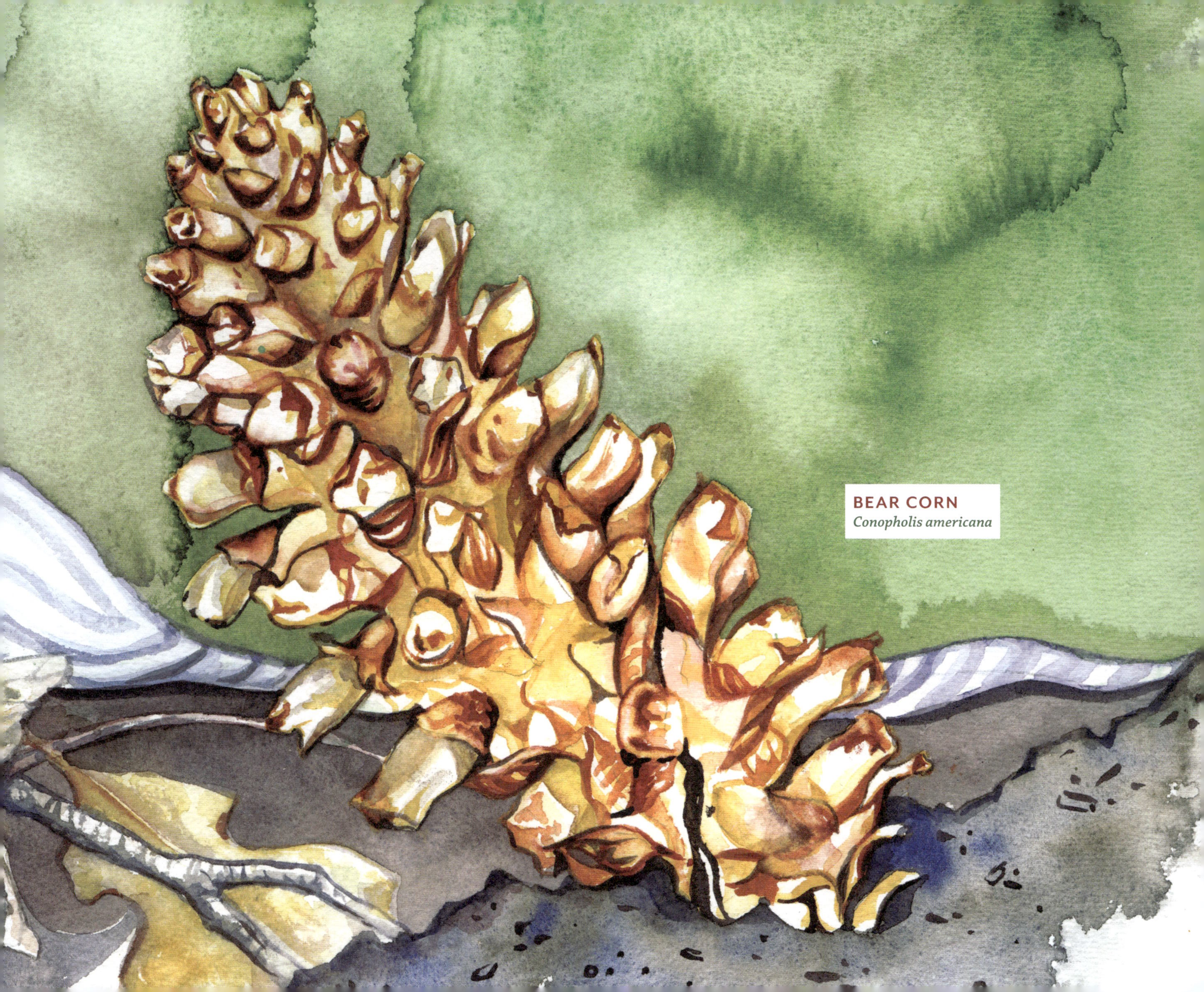

BEAR CORN
Conopholis americana

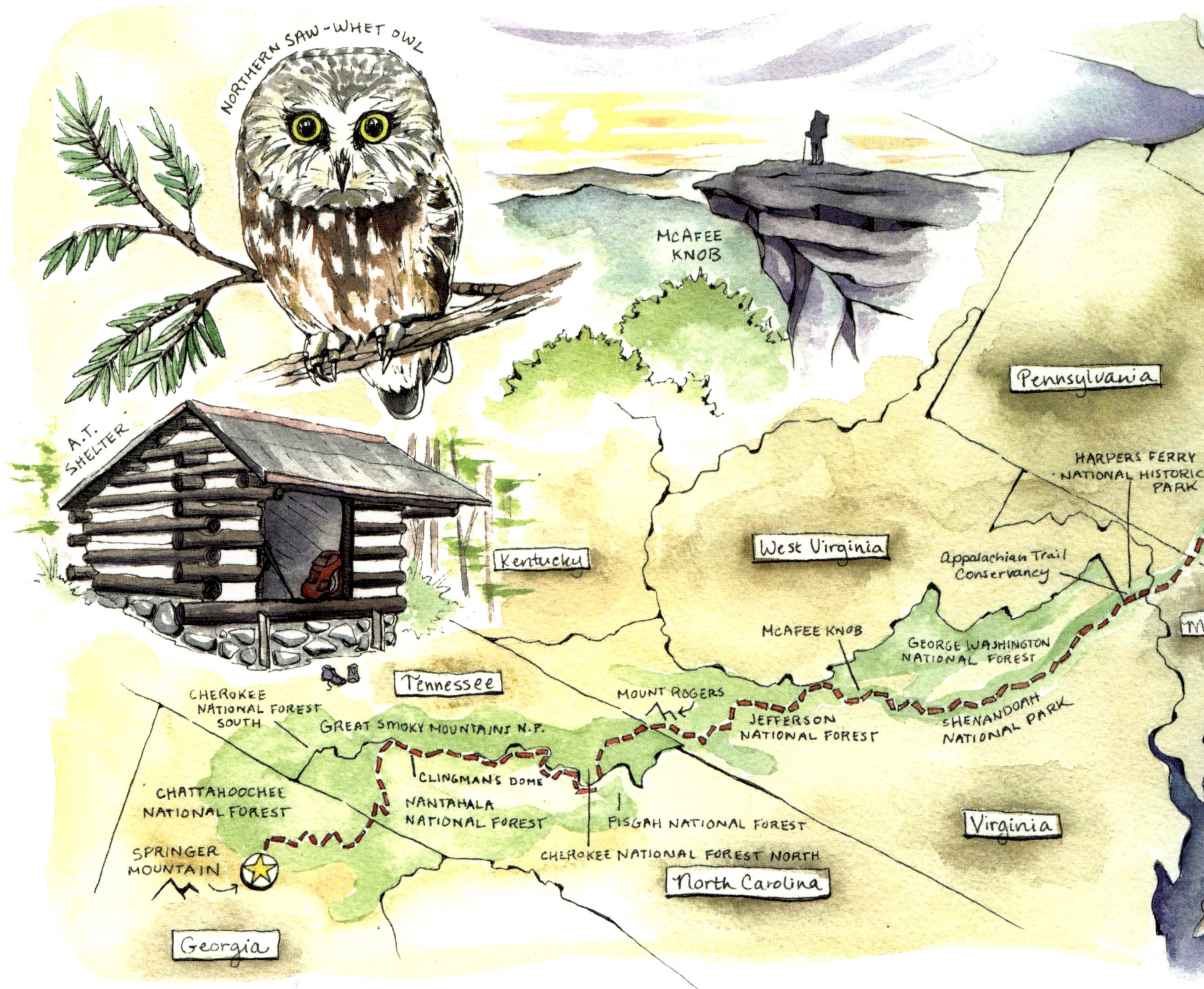

NORTHERN SAW-WHET OWL
A.T. SHELTER
McAFEE KNOB
Pennsylvania
HARPERS FERRY NATIONAL HISTORICAL PARK
West Virginia
Appalachian Trail Conservancy
Kentucky
McAFEE KNOB
GEORGE WASHINGTON NATIONAL FOREST
Man
Tennessee
MOUNT ROGERS
JEFFERSON NATIONAL FOREST
SHENANDOAH NATIONAL PARK
CHEROKEE NATIONAL FOREST SOUTH
GREAT SMOKY MOUNTAINS N.P.
CHATTAHOOCHEE NATIONAL FOREST
CLINGMAN'S DOME
NANTAHALA NATIONAL FOREST
PISGAH NATIONAL FOREST
Virginia
SPRINGER MOUNTAIN
CHEROKEE NATIONAL FOREST NORTH
North Carolina
Georgia

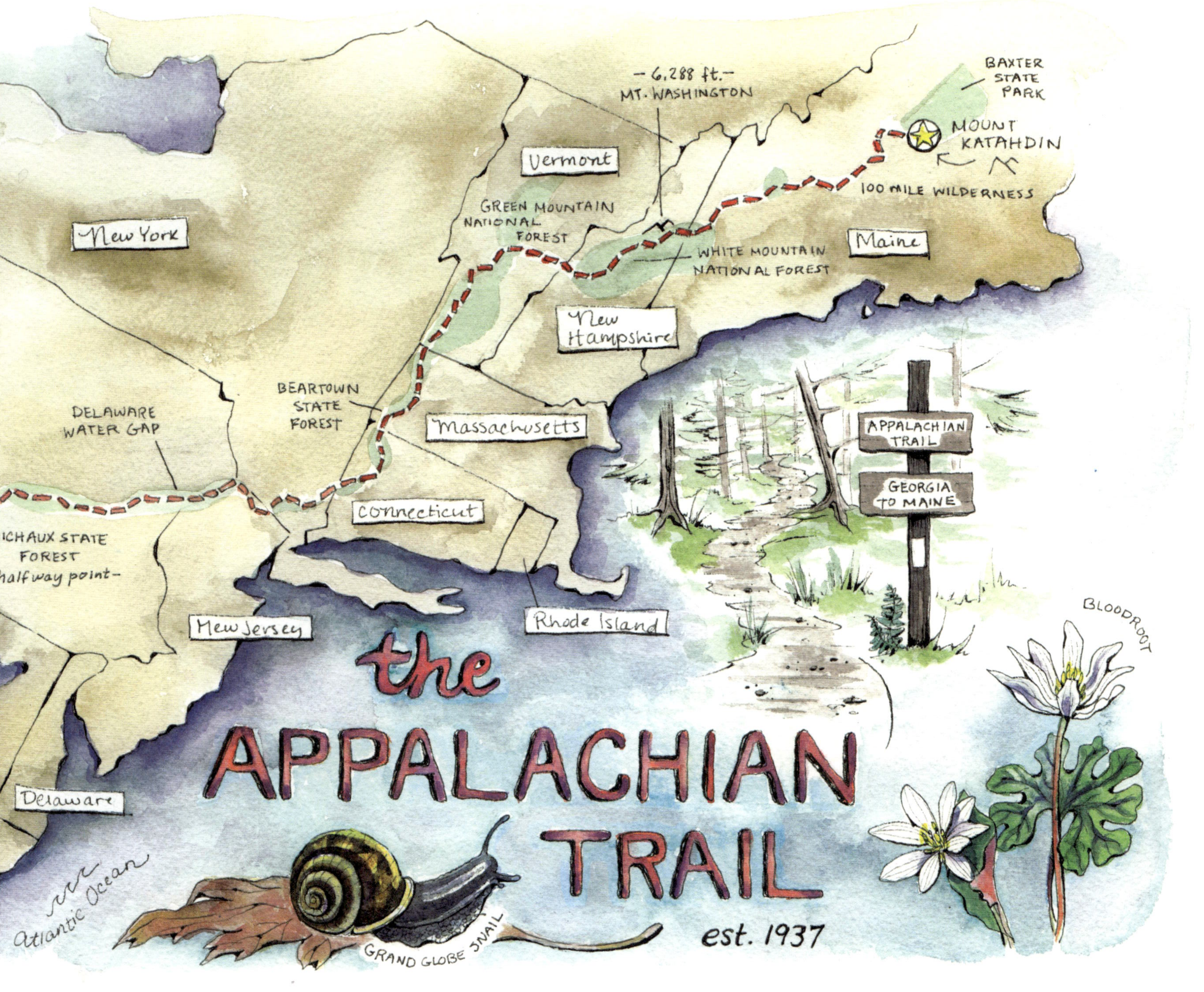

the APPALACHIAN TRAIL
est. 1937
BAXTER STATE PARK
MOUNT KATAHDIN
100 MILE WILDERNESS
~6,288 ft.~ MT. WASHINGTON
Vermont
GREEN MOUNTAIN NATIONAL FOREST
WHITE MOUNTAIN NATIONAL FOREST
Maine
New York
New Hampshire
BEARTOWN STATE FOREST
Massachusetts
DELAWARE WATER GAP
Connecticut
MICHAUX STATE FOREST ~halfway point~
Rhode Island
New Jersey
Delaware
Atlantic Ocean
GRAND GLOBE SNAIL
APPALACHIAN TRAIL
GEORGIA TO MAINE
BLOODROOT

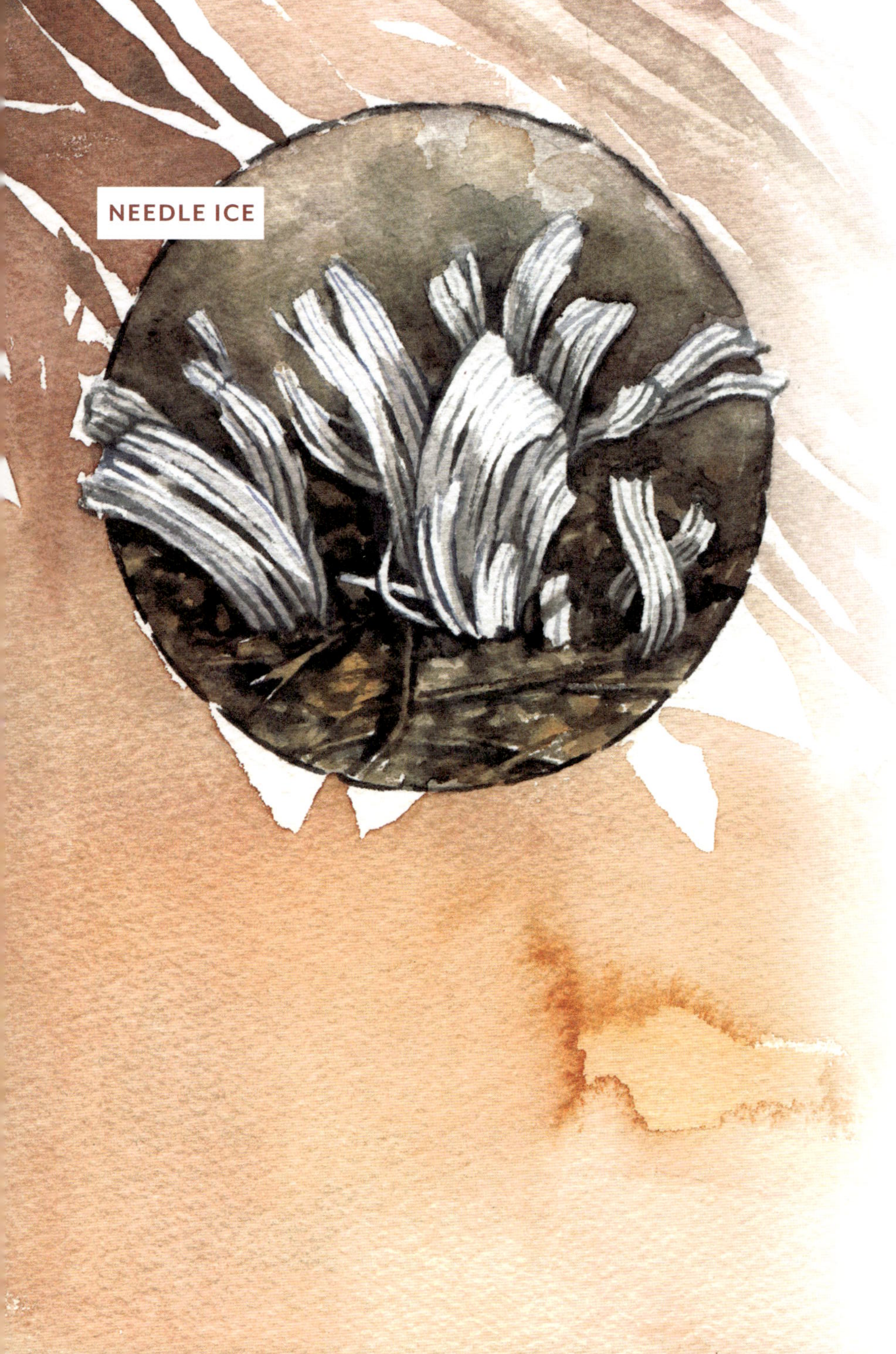

Spring Time Traveling

APRIL 19 | WAYAH BALD, NC

Bundled in hats and mittens to face the 28°F weather, we climb the old stone fire tower for a panoramic scene of powdery blue-and-purple mountains, which ripple outward into the Great Smoky Mountains to the north and the North Georgia mountains to the south. At this elevation—5,300 feet—there isn't any spring growth yet and it feels like we've gone back in time by a full month. After taking in the view, we hop onto the Appalachian Trail, which passes below the tower, and begin descending the steep footpath, our boots crunching on needle ice with each step. Have you ever seen strange little ice tubes emerging from the muddy ground after a cold night? This occurs when the temperature of the soil is above freezing (32°F) and the temperature of the air is below freezing. Water in the ground rises to the surface, where it freezes and gradually builds a needle-like column of ice.

As we hike down the mountain, the forest takes on dabs of green and we see a patch of pungent ramps sprouting beside the trail. These wild onions hold a prominent place in Appalachian cuisine, their elegant, elongated leaves a proclamation that spring has arrived in the mountains. A little farther down, we pass a bed of violet woodsorrel. It's not yet in bloom, but the leaves alone create a lively, colorful pattern. A few large-flowered bellwort stand timidly nearby, their twisted yellow petals peeking out from behind leafy cloaks.

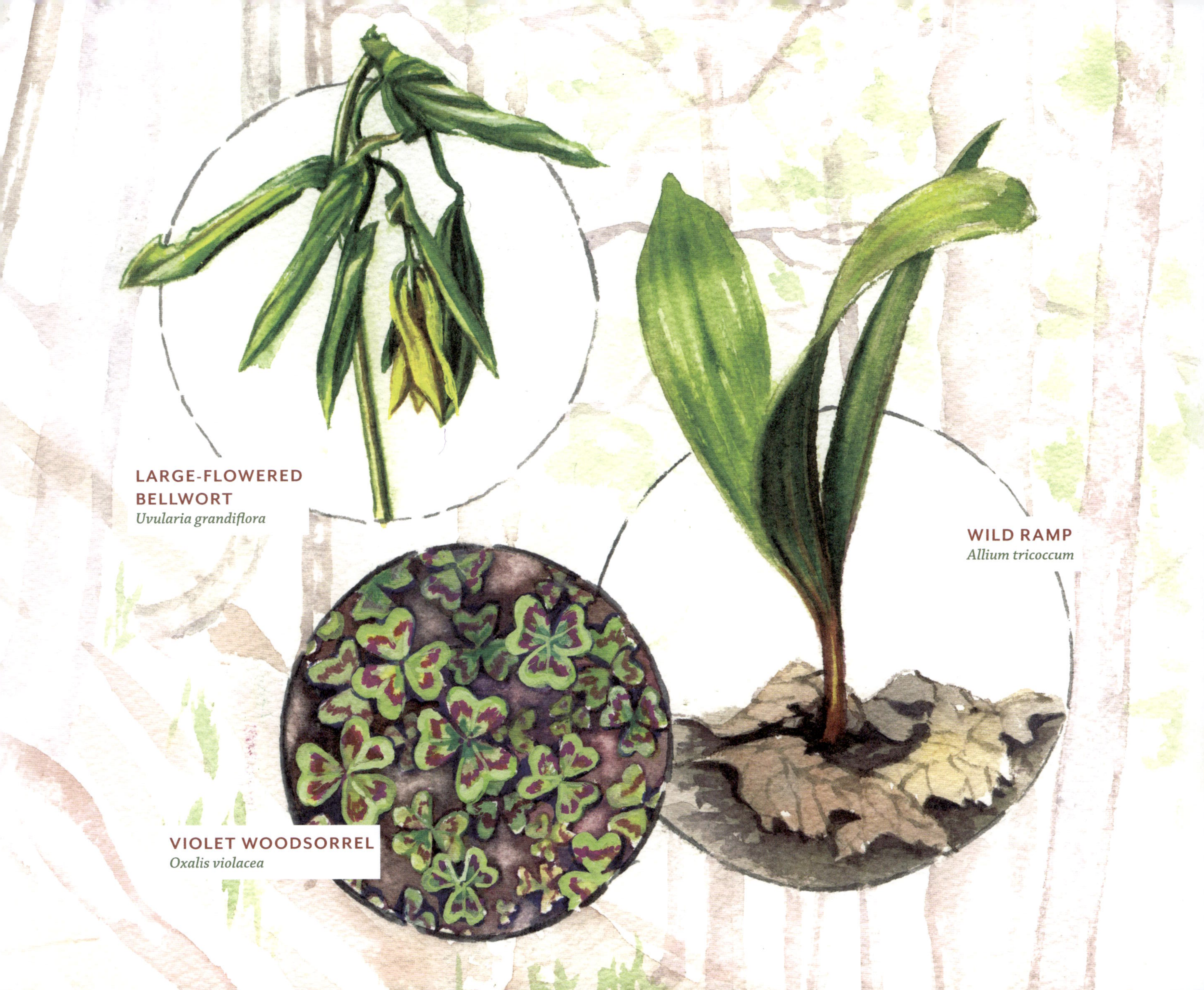

LARGE-FLOWERED
BELLWORT
Uvularia grandiflora
WILD RAMP
Allium tricoccum
VIOLET WOODSORREL
Oxalis violacea

Small Things That Have No Words

"Rosie, I have some sad news for you," my mom told me in a somber tone as I jumped into the back seat of the car after a sleepover at my grandma's house. "I'll tell you when we get home." In an instant, my heart dropped and my eleven-year-old mind went wild with all the possibilities of what might have happened. By the time we completed the five-minute drive to our house, I had concluded that my dad had died.

Mom led me inside and sat me down at the table. Then she brought out my hamster's cage and I immediately noticed that one of the plastic play tubes was broken and there was no hamster to be found. I'd recently spent all my money ($12) on a fancy pack of tubes that had turned Goldie's ordinary cage into a rodent jungle gym, something resembling a miniature McDonald's PlayPlace. Our cat, my mom informed me, had broken the tubes off the cage and had eaten my beloved Goldie.

I sobbed uncontrollably. My mom held me and cried, too, knowing the grief that was tearing through my little heart. "This was the worst day of my entire life," I scribbled in my journal that night. Looking back at how badly this small tragedy rocked my world, it's obvious that I had it pretty good.

My affinity for the smaller creatures of the world started at a young age. First there was Goldie (full name: Golden Red Haizlett), a Syrian golden hamster I received for Christmas when I was nine years old. We were soul sisters, and I adored all four ounces of her fluffy self. I would put her in my shirt and she'd run laps around my belly, her tiny claws tickling me as I laughed and laughed. When I went on summer vacation and entrusted her to a pet sitter, I missed her so badly that I began writing daily journal entries from her perspective.

Goldie inspired me to establish the Brooke County Rodent Club, a collective composed of me, my little sister, and a few neighborhood friends who also owned hamsters or gerbils. I made flyers for my club, and my ever-supportive mother drove me to the local elementary school to pin them up. I was homeschooled until I was twelve, and I'm now well aware that the students were most definitely weirded out by this strange girl showing up to advertise her rodent club. I even rented out a church basement for

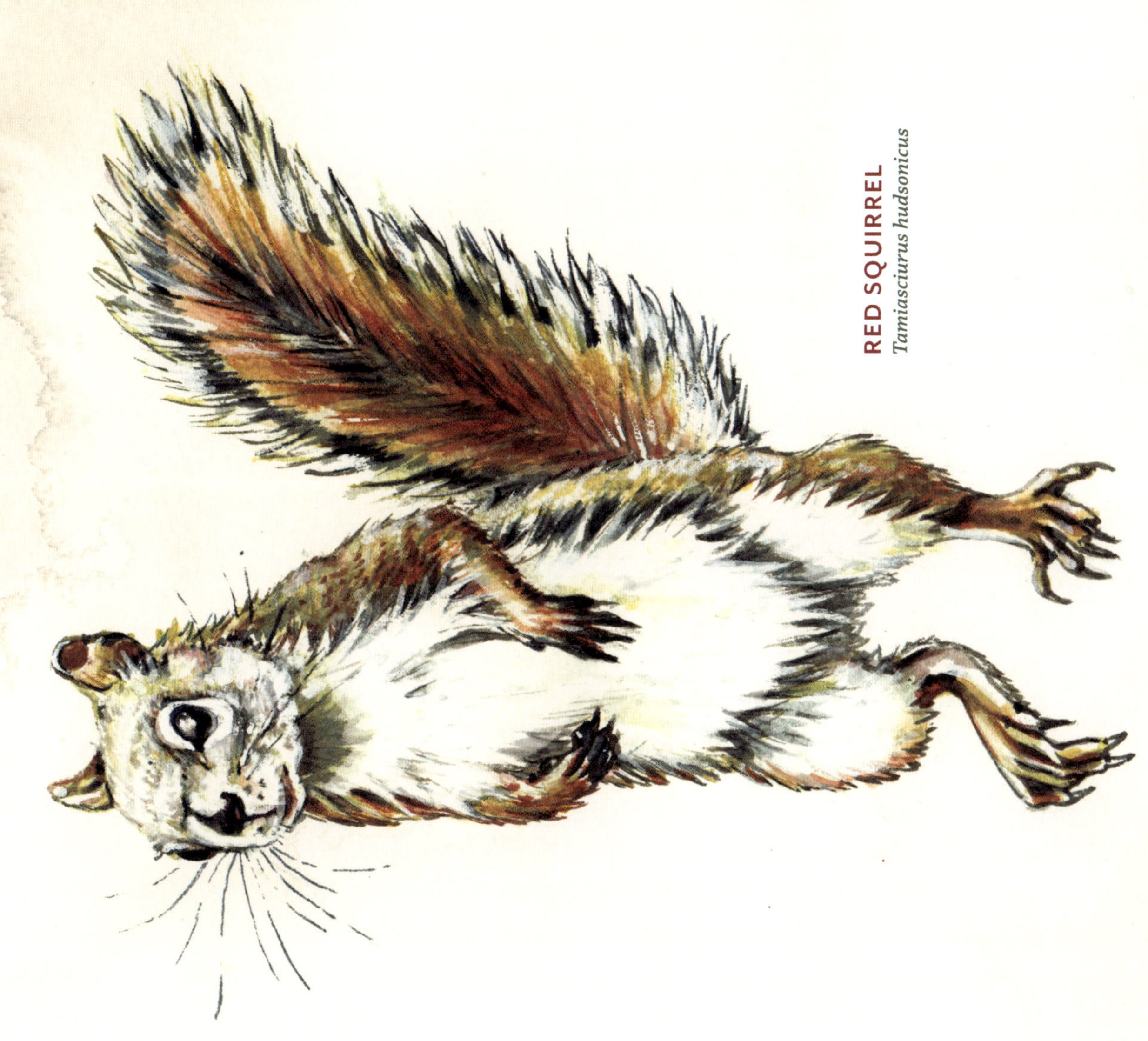

RED SQUIRREL
Tamiasciurus hudsonicus

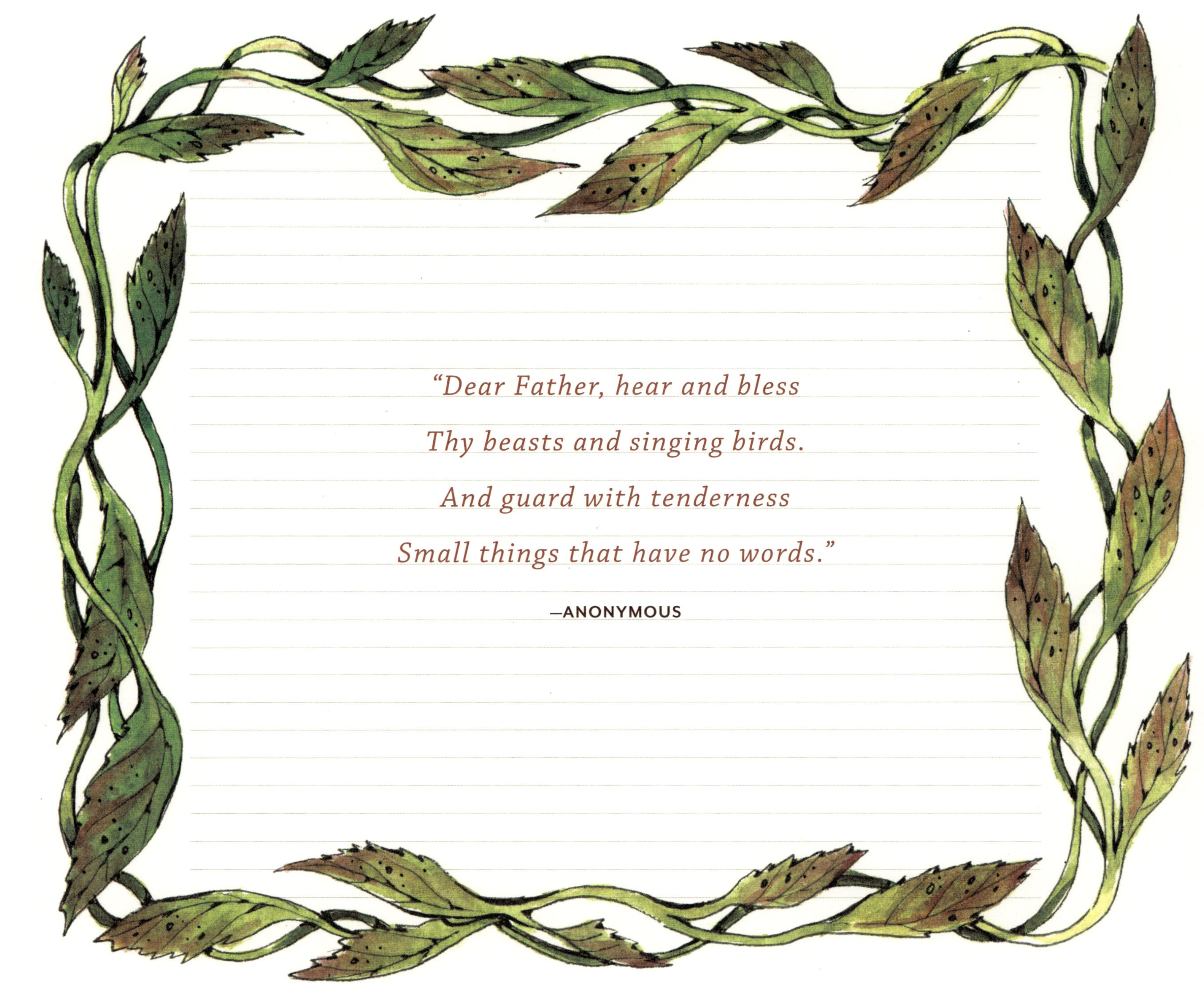

"Dear Father, hear and bless

Thy beasts and singing birds.

And guard with tenderness

Small things that have no words."

—ANONYMOUS

the inaugural meeting, during which we spent a blissful few hours completing my rodent-themed worksheets. I recall that one of the prompts was "Draw you and your rodent on your dream tropical vacation." At the end of every club meeting, we'd have a dance party to the tracks of my *Hamsterdance* CD, our hamsters, gerbils, and guinea pigs nestled against our chests.

When Goldie died, my older brother hand-crafted a six-inch pine coffin and used his wood-burner to carve her name into the lid. My mom shaved a few bits of sweet-scented soap into the coffin to make her final resting place smell nice. We went out into the yard and I said my tearful goodbyes at our sacred little pet graveyard that was positioned behind the trampoline.

After Goldie, a whirlwind of other rodents burrowed their way into my heart. "Tiny" was a baby mouse that I rescued from our yard when the propane tank lid had to be cleaned and her nest was disturbed. Her eyes were still shut when I found her, and I faithfully set my purple alarm clock for 2:00 a.m. every night to feed her eyedroppers full of puppy milk formula. Champ and Bear were other hamsters that I tended to, each bringing their own endearing person-

alities. I loved their wiggling noses, perky ears, and minuscule hands and feet. I loved watching them stuff their stretchy cheeks with enormous amounts of seeds to bury in caches around their woodchip-filled cages.

These days I move around too frequently to have a pet rodent, but I get my tiny-creature fix whenever I spot them in the wild. I am not embarrassed to admit that endorphins flood my body each time I cross paths with a vole, squirrel, or chipmunk. On the flip side, when I see one squished in the road while I'm driving, a deep wave of sadness hits me, leaving me with a queasy feeling in my gut.

It would be easy to dismiss this heightened sensitivity as frivolous, or to view it as something to be hidden. But I observed at an early age that my pets, like all living beings, were unique critters with quirky personalities, exercise habits, sleeping schedules, and food preferences. And although it probably isn't realistic for everyone to feel such intense affection for the littlest creatures among us, my mom's empathetic tears and my brother's hand-hewn hamster coffin communicated to me that it is a blessing to see value in every life, no matter how small.

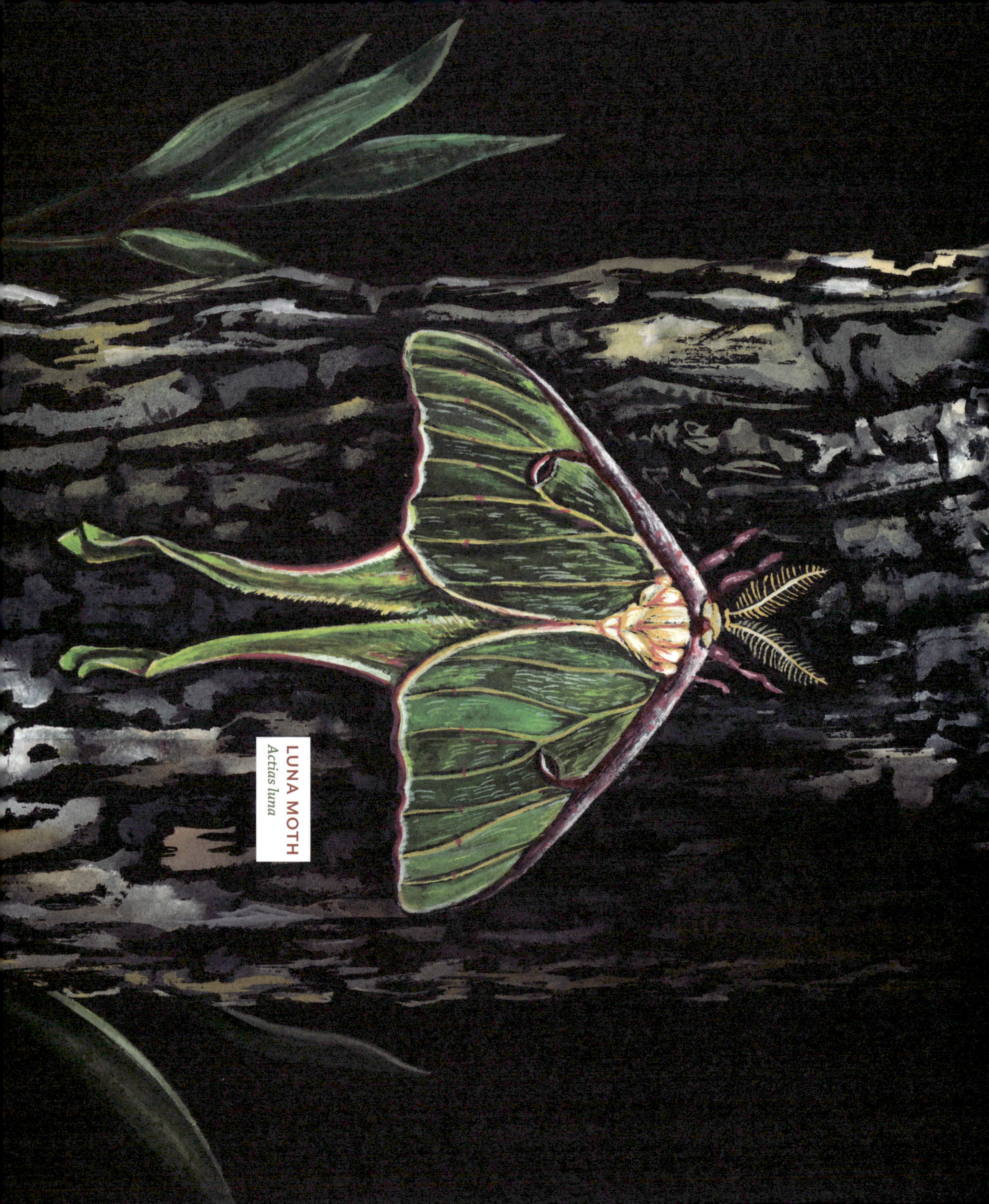

LUNA MOTH
Actias luna

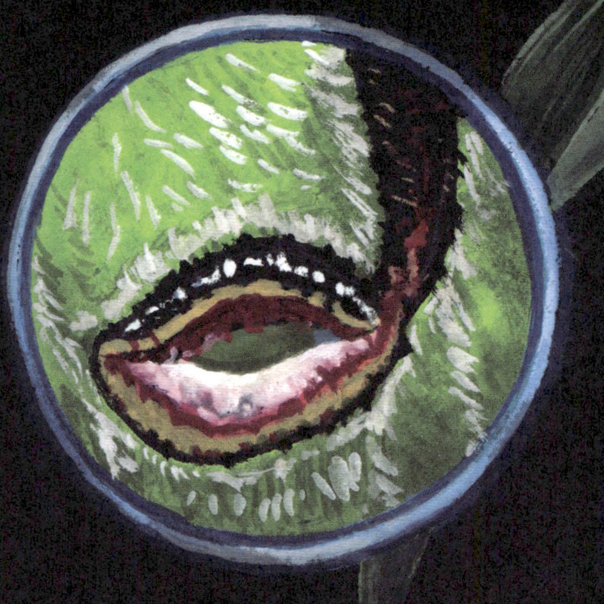 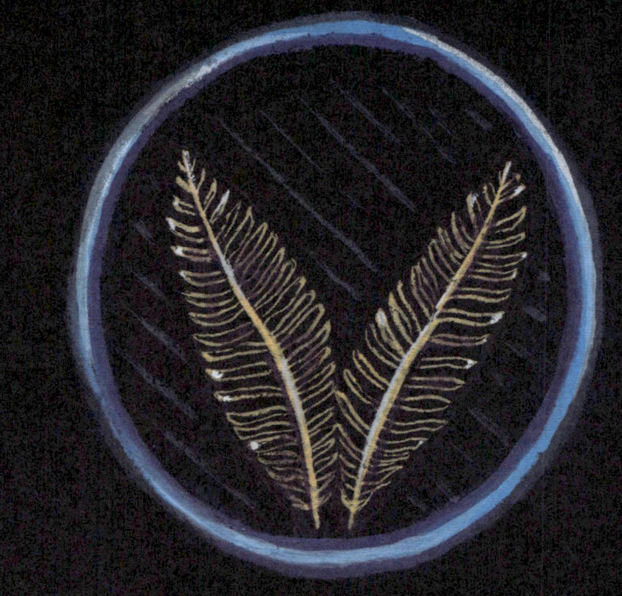

Orchard Angel

APRIL 21 | TESSENTEE BOTTOMLAND PRESERVE, NC

I walk through an overgrown apple orchard near an eerie abandoned house, both remnants of an old homestead that has recently been turned into a nature preserve. I inhale the sweet purple-grape scent of Carolina allspice flowers, then scan the meadow floor in the hope of finding some morel mushrooms, an edible Appalachian delicacy. But a treasure of a different kind is in store for me today: a lime-green luna moth hangs from the bark at the base of an apple tree!

Luna moths live as adults for a maximum of ten days; I'd previously found them only after they had died. But this one is flawless, with not a single tear in its papery wings. It doesn't move or react to my presence, so I sit next to it and take in every detail: the fine hairs covering the wings, the striking pink-and-black eyespots, and the long hind wings, poised like an elegant ballerina standing on her toes.

TOP LEFT: *Feather-like antennae to detect food and potential mates* **TOP RIGHT:** *Bold eyespots to intimidate predators*

Fungal Friends on the Appalachian Trail

APRIL 22 | SILER BALD, NC

If you're into mushroom hunting, the southern Appalachians are the place for you. Scientists have identified nearly 2,300 species of fungi and estimate that there could be as many as 20,000 species hiding within these mountains. While mushroom growth really takes off in May and June in the region, I find a few specimens camouflaged against the leafy forest floor. The first ones to catch my eye are a pair of earthstar mushrooms tucked under a root in the middle of the trail, looking like two baby octopuses playing hide-and-seek.

As I hunch down to take photos and get a closer look, a thru-hiker approaches and together we marvel at these strange fungal structures. We gently tap the center puffballs and watch a cloud of spores erupt from each. I'm naturally shy and tend to keep to myself while hiking, but I'm learning that beautiful moments of connection can arise when I speak up about something interesting. Noticing is contagious!

The next mushroom to capture my attention is a four-inch-tall morel. These brain-like fungi are a harbinger of spring in many parts of the Appalachian Mountains. When I was growing up, we'd sometimes hunt for morels on Easter Sunday instead of hunting for eggs. This specimen is standing right beside the

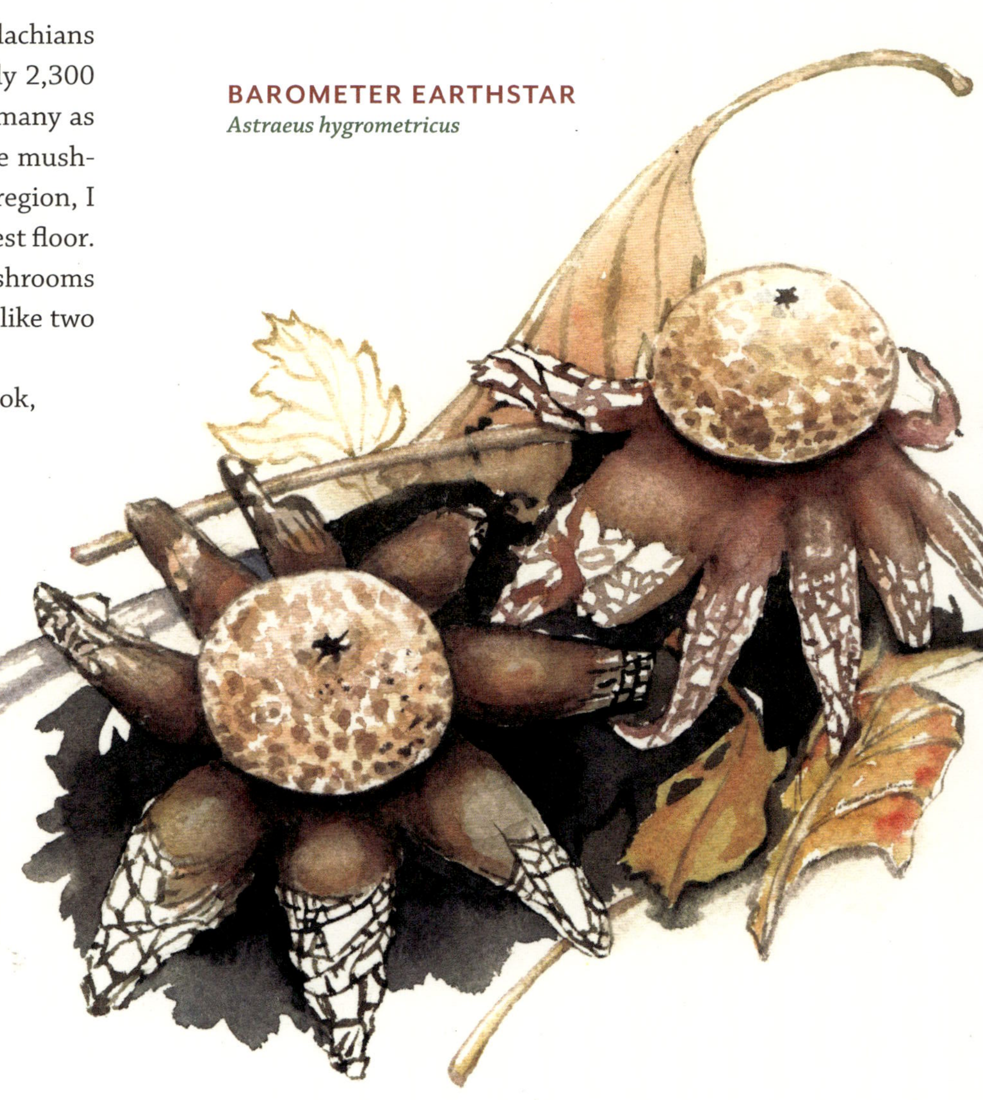

BAROMETER EARTHSTAR
Astraeus hygrometricus

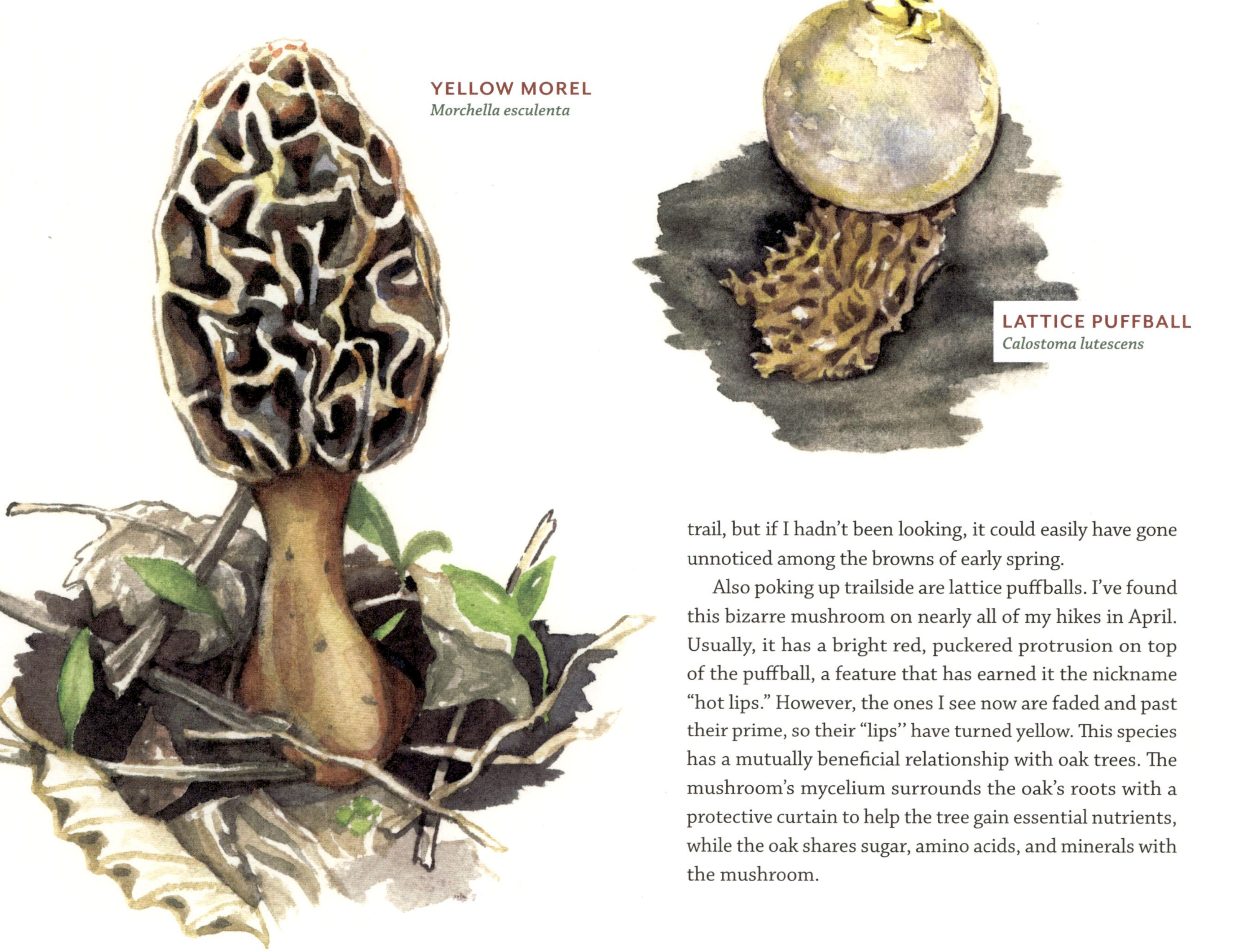

YELLOW MOREL
Morchella esculenta

LATTICE PUFFBALL
Calostoma lutescens

trail, but if I hadn't been looking, it could easily have gone unnoticed among the browns of early spring.

Also poking up trailside are lattice puffballs. I've found this bizarre mushroom on nearly all of my hikes in April. Usually, it has a bright red, puckered protrusion on top of the puffball, a feature that has earned it the nickname "hot lips." However, the ones I see now are faded and past their prime, so their "lips" have turned yellow. This species has a mutually beneficial relationship with oak trees. The mushroom's mycelium surrounds the oak's roots with a protective curtain to help the tree gain essential nutrients, while the oak shares sugar, amino acids, and minerals with the mushroom.

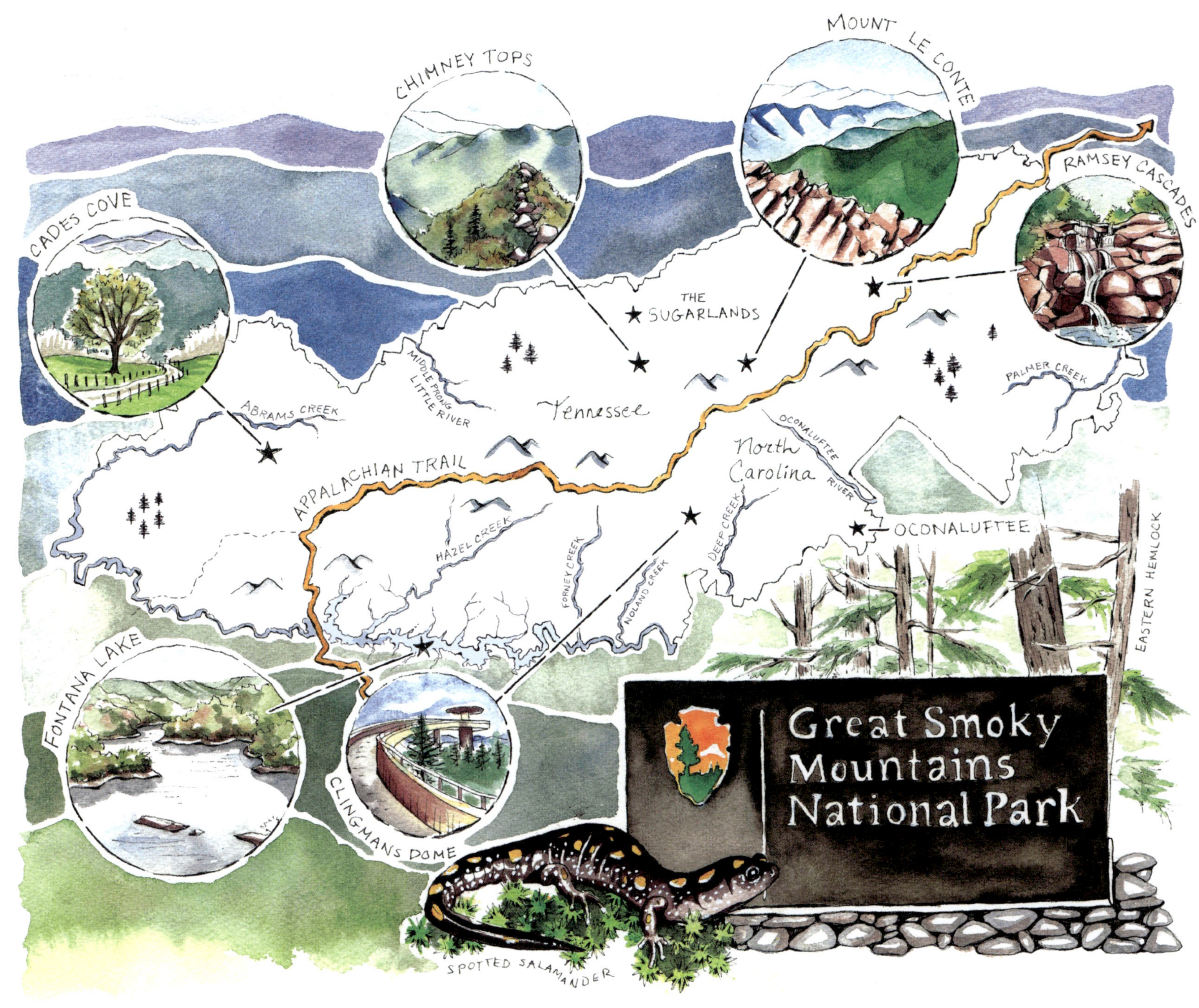
CHIMNEY TOPS
MOUNT LE CONTE
RAMSEY CASCADES
CADES COVE
THE SUGARLANDS
MIDDLE PRONG LITTLE RIVER
Tennessee
PALMER CREEK
ABRAMS CREEK
OCONALUFTEE RIVER
North Carolina
APPALACHIAN TRAIL
HAZEL CREEK
FORNEY CREEK
NOLAND CREEK
DEEP CREEK
OCONALUFTEE
EASTERN HEMLOCK
FONTANA LAKE
CLINGMANS DOME
Great Smoky Mountains National Park
SPOTTED SALAMANDER

GREAT SMOKY MOUNTAINS NATIONAL PARK

Tucked tenderly into the lush southern Appalachians, where it spans the border of Tennessee and North Carolina, Great Smoky Mountains National Park encompasses most of its namesake subrange. This national park, the most visited in the United States, boasts expansive views from high points like Clingmans Dome and Mount Le Conte, but it is also well worth the pilgrimage for those who delight in tiny worlds. The park is believed to be the most biodiverse in the entire national parks system, with over 19,000 described species of plants and animals. It remains a safe haven for synchronous fireflies, spring ephemerals, a treasure trove of fungi, and thirty different species of adorable salamanders—enough to earn it the nickname "Salamander Capital of the World."

In 2018, I was fortunate enough to spend six weeks living within the park as the artist-in-residence. My residency lasted through the month of June, which is when the synchronous fireflies perform their annual mating ritual—a spectacular display. This meant that I got to volunteer for the firefly-viewing events. Over two weeks each summer, people from all over the world flock to the Smokies to sit on picnic blankets in the hardwood forest and watch the male insects flash their lights as they zoom between the trees, their flashes becoming progressively more in sync as it gets closer to midnight. Once the females are thoroughly impressed by the synchronicity, they will flash twice from their places on the ground to signal that they are ready to mate. This phenomenon is known to occur in only a few parts of the world, and as I glimpsed the faces of each visitor in the moonlight, our shared smiles and hushed voices reflected our knowledge that this was truly a sacred experience.

Smoky Mountain Stunners

APRIL 23 | GREAT SMOKY MOUNTAINS NATIONAL PARK, TN

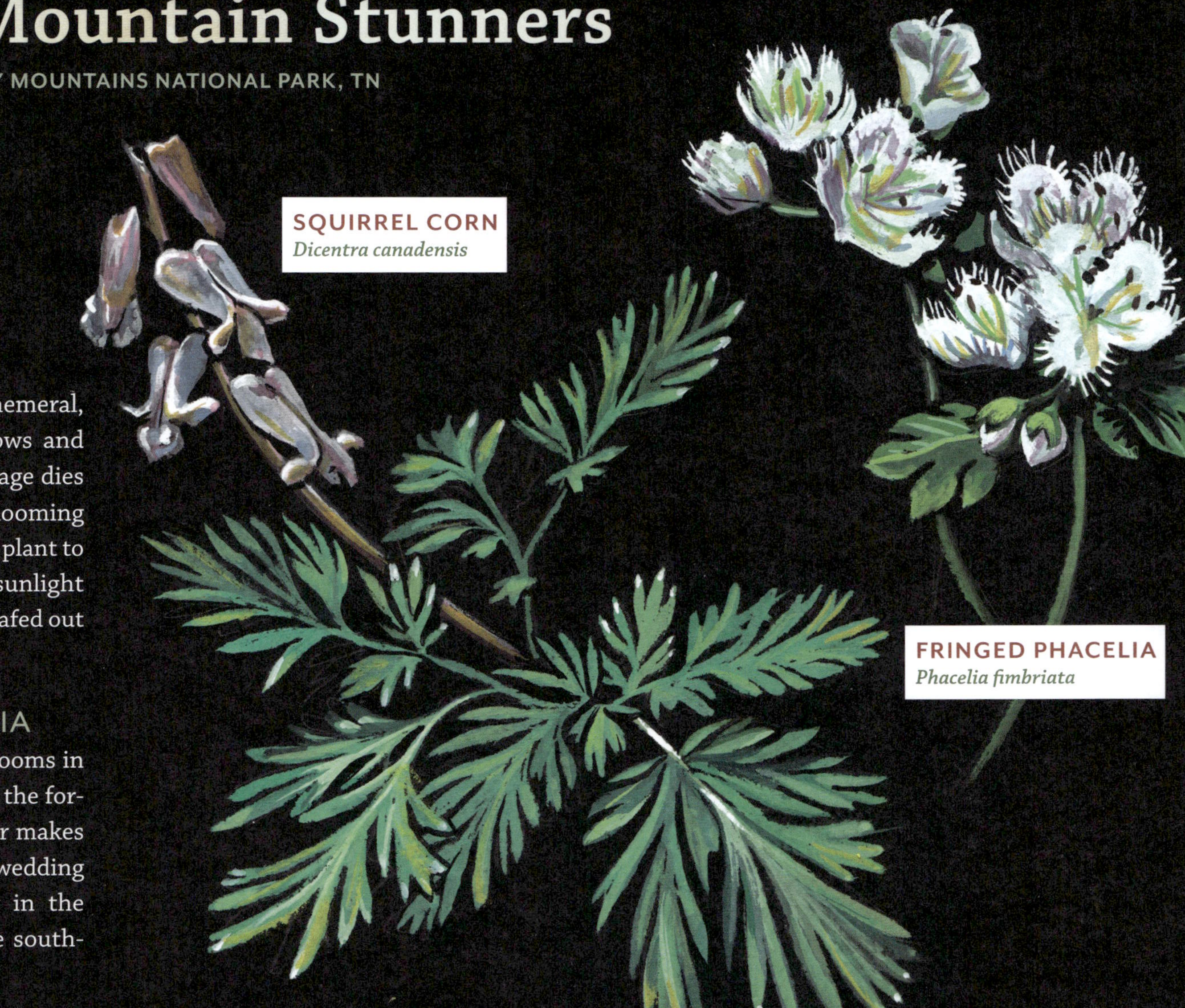

SQUIRREL CORN

This plant is a spring ephemeral, which means that it grows and blooms early and the foliage dies back in early spring. Blooming early allows this perennial plant to access huge amounts of sunlight before all the trees have leafed out and created shade.

FRINGED PHACELIA

When fringed phacelia blooms in thick white blankets over the forest floor, this fancy flower makes the landscape look like a wedding venue. It is found only in the wooded highlands of the southeastern United States.

SHOWY ORCHID

Juvenile showy orchids don't have their own food reserves, so they must form a mycorrhizal partnership with fungi in the genus *Ceratobasidium*. The fungi provide carbohydrates to the plants, while the plants provide moisture and organic matter to the fungi. Without this partnership, the showy orchid's seeds couldn't establish.

YELLOW TROUT LILY

A spring ephemeral that grows in large colonies, this flower gets its common name from the markings on its leaves, which are similar to those of brook and brown trout.

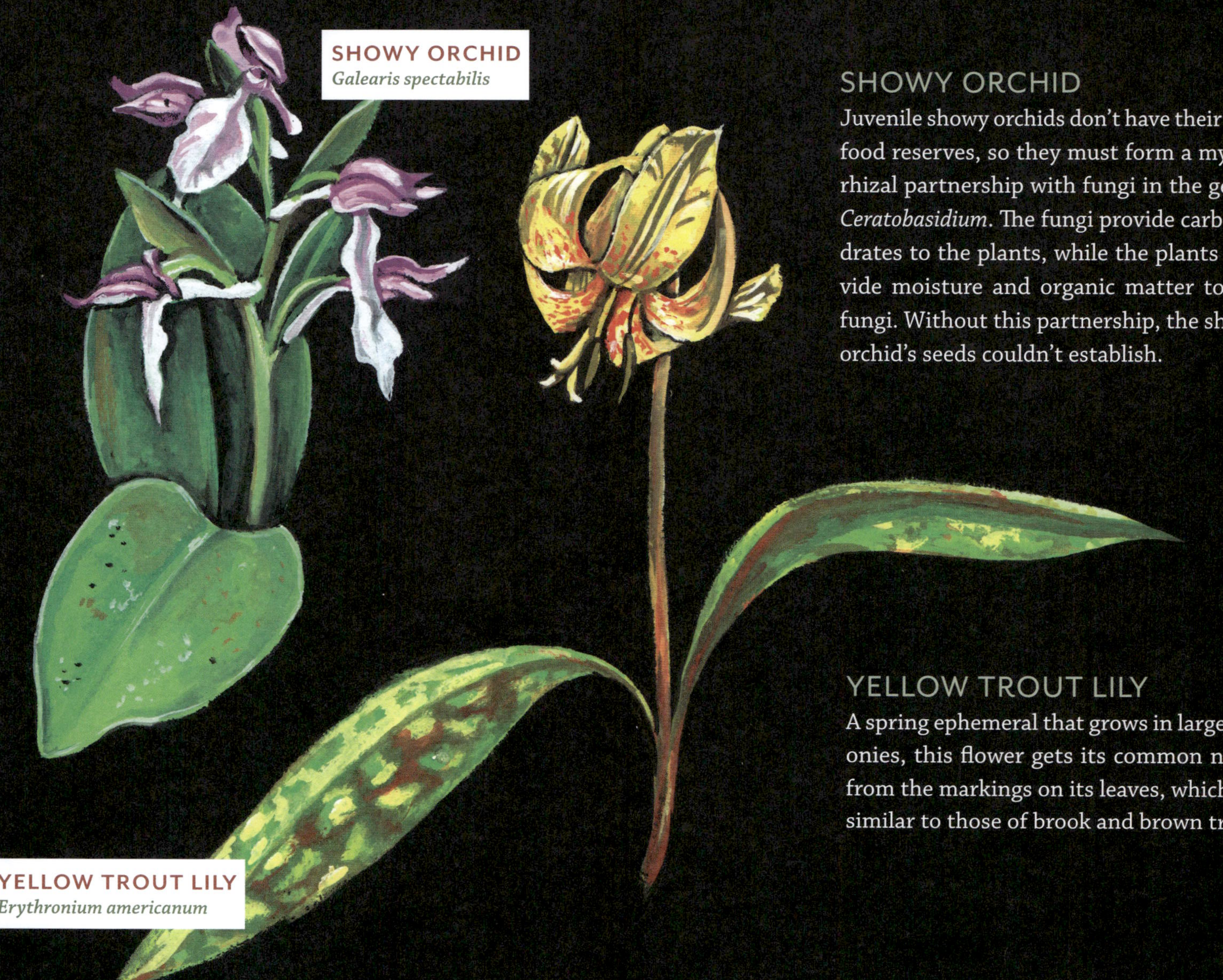

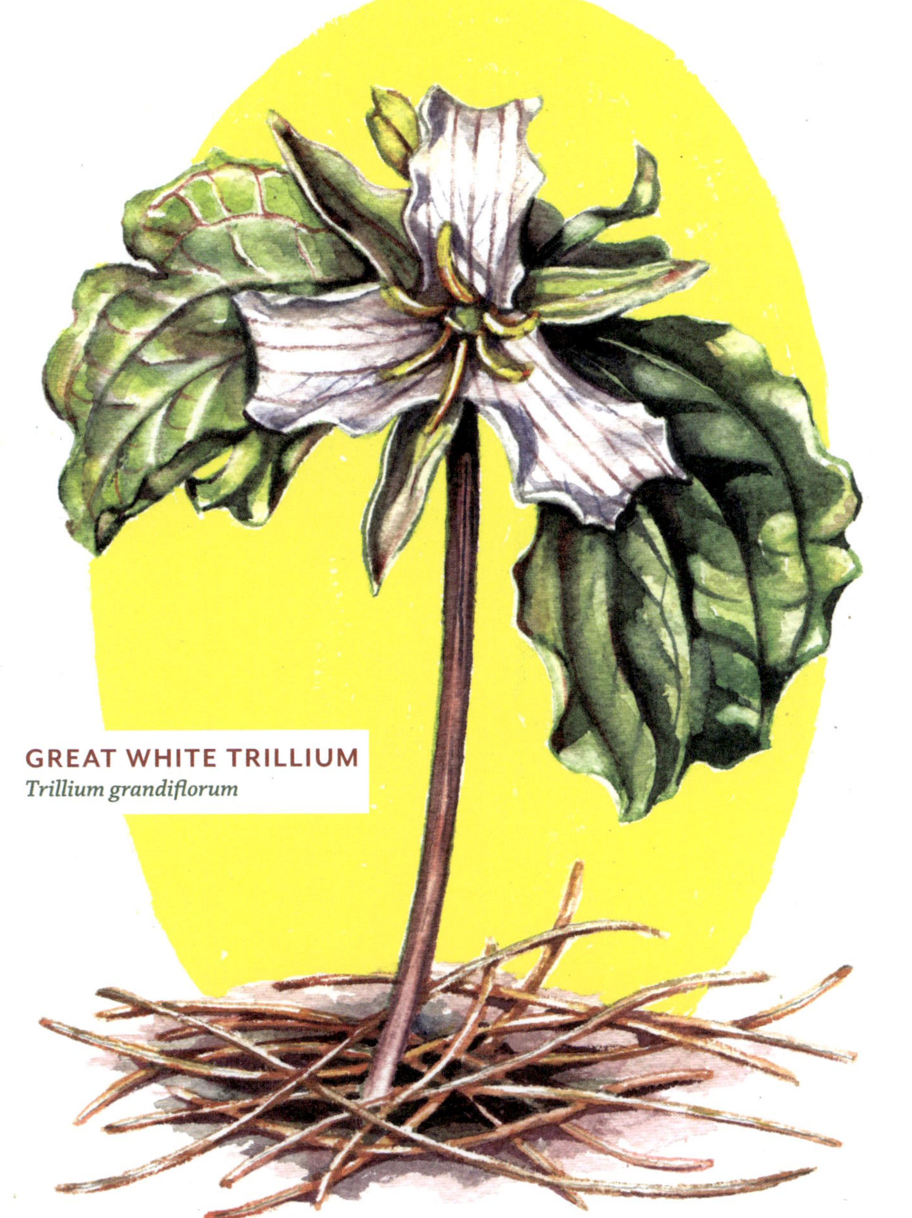

Seed Scatterers

APRIL 24 | GREAT SMOKY MOUNTAINS NATIONAL PARK, TN

We pass through a field of great white trilliums in a partially shaded forest cove. These flowers initially bloom with stark white petals that gradually turn light pink with age. Later in the summer, a single bright red fruit filled with seeds will appear, and ants will play an important role in dispersing the seeds from this fruit. Each trillium seed holds an elaiosome, a small lipid-and-protein-filled packet that represents an excellent food source for the ants' young. The ants carry the fruit into their underground kingdoms, and after the babies are all fed, they then discard the seeds in the colony's dedicated waste pile (ants are very organized!). This waste pile, filled as it is with rich ant excrement, is the perfect place for seed germination. So the next time you see a lovely trillium on an early spring walk, thank an ant.

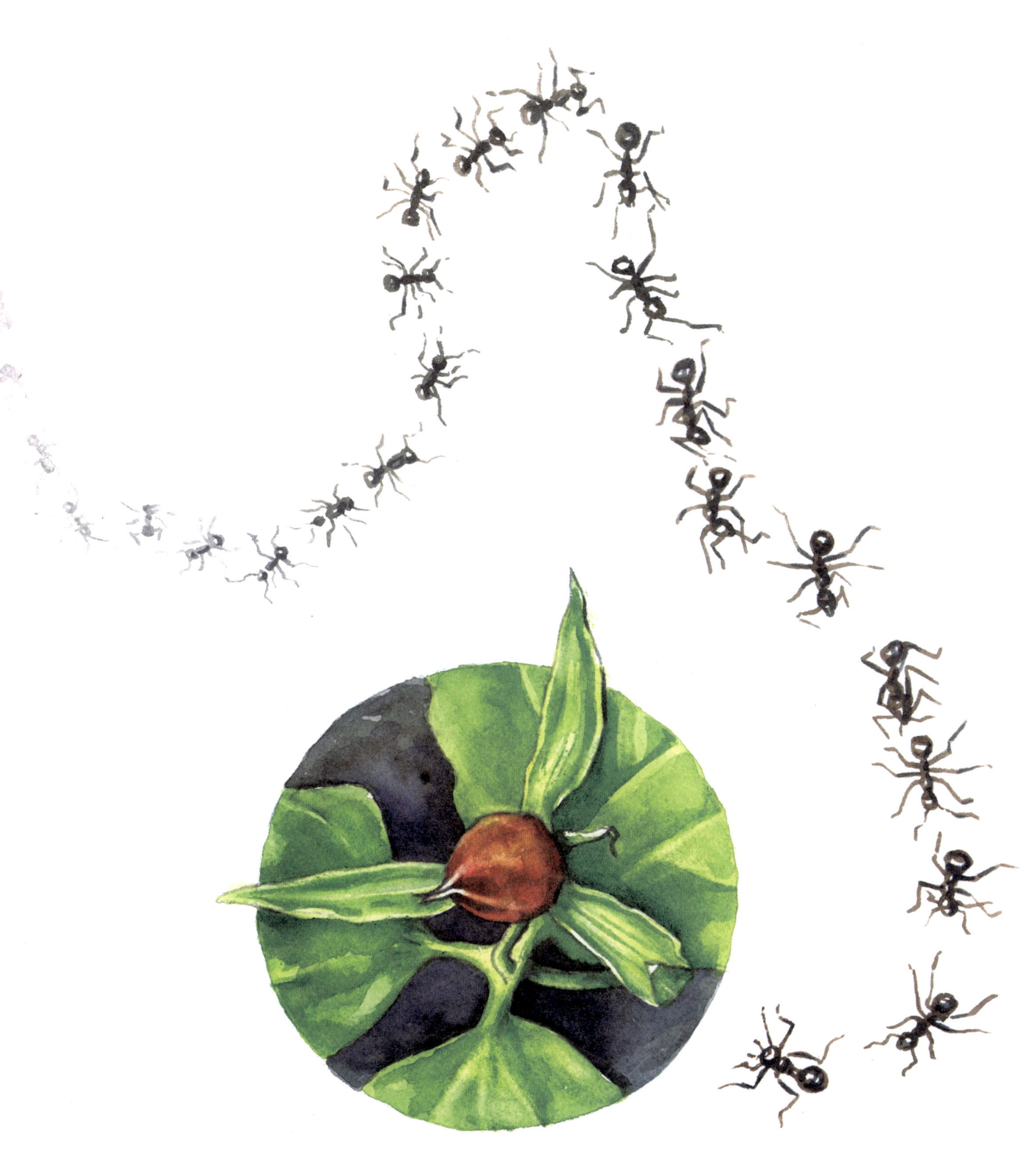

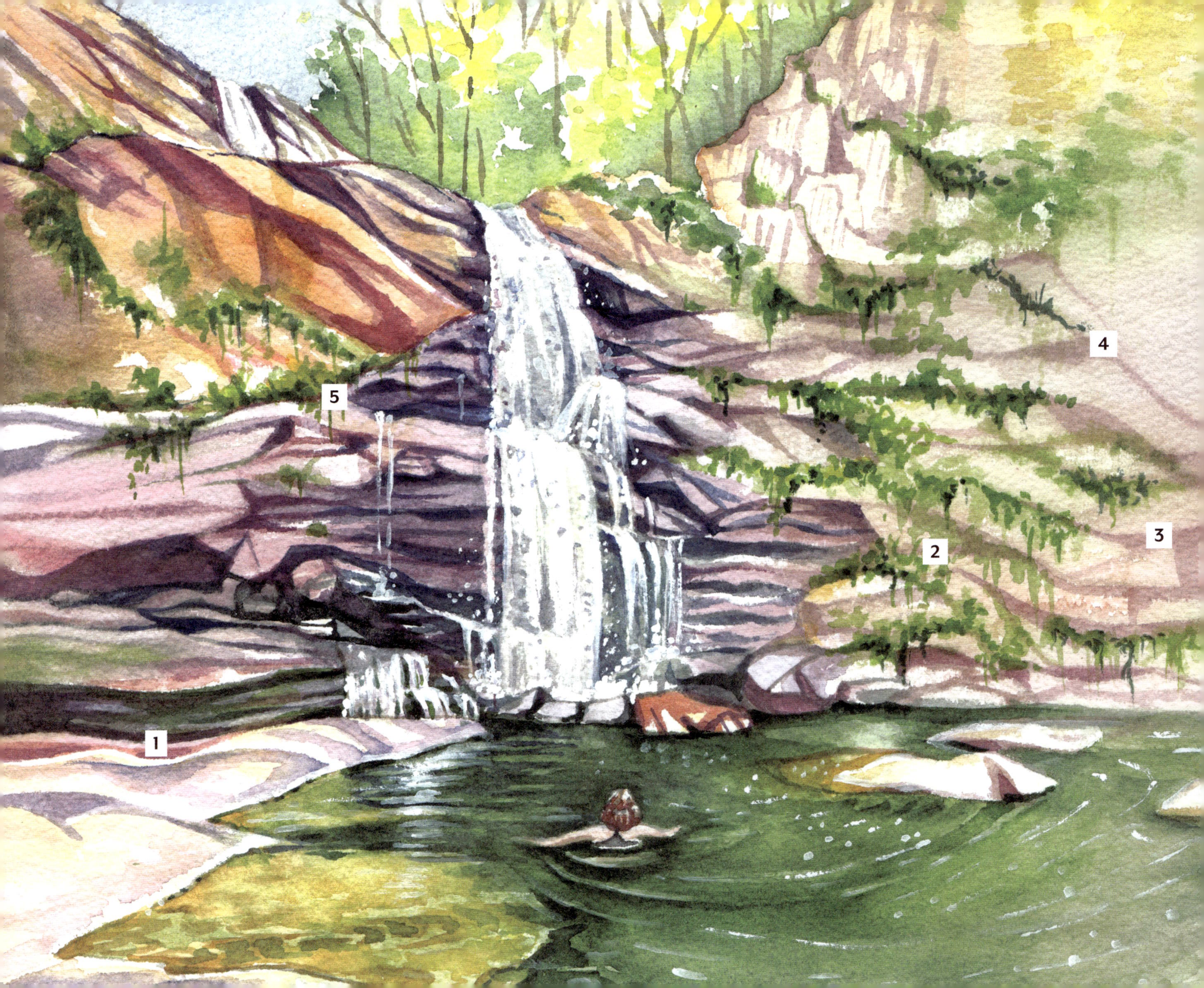

1
2
3
4
5

Chasing Vertical Bogs

APRIL 25 | CULLOWHEE, NC

Oh, the intoxicating sound of water tumbling down a cliff! I've always enjoyed soaking up a waterfall's misty magic, but this is my first time looking intently for the plants and animals that make their homes nearby. Surrounding this waterfall are boulders and rock ledges teeming with various shades of green growth. Over time, damp soil has built up on these seepy ledges, inviting moisture-loving plants like mosses, liverworts, club mosses, and the carnivorous roundleaf sundew to take root. These crevices also make nice homes for critters like salamanders.

I climb the slick rocks alongside a local botanist to inspect the plants, moving very carefully so as not to slip or step on any delicate vegetation. Using a hand lens to examine the tiniest specimens, we identify snakeskin liverwort, dogtooth lichen, and rock club moss. Many of the plants have a slight reddish tinge to them, which is due to the presence of a chemical compound called anthocyanin. This pigmentation protects the plants against UV damage while they are still extra sensitive in early spring.

The next few pages offer a zoomed-in look at some of the little wonders hiding out beside this impressive waterfall.

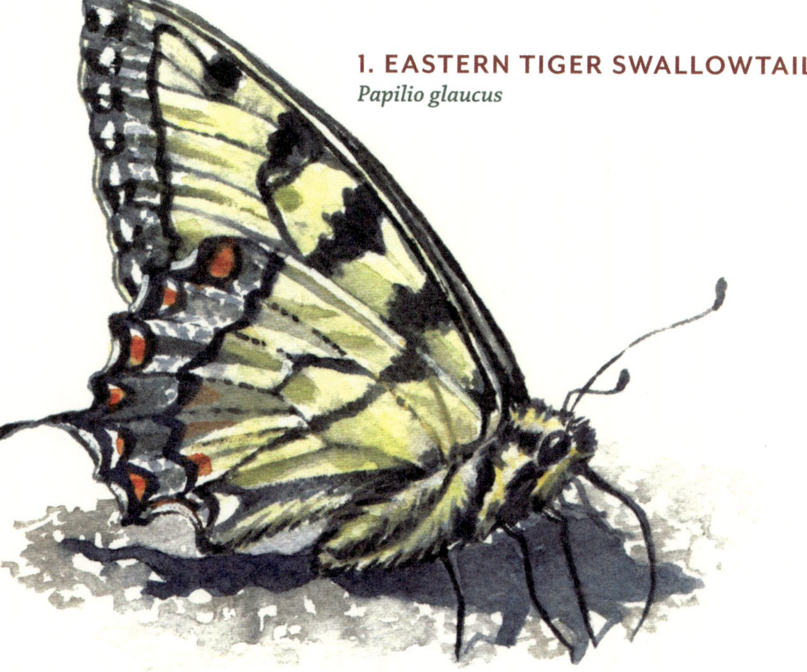

1. EASTERN TIGER SWALLOWTAIL
Papilio glaucus

EASTERN TIGER SWALLOWTAIL

A cluster of five eastern tiger swallowtail butterflies is gathered to engage in an activity called "puddling," in which a butterfly uses its proboscis—a tubelike mouth structure—to consume salts and other nutrients that have leached from the soil into a puddle on the ground.

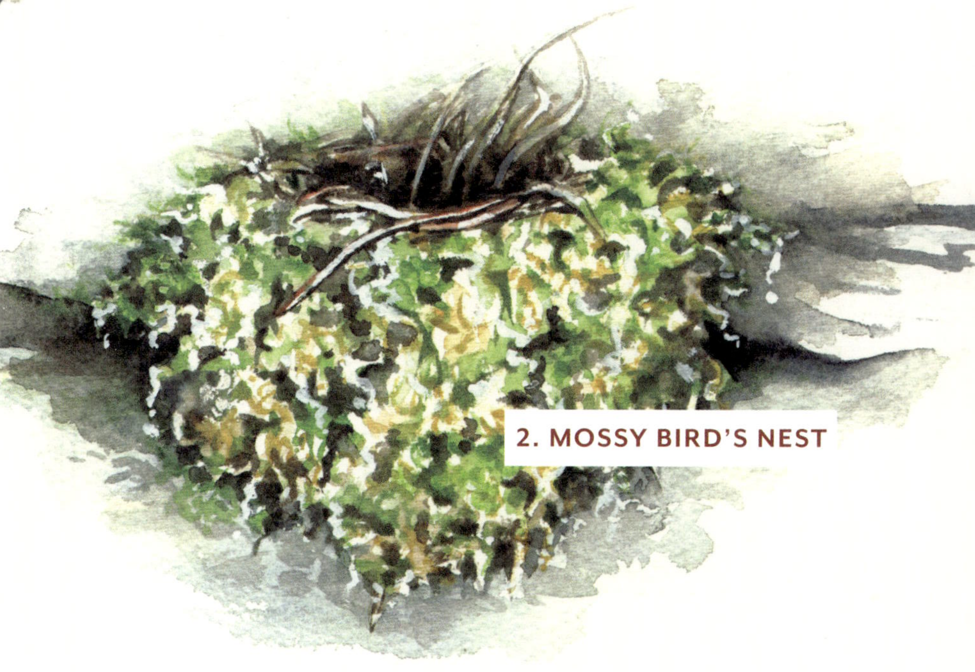

2. MOSSY BIRD'S NEST

MOSSY BIRD'S NEST

I find a well-hidden bird's nest made up of several colors of moss and pine needles tucked away under a shady rock ledge. While I'm not sure what kind of bird the nest belongs to, it's definitely a small species since the nest is only about three inches wide.

ROCK CLUB MOSS

This fuzzy club moss, an imperiled species in North Carolina, is perched high up on the wet cliffs. Club mosses are small evergreen herbs that grow close to the ground, reaching only a few inches tall.

DUSKY SALAMANDER

In the same kind of shadowy crevice where I find the mossy bird's nest, I catch the slightest movement as this salamander positions its body right up against the rock wall for a perfect hiding spot. When summer arrives, the female dusky will lay her eggs in moss or secure them to the underside of a rock or in another secret spot near the water. She will keep watch over her eggs until they are ready to hatch.

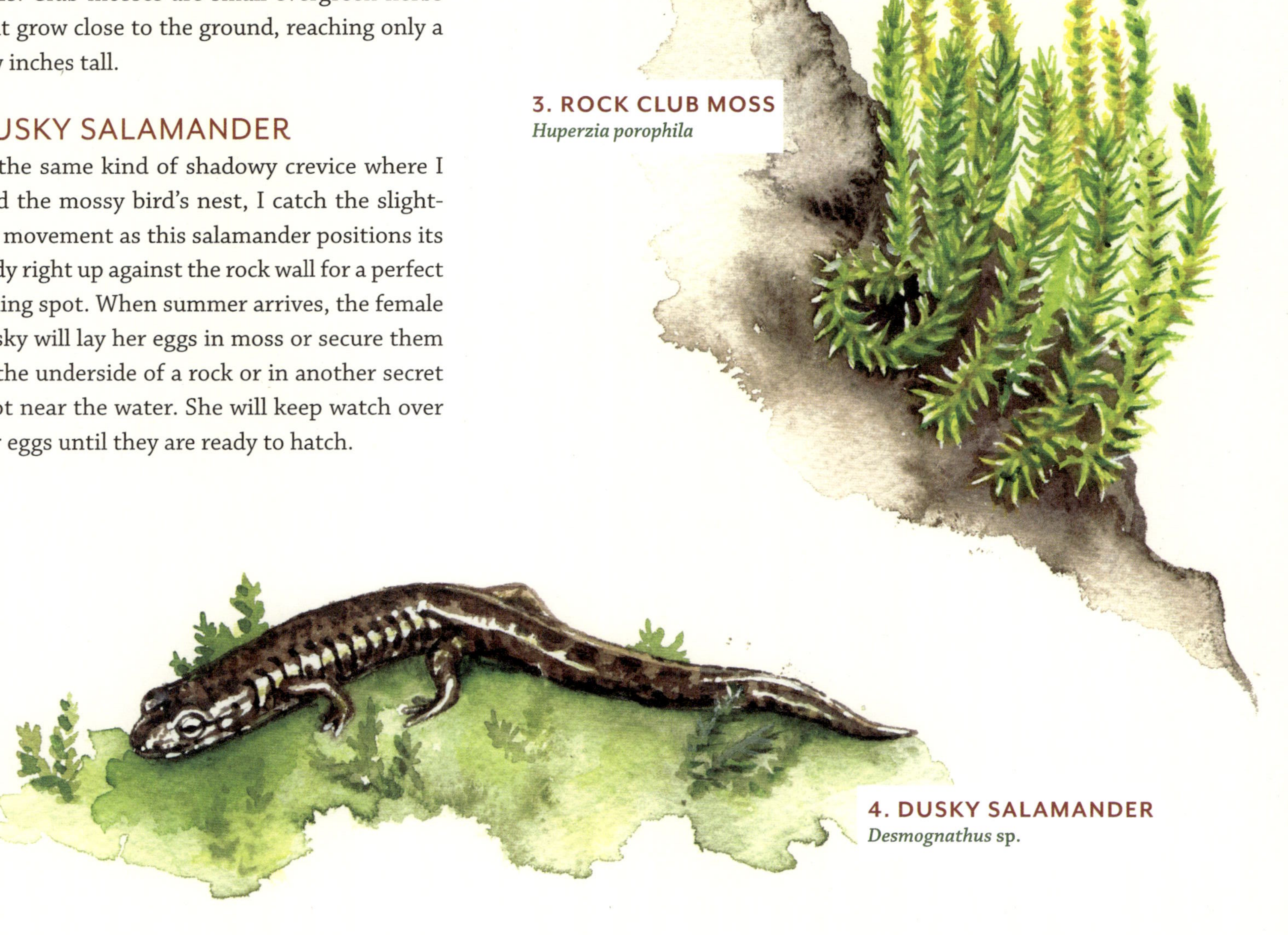

3. ROCK CLUB MOSS
Huperzia porophila

4. DUSKY SALAMANDER
Desmognathus sp.

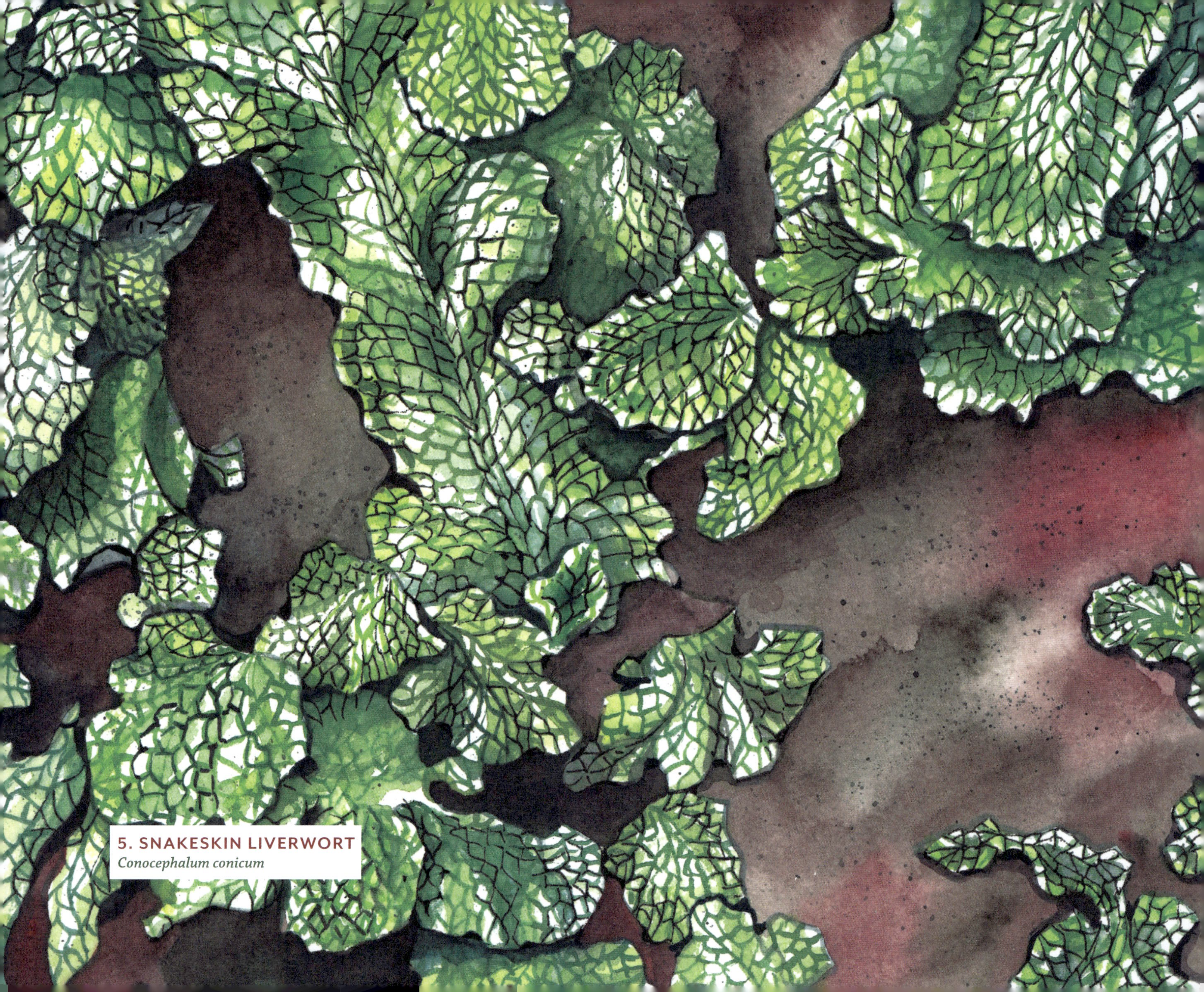
5. SNAKESKIN LIVERWORT
Conocephalum conicum

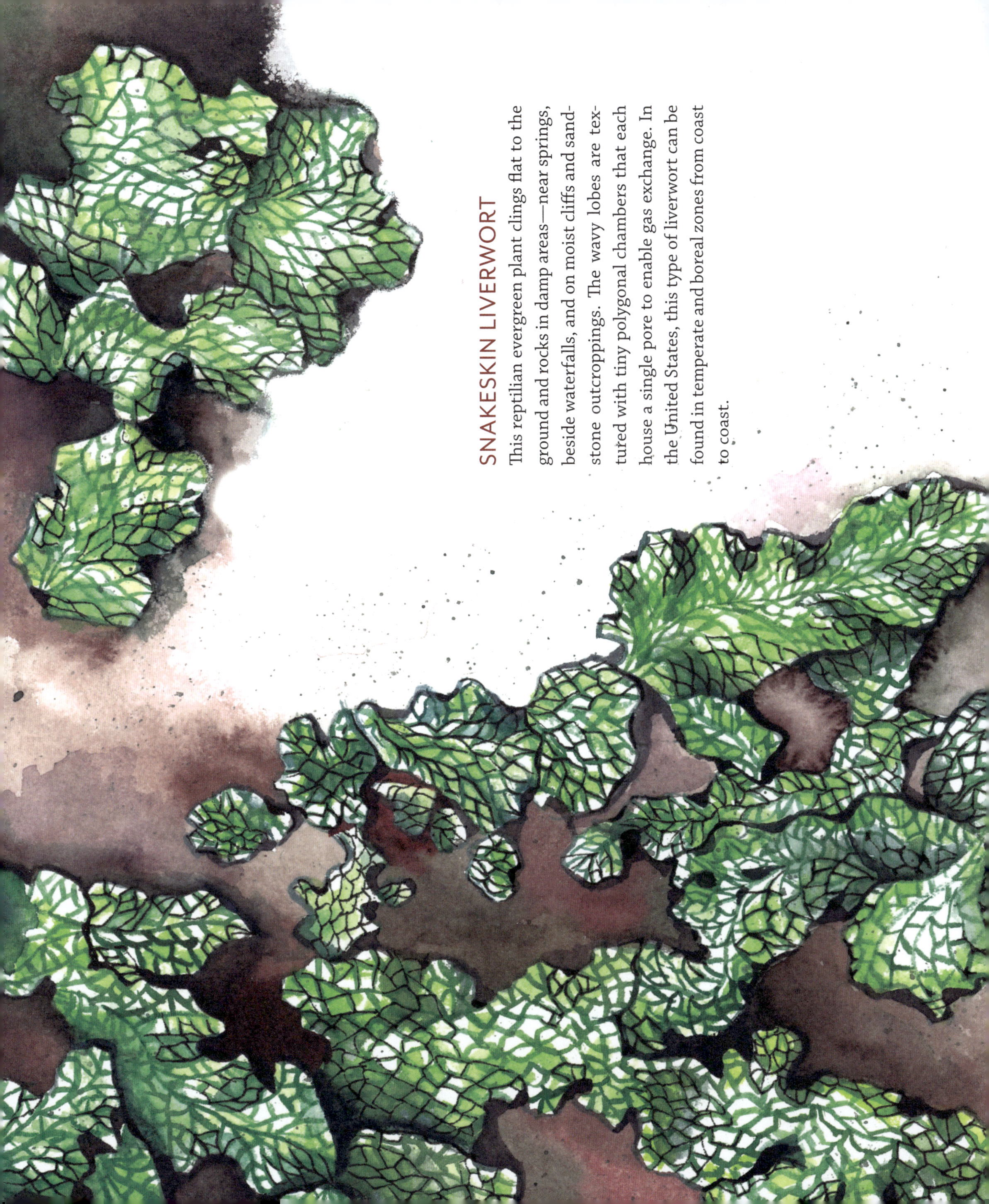

SNAKESKIN LIVERWORT

This reptilian evergreen plant clings flat to the ground and rocks in damp areas—near springs, beside waterfalls, and on moist cliffs and sandstone outcroppings. The wavy lobes are textured with tiny polygonal chambers that each house a single pore to enable gas exchange. In the United States, this type of liverwort can be found in temperate and boreal zones from coast to coast.

ENTERING THE CENTRAL BLUE RIDGE

MAY 1

If I had to describe this stretch of the Appalachians in one word, it would be *cozy*. The temperate rainforest ecosystem creates a lush understory so dense that sometimes the trails transform into tunnels through thickets of rhododendron and mountain laurel. Lush layers of moss blanket rocks and logs, and the steep, partially shaded banks of stream hollows create the perfect environment for delicate spring ephemerals. A wealth of spring-fed rivers, streams, and wetlands are home to freshwater critters of all colors, shapes, and sizes, including the famously awkward-looking hellbender salamander, as well as Seussian flora like swamp pink. And when you view the mountains from a distance, they give off a hazy, bluish hue, the result of the interaction of sunlight with an organic chemical released by conifer trees, which scatters the blue light.

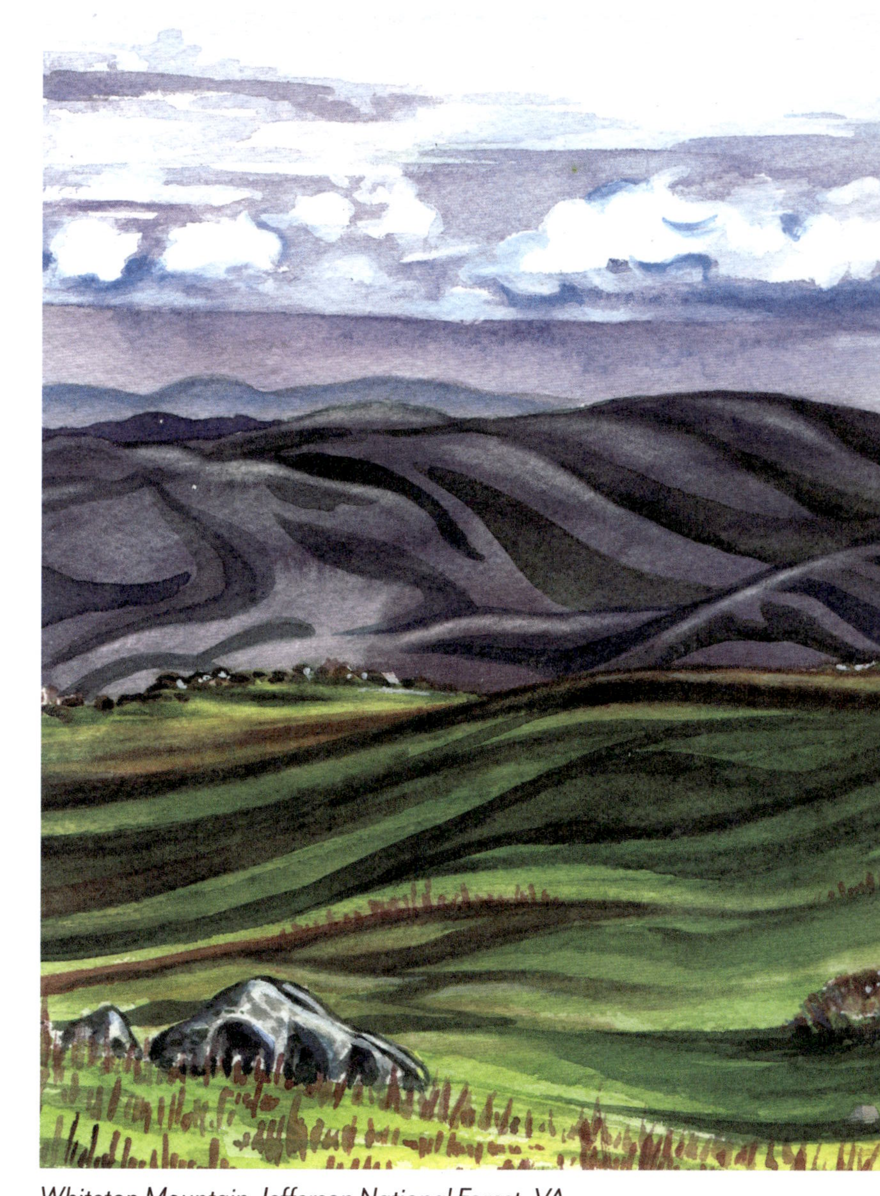

Whitetop Mountain, Jefferson National Forest, VA

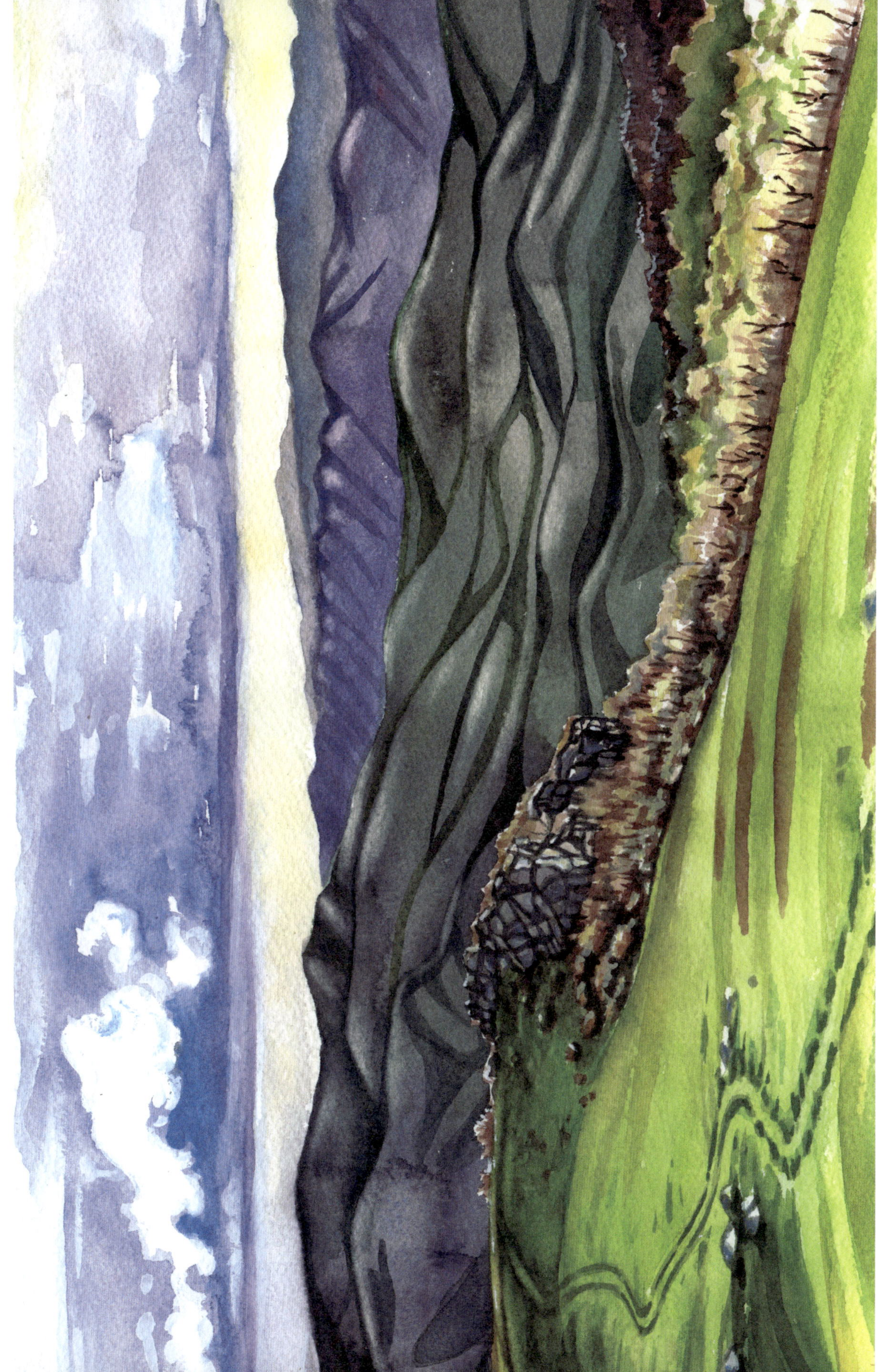

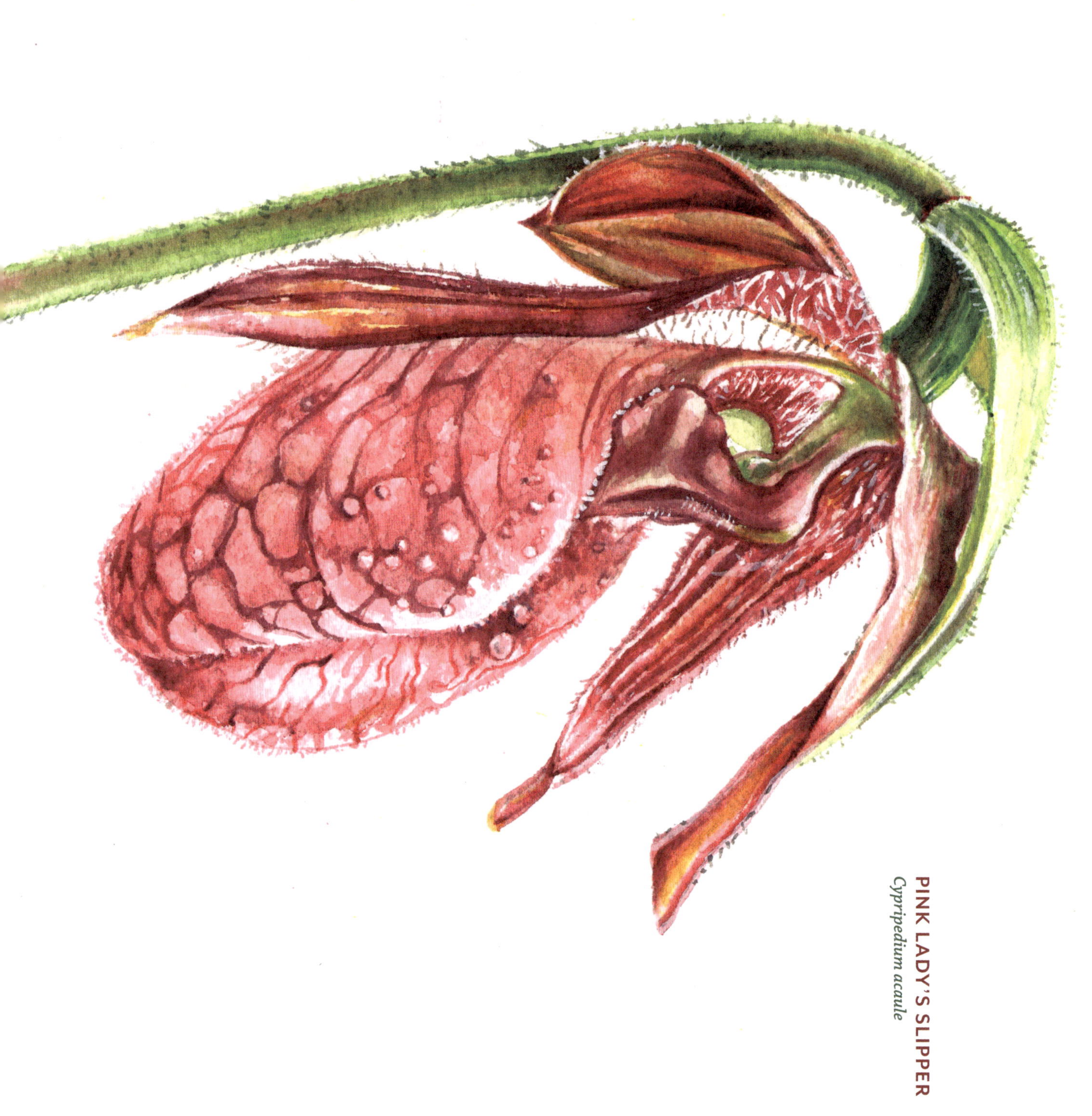

PINK LADY'S SLIPPER
Cypripedium acaule

Delicate Dancer

MAY 6 | PRIEST WILDERNESS, VA

The trail carries us through a grove of pines. Perched on the steep bank above the path is a single pink lady's slipper flower, a perfectly poised dancer performing a solo recital. Drops of dew sit on the leaves and sparkle in the morning light. These native orchids thrive only in acidic soil and in limited climates, which means that finding one in the woods is a somewhat rare treat. A symbol of femininity in many cultures around the world, the sweet-smelling lady's slipper has historically been used to create floral essences to enhance desire.

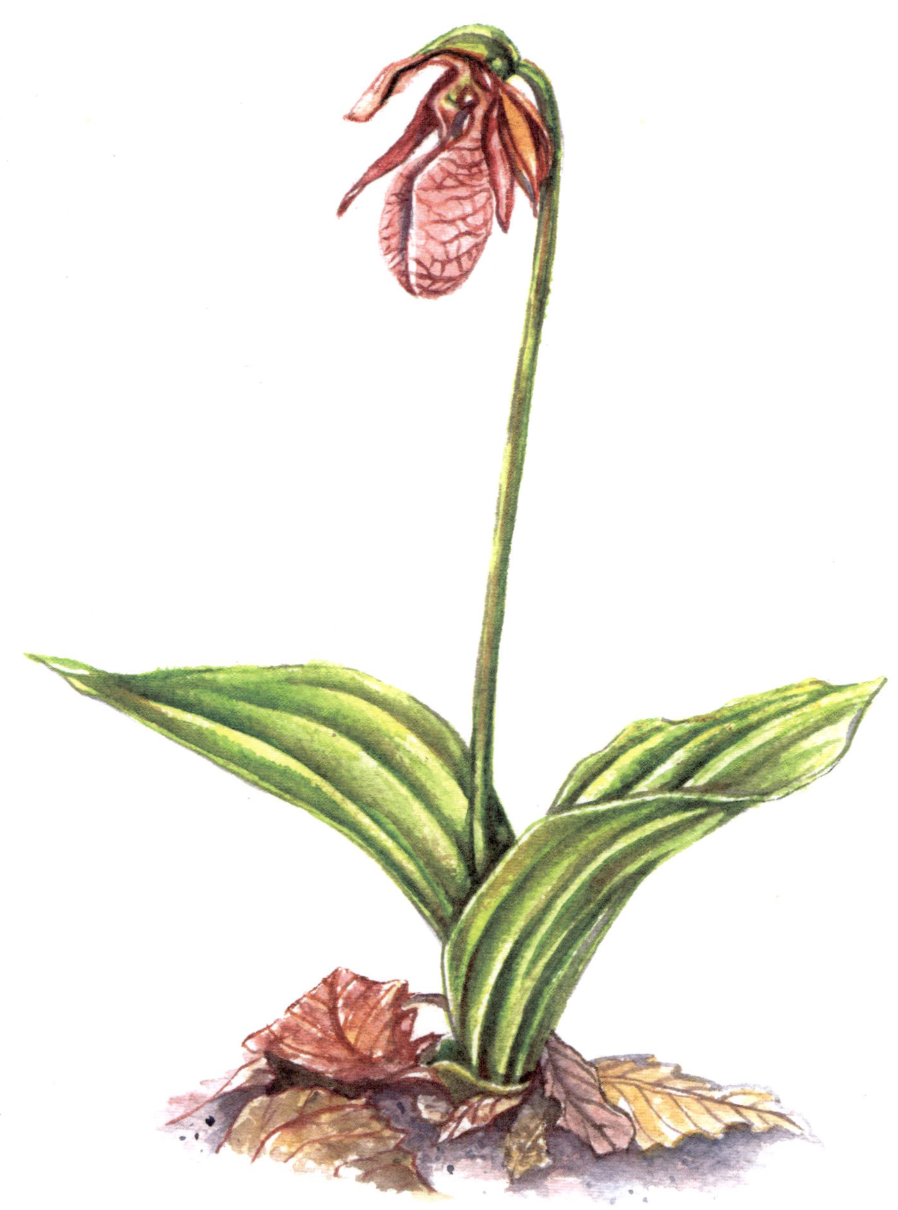

A Beauty All Your Own

My reflection finds me wherever I go. I see it above the sink in the coffee shop bathroom, in the glassy surface of storefront windows. It's there in the rearview mirror or in the corner of my computer screen during a video call. And each time I meet my own eyes in my reflection, I find myself assessing the way I look.

I've had a tumultuous relationship with my appearance throughout my life. My exterior shell features red hair, acne-scarred skin, a bumpy nose that runs in our family, thin lips, and a scrawny body that won me the role of Village Boy in the school play and hurtful nicknames from boys in my class. Like a butterfly stuck in a chrysalis purgatory, I kept waiting for my body to transform into the voluptuous woman who I thought would finally be seen as beautiful. But that day never came.

As adults, we collectively refer to our teenage years as that horrendous chapter where we were all so insecure. As we get older, however, we're expected to shed the bulk of those worries, to become self-confident and self-assured. But in my experience, those feelings of insecurity don't just magically disappear. I'm in my late twenties now and have a fulfilling career and a loving partner, yet on many days I still feel bogged down with insecurity. And even though my acne has faded and I feel more at home in my boyish frame, I find that I'm hyperaware of every tiny new line that carves its way deeper into my flesh with each passing year.

In the woods, there's freedom from this relentless self-scrutiny. After a glorious day and night backpacking through Virginia's Priest Wilderness, I stop thinking about my physical appearance altogether. My muscles feel sturdy as I hike the steep footpath, a steady drizzle soaking through my clothes, my whole body surging with life.

I wonder if everyone feels more beautiful when they are outside, immersed in the organic messiness of nature. As I walk beneath crooked branches, admiring a moth with a ripped wing or the flaming-red head of the pileated woodpecker, my own natural imperfections feel normal, even lovely. Our cities, however, are constructed with symmetry, clean angles, matching colors, and machine-fabricated textures. Whenever I find myself in an urban setting, I feel an urge to match this engineered level of perfection just to belong. I'm tempted to smooth the texture of my pores with concealer, wear clothes that shape and enhance, smear on eye makeup to make my face look more symmetrical, and whiten my teeth to dazzle under fluorescent light.

And sure enough, when I step into a gas station bathroom on the drive back from the trail, the confidence I felt during the hike vanishes as I catch my reflection in the mirror. I am reminded of those hard-to-shake insecurities once again and am suddenly disgusted by my greasy hair and dirt-speckled skin.

As an artist, I deliberately seek out aspects of the natural world that aren't seen as conventionally beautiful—slime molds, moths, salamanders. In fact, the more unusual the plant's or animal's appearance is, the more compelled I am to capture it with my paintbrush. So why, then, is it so hard to appreciate the subtle beauty of my own body? I have trained myself to find beauty in every facet of my surroundings. Why am I so blind to my own?

I take a longer look in the mirror and try to imagine that I am the next tiny creature to be painted for this book. When I examine a subject closely before I start to paint it, I think about its texture, form, and color. Now I try my hardest to see my features through the objective eyes of an artist. It is difficult at first, but slowly a new perspective takes hold:

The gentle waves and copper tint of the hair, which appears brown in shadow but vibrant orange when it hits the light, will be perfect for adding movement to the painting.

The nose's ridge, its curve and bend, has an elegance all its own. And don't forget the little divot at its tip. Oh, I do love the challenge of painting a unique shape!

The two parallel creases between the eyebrows provide a helpful indicator of the subject's age—youthful but slowly maturing, with lines that arise from years of laughing and squinting into the afternoon sun. I'll mix up some darker paint for those shadows.

A pleasing thought occurs to me: I wonder if, the more time I spend in the woods, the more I'll look like the woods. After we've spent years and years together, will my skin have the delicate, wind-blown texture of paper birch bark? Will my eyes reflect the sky the way that ponds and puddles echo the shapes of clouds? Will the veins of my hands resemble the branches of all the trees I've stood beneath? I'd like to think so.

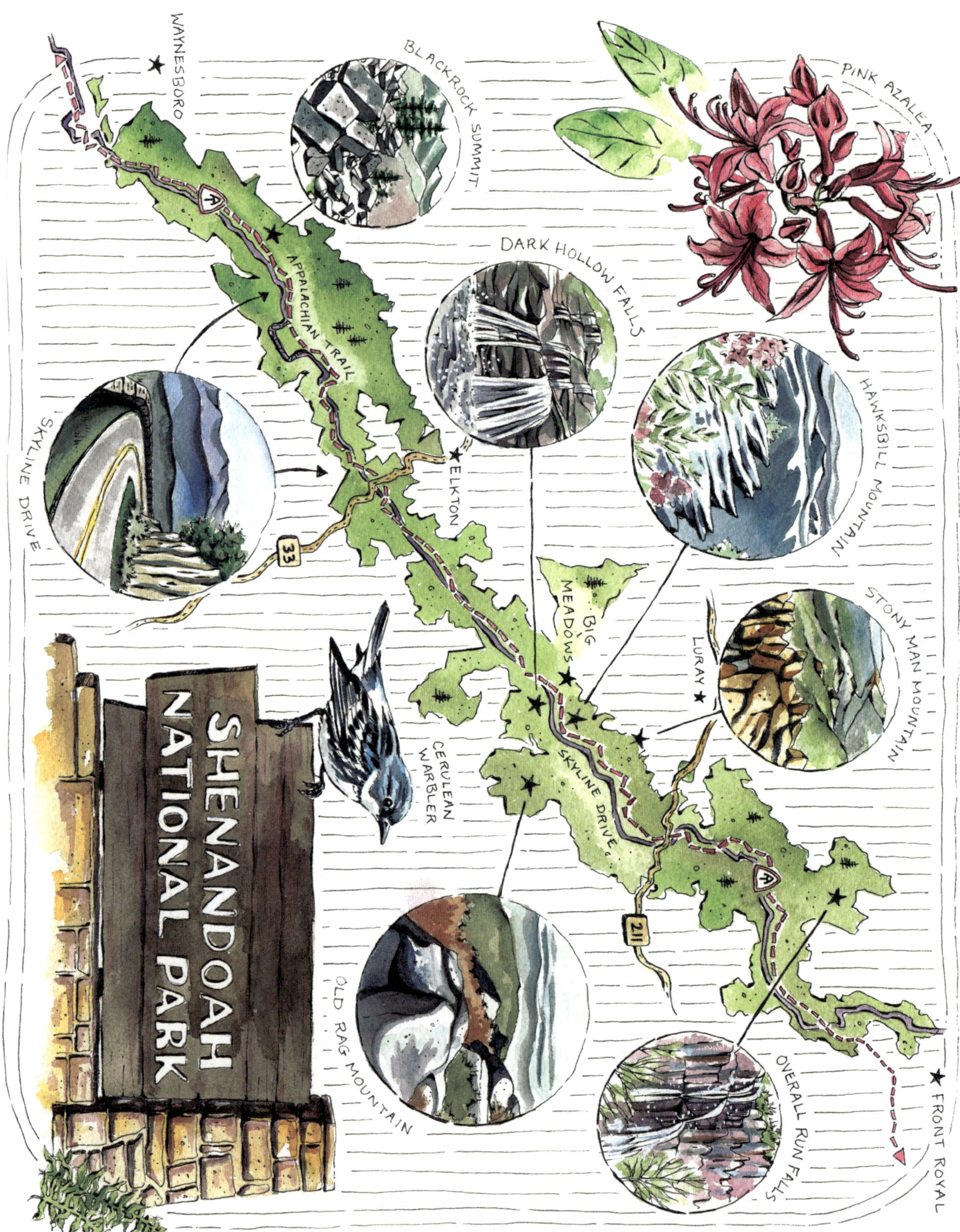

WAYNESBORO
BLACKROCK SUMMIT
PINK AZALEA
DARK HOLLOW FALLS
APPALACHIAN TRAIL
SKYLINE DRIVE
HAWKSBILL MOUNTAIN
ELKTON
33
BIG MEADOWS
STONY MAN MOUNTAIN
LURAY
SHENANDOAH NATIONAL PARK
CERULEAN WARBLER
SKYLINE DRIVE
OLD RAG MOUNTAIN
211
OVERALL RUN FALLS
FRONT ROYAL

SHENANDOAH NATIONAL PARK

Comprising 200,000 forested acres blanketing either side of a narrow mountain crest, Shenandoah National Park is a jewel of the Blue Ridge Mountains. This area is the ancestral home of several Indigenous tribes including the Manahoac, Monacan, and Massawomeck. Both the Appalachian Trail and the Skyline Drive scenic highway wind along the ridge through the middle of the park, offering sweeping views of the mountains and valleys from numerous overlooks. In May, pink azaleas adorn the ridge, followed by stunning swaths of white mountain laurel in June. If you're lucky, you might also spot a black bear meandering near the road.

While the scenic drive provides a great overview, the park's 500 miles of hiking trails allow a less hurried, more intimate look at the lush hollows, waterfalls, and rocky outcroppings that make this area so much fun to explore. The forests are dominated by deciduous trees, primarily oak and hickory. Among the rare plants that are tucked away in these woods is American ginseng, an herb that has long been an important component of traditional medicine in cultures around the world.

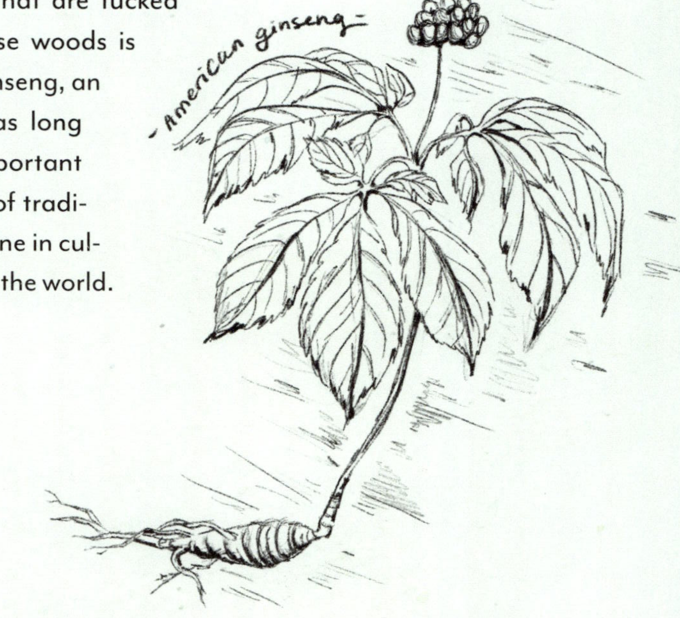

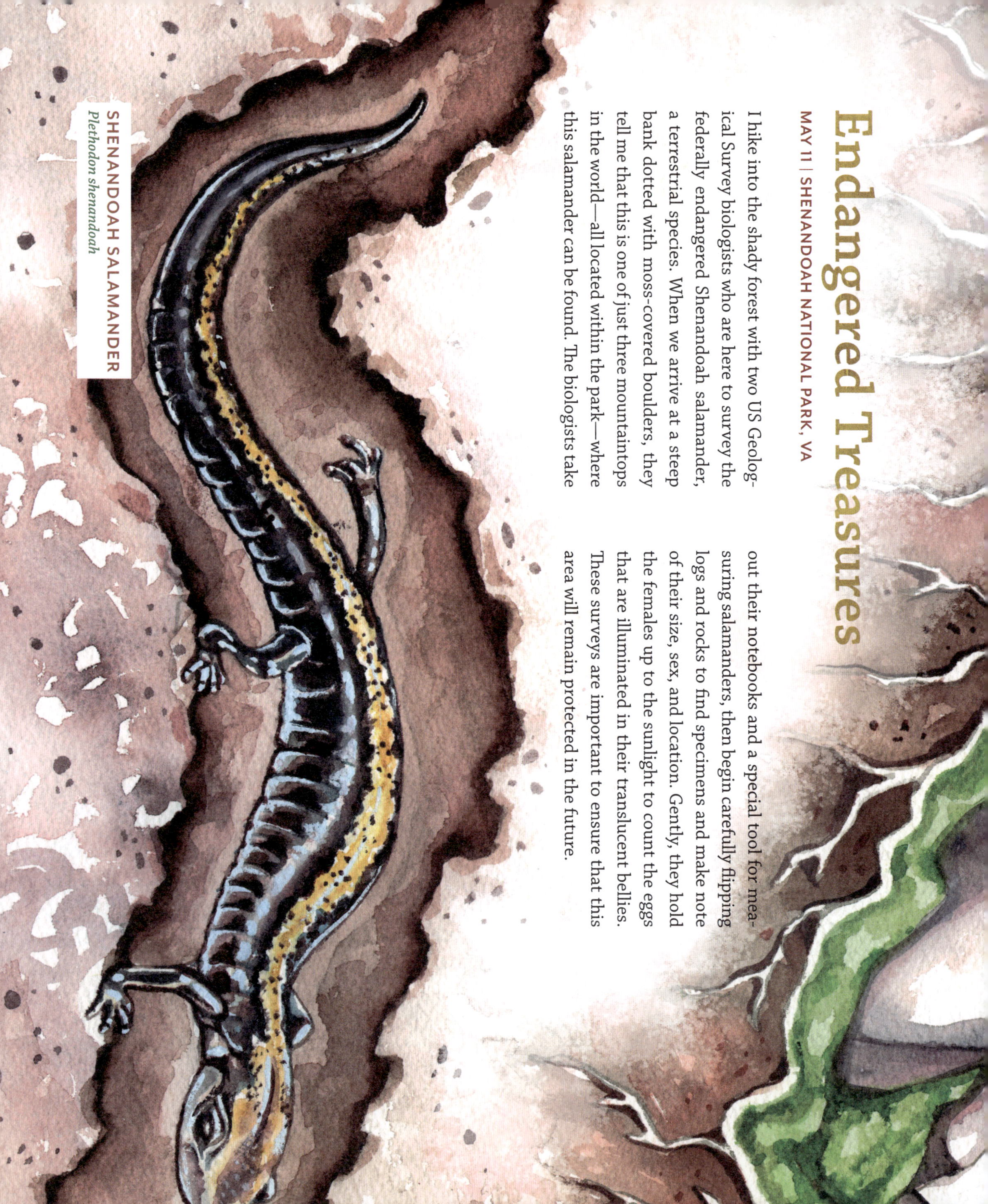

Endangered Treasures

MAY 11 | SHENANDOAH NATIONAL PARK, VA

I hike into the shady forest with two US Geological Survey biologists who are here to survey the federally endangered Shenandoah salamander, a terrestrial species. When we arrive at a steep bank dotted with moss-covered boulders, they tell me that this is one of just three mountaintops in the world—all located within the park—where this salamander can be found. The biologists take out their notebooks and a special tool for measuring salamanders, then begin carefully flipping logs and rocks to find specimens and make note of their size, sex, and location. Gently, they hold the females up to the sunlight to count the eggs that are illuminated in their translucent bellies. These surveys are important to ensure that this area will remain protected in the future.

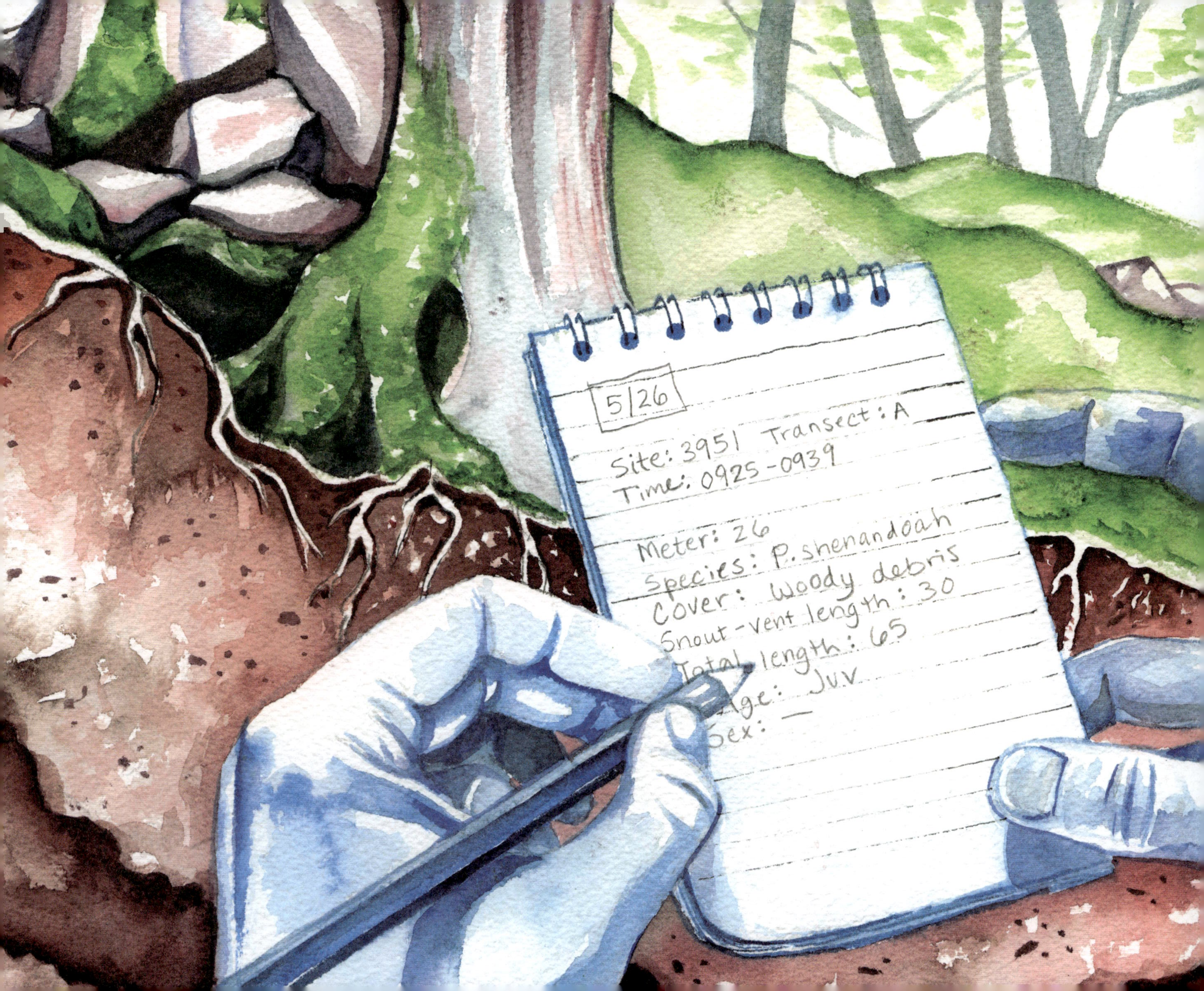

5/26
Site: 3951   Transect: A
Time: 0925-0939

Meter: 26
Species: P. shenandoah
Cover: Woody debris
Snout-vent length: 30
Total length: 65
Age: Juv
Sex: —

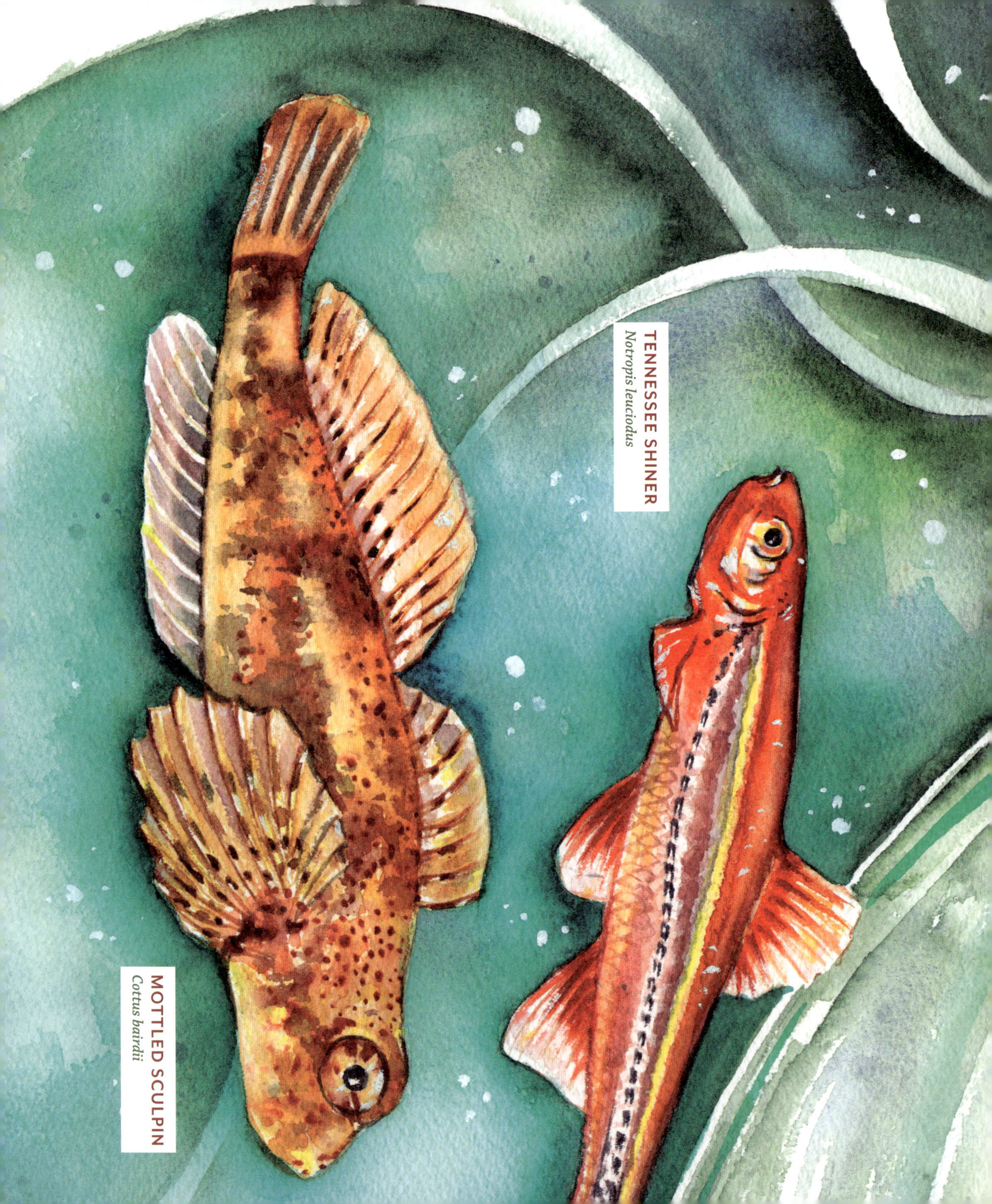

TENNESSEE SHINER
Notropis leuciodus
MOTTLED SCULPIN
Cottus bairdii

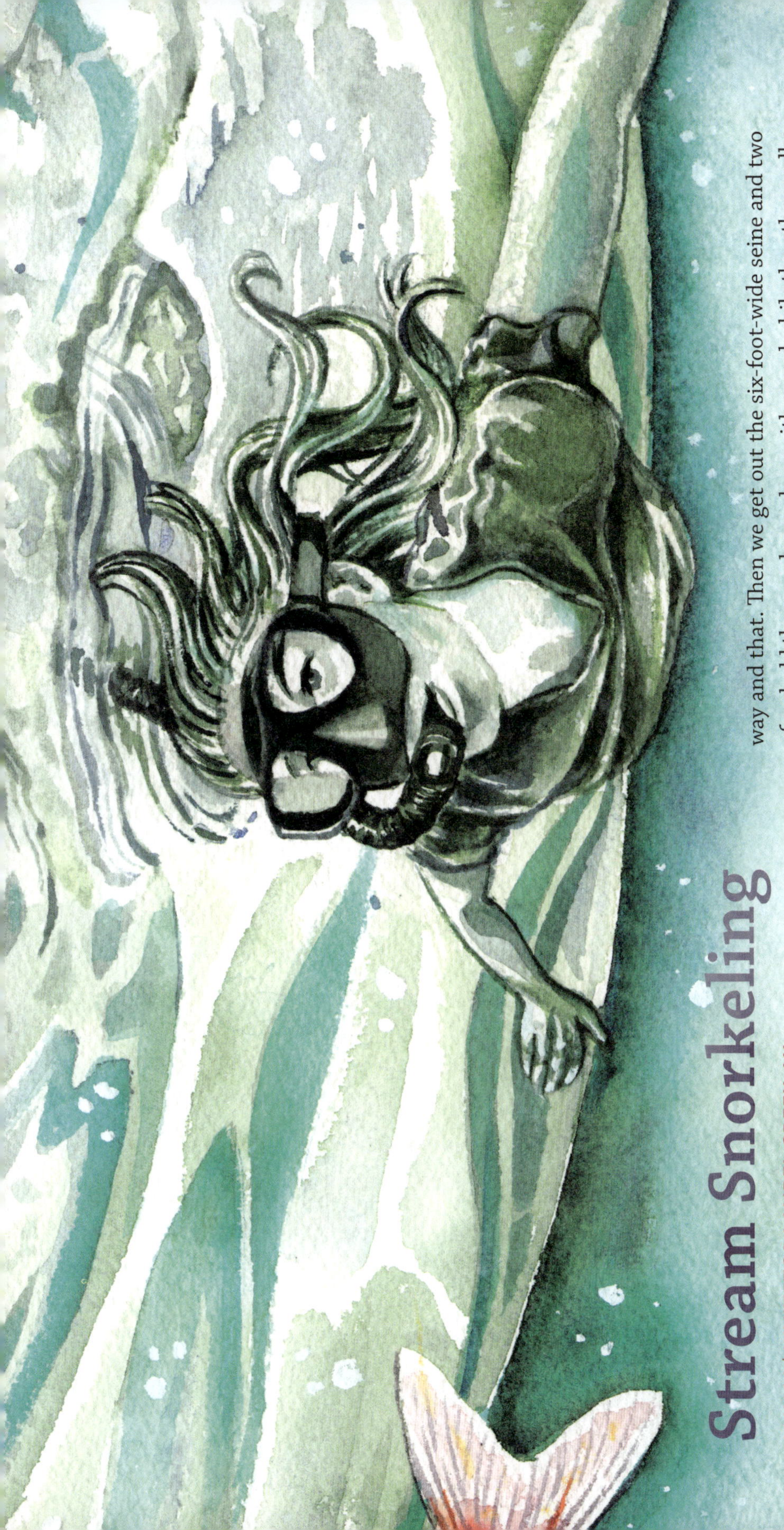

Stream Snorkeling

Accompanied by a local naturalist, we walk past a few fly-fishers to a stretch of stream that's unoccupied, where we set up a station with a glass fish tank, hand nets, a seine, and field guides. I notice a fat water snake keeping watch on the bank about ten feet away and suddenly I'm very grateful for the protection of my thick wetsuit. Once we've plunged into the freezing water, we practice treading water instead of standing up so that we don't kick up sediment that will cloud our view. We float on our bellies, watching through our goggles as the tiny fish dart this way and that. Then we get out the six-foot-wide seine and two of us hold the wooden rods on either end while the other walks quickly toward the net to drive fish and other critters our way.

Each time we lift the seine, we gently place our findings in the fish tank. I remember going out with my dad a few times as a kid to help him catch minnows for fishing with the same sort of net. We referred to all the little fish that we caught as "minnows" back then, but as we flip through the field guides to identify each small fish, we're amazed at how many different species there really are. Each of the darters, shiners, chubs, and sculpins that we find reveals unique colors and markings upon closer study.

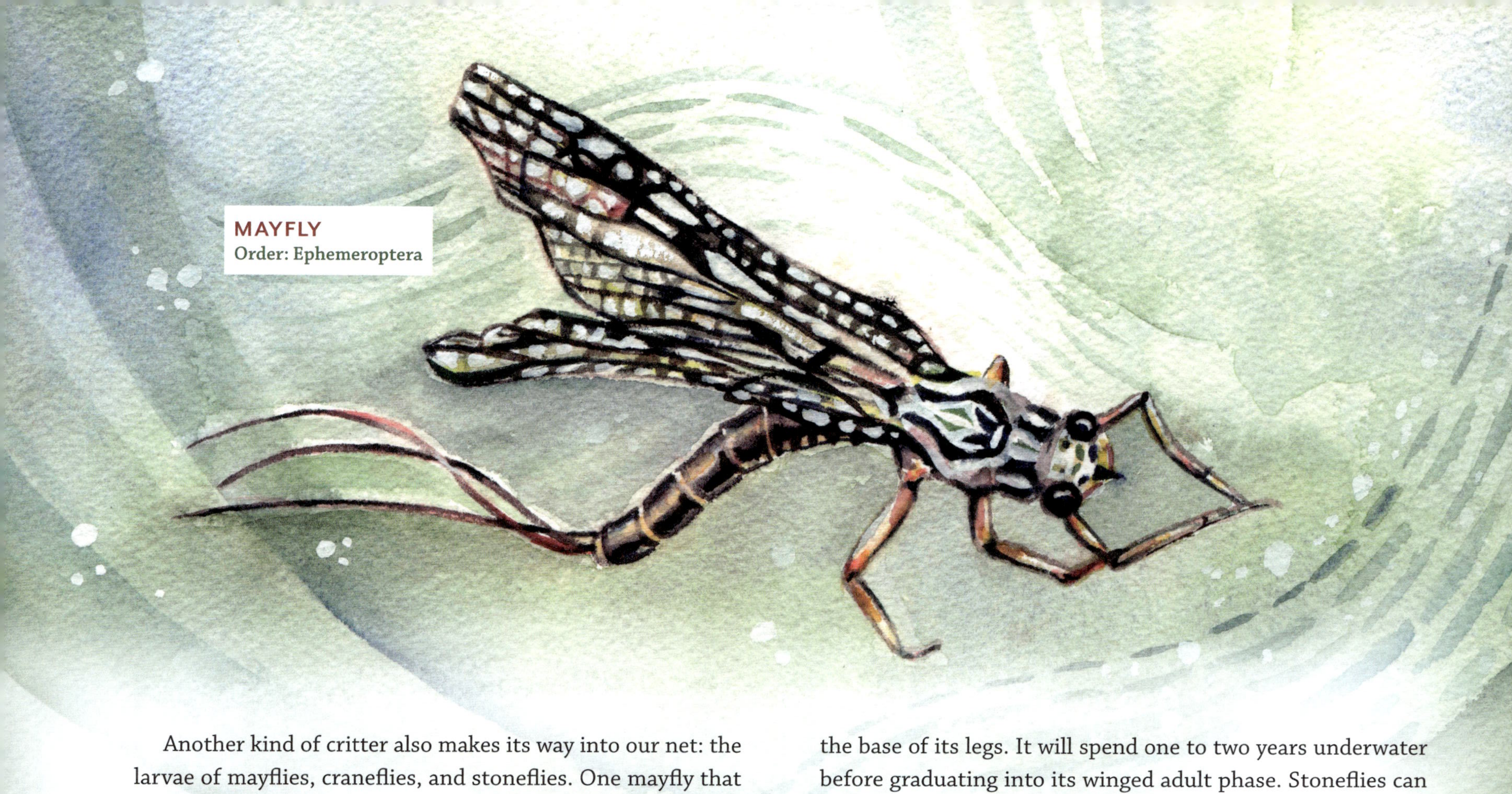

Another kind of critter also makes its way into our net: the larvae of mayflies, craneflies, and stoneflies. One mayfly that swims into our net has already sprouted wings; soon it will enter its terrestrial phase of life above water. A funky-looking stonefly nymph winds up in our fish tank, too, with feathery gills on the base of its legs. It will spend one to two years underwater before graduating into its winged adult phase. Stoneflies can only survive in clear, highly oxygenated streams, which makes them a good indicator of stream health and a tasty meal for the brook trout that also inhabit these high-elevation waterways.

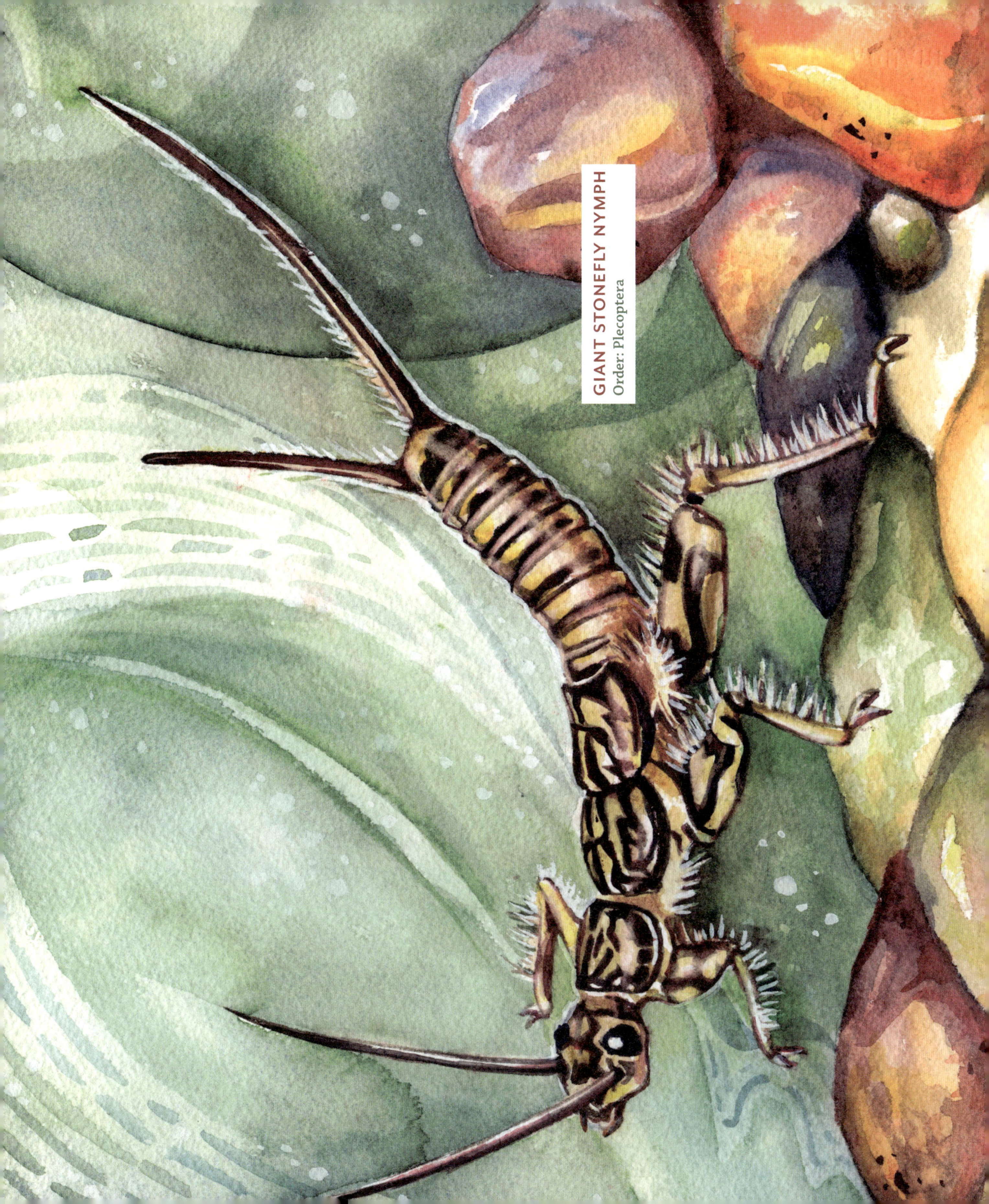

GIANT STONEFLY NYMPH
Order: Plecoptera

Expanding the Circle

I'm almost to the summit of Virginia's Mount Rogers when a giant, moss-covered tree a short distance from the trail captures my attention. It looks like it's been here for many centuries. As I inspect its gnarled form more closely, the layered discoveries remind me of opening a Russian nesting doll. First, I spot a perfectly camouflaged leopard slug slowly cruising across the lichen-covered bark. While examining the slug's distinctive markings, I notice the tiniest white spider sitting on the slug's back, taking a slow-motion joy ride. The spider is only about two millimeters long and I can barely see it, but there it is—a little life, just like mine.

I can only speculate about the relationship between these two critters, but the observation gets me thinking about friendship. Last June, I stood next to my husband as we read our vows to one another against a backdrop of rolling West Virginia hills. With the pandemic still in full swing, we had opted for a small wedding with only our immediate families. Before Covid, back when we thought we might have a full-size ceremony, I brainstormed the friends I would invite. My heart sank as I realized just how short my list of close, current, non-family friends truly was. What was wrong with me?

I've moved from place to place almost nonstop as an adult, immersing myself in a variety of natural landscapes and cultures that provide inspiration for my art and opportunities for personal growth. In fact, I've moved more than a dozen times since graduating from college in 2016, if you count all the places I've lived in for a month or more. One of the by-products of my meanderings has been a lack of long-lasting friendships. I meet wonderful humans wherever I go, but my lifestyle has made it hard to get really close to any of them.

Yet even when I have lived somewhere for a longer stretch, I find it difficult to build and sustain large social circles. My introverted disposition allows me to have a fabulous time by myself for several weeks at a time. I hike, I read, I journal, I dream, I play my fiddle, I do puzzles. But then a sudden wave of loneliness will catch up to me, like in that moment when I took a hard look at my short list of close friends.

Recently, I was put in touch with a friend of a friend who, upon hearing that I was in Virginia for the month, invited me to join her for a day of nature sketching in Shenandoah National Park. Betty is in her seventies, a retired Shenandoah park ranger and an avid painter. When we met up, she showered me with gifts: a field guide for the park, copies of her children's books that she illustrated, and an assortment of yummy trail snacks. Although she's forty years my senior, Betty is exuberant to the point that many of her words trail off into a birdlike trill.

We hiked to several of her favorite spots in the park. Betty squealed with delight when we stumbled upon two yellow morels, and in a quieter moment we unpacked our art supplies and split off from one another for about forty-five minutes to sketch our surroundings in silence. Betty has a group of friends called the "Sketchy Ladies" with whom she regularly meets up for hiking and painting excursions. I went home that day with a full heart, thankful to have been an honorary Sketchy Lady. Betty and I went adventuring together once more and I felt such a beautiful, age-defying sense of kinship with her, this fellow nature-lover and artist.

Inspired by the spider riding the leopard slug, I begin to rack my memory for other unlikely friendships. What if, instead of trying to squeeze my idea of friendship into a limiting, narrow mold, I expanded that definition to include all the people in my life who love and have loved me over the years? My sisters have been my closest companions throughout my entire life, but I've often thought that I can't count them as being "friends," since we're related.

But now I'm starting to view it all through a different lens. Though most of the friendships I've formed throughout this decade of frequent relocation have been short-lived, these connections have shown me that the world is full of kindred spirits. And as silly as it may sound, I am a friend to myself. I have worked to cultivate an internal voice that is gentle and kind, and this self-love is a form of friendship not to be ignored.

The friendships that dot my life are like twinkling stars in the night sky. Some are steadfast constellations, there since the beginning: my parents, my siblings, a few friends whom I've known since early childhood. Others are irregular and sometimes unexpected, like a shooting star that burns bright and fast but leaves an impression long after it has faded. And if I keep my heart open to the possibility of intergenerational and unconventional friendships, there's no limit to the connections that will find their way into my orbit.

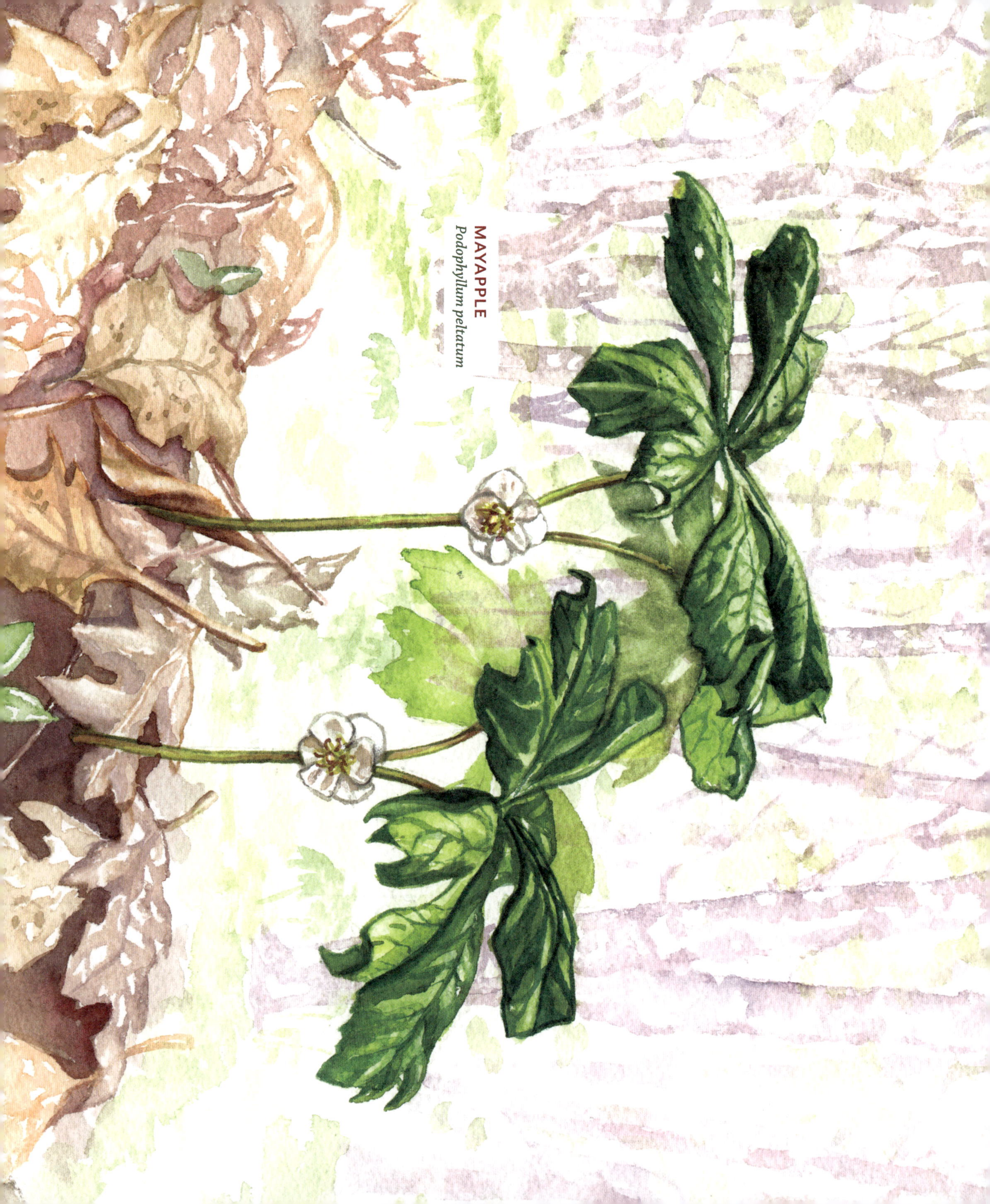

MAYAPPLE
Podophyllum peltatum

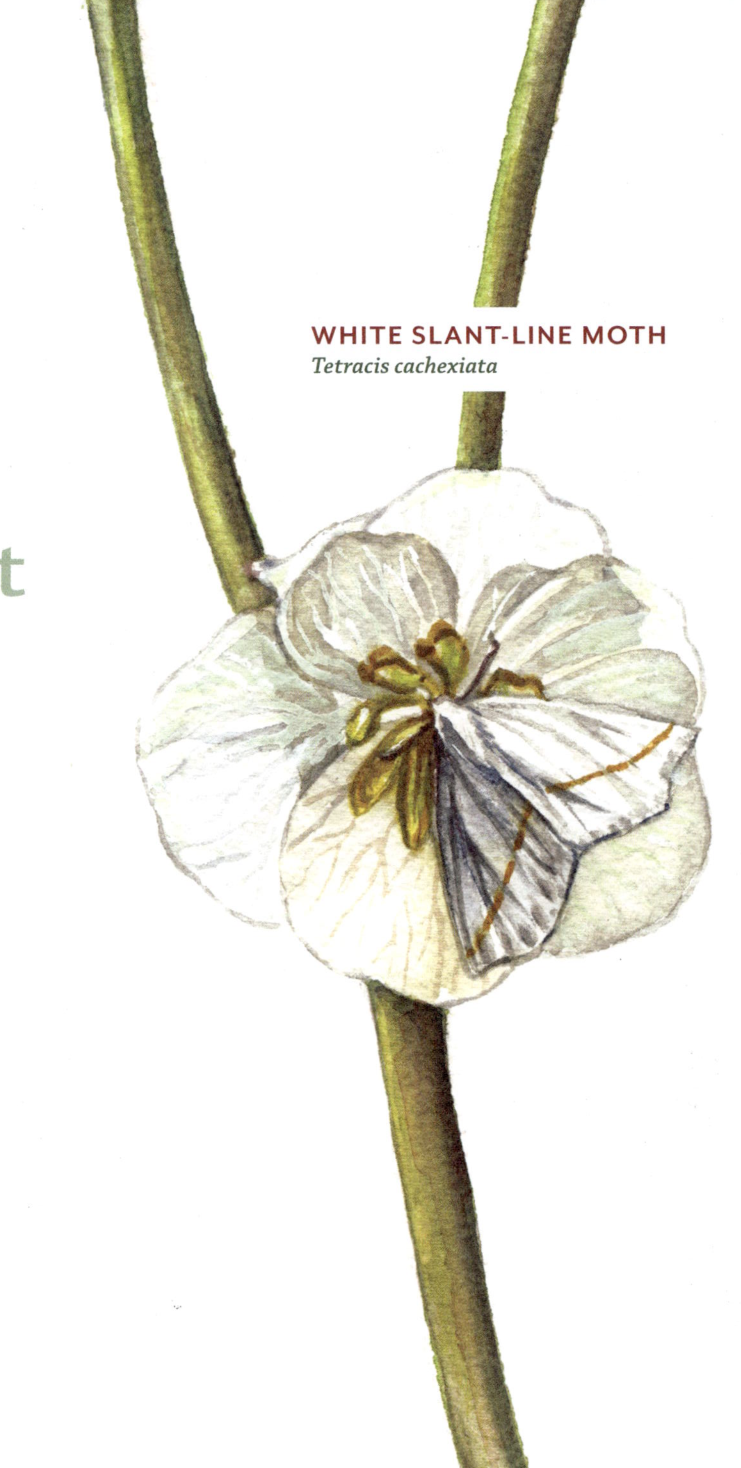

Hidden in Plain Sight

MAY 15 | GRAYSON HIGHLANDS STATE PARK, VA

"Two hours before the rain!" I say as I check my weather app. We've just pulled into Grayson Highlands State Park and we hustle to get all our gear together before setting out for a wetland area that we'd read about. Along the way, we pass through a lively meadow filled with hundreds of mayapples in bloom. These woodland wildflowers hold fond memories for me: when I was a kid, my mom taught me how to transform the umbrella-shaped plants into a fairylike doll.

As we walk through the meadow, I spot a tiny white moth with a distinct orange line running across each wing. This moth is commonly called the white slant-line moth or the "mayapple moth" because it sometimes camouflages itself against pearly white mayapple flowers for a foolproof hiding spot.

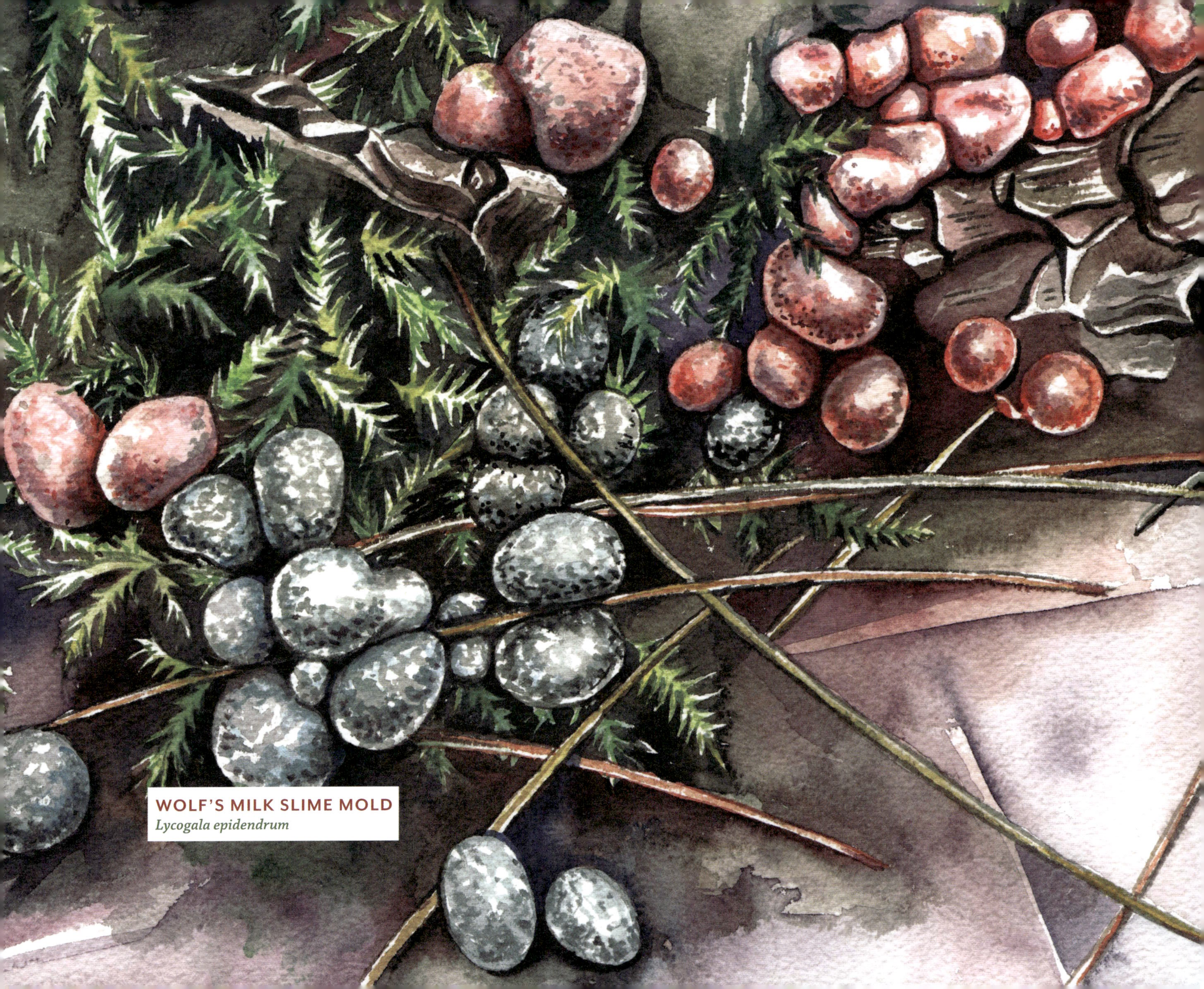

WOLF'S MILK SLIME MOLD
Lycogala epidendrum

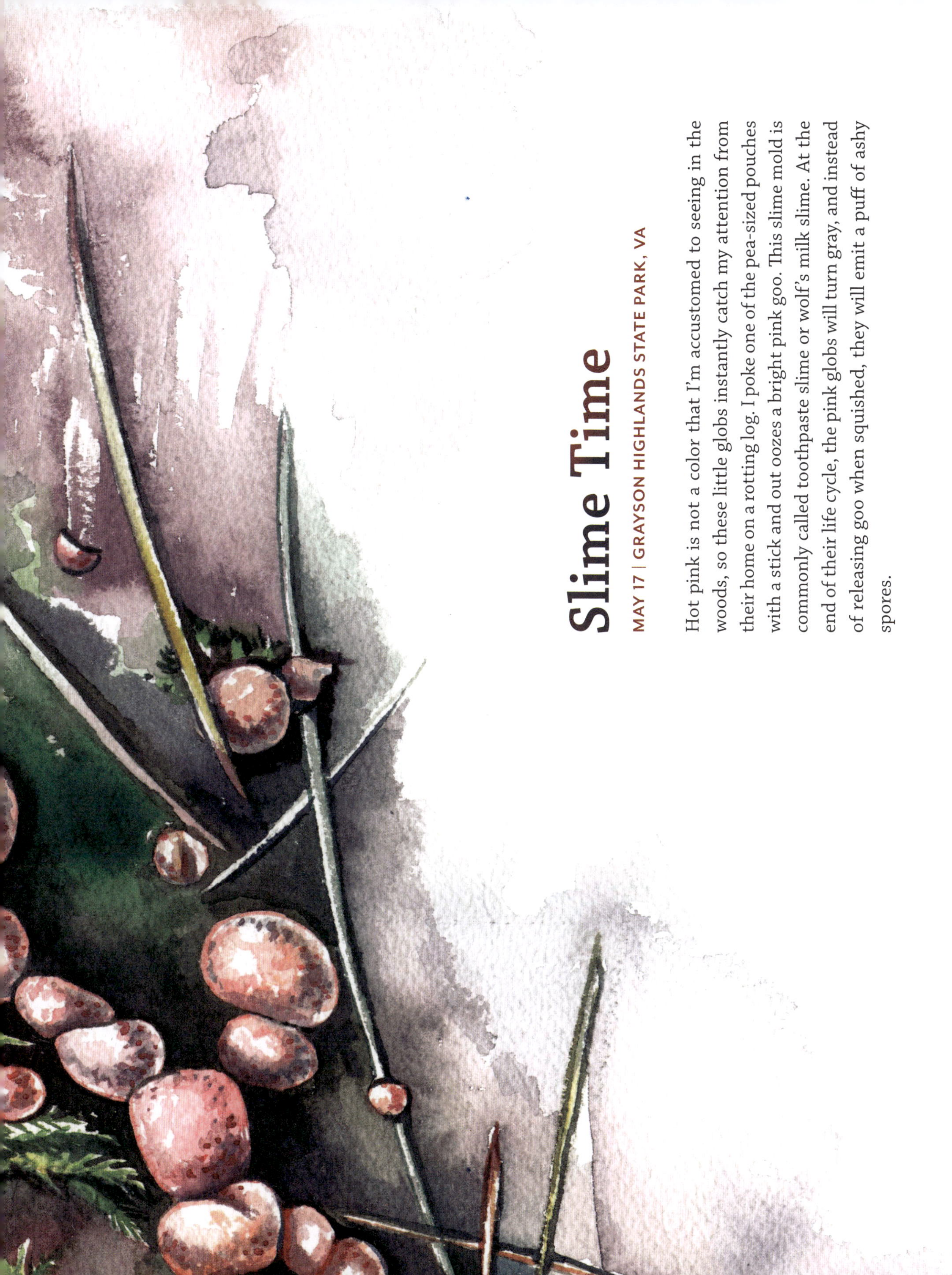

Slime Time

Hot pink is not a color that I'm accustomed to seeing in the woods, so these little globs instantly catch my attention from their home on a rotting log. I poke one of the pea-sized pouches with a stick and out oozes a bright pink goo. This slime mold is commonly called toothpaste slime or wolf's milk slime. At the end of their life cycle, the pink globs will turn gray, and instead of releasing goo when squished, they will emit a puff of ashy spores.

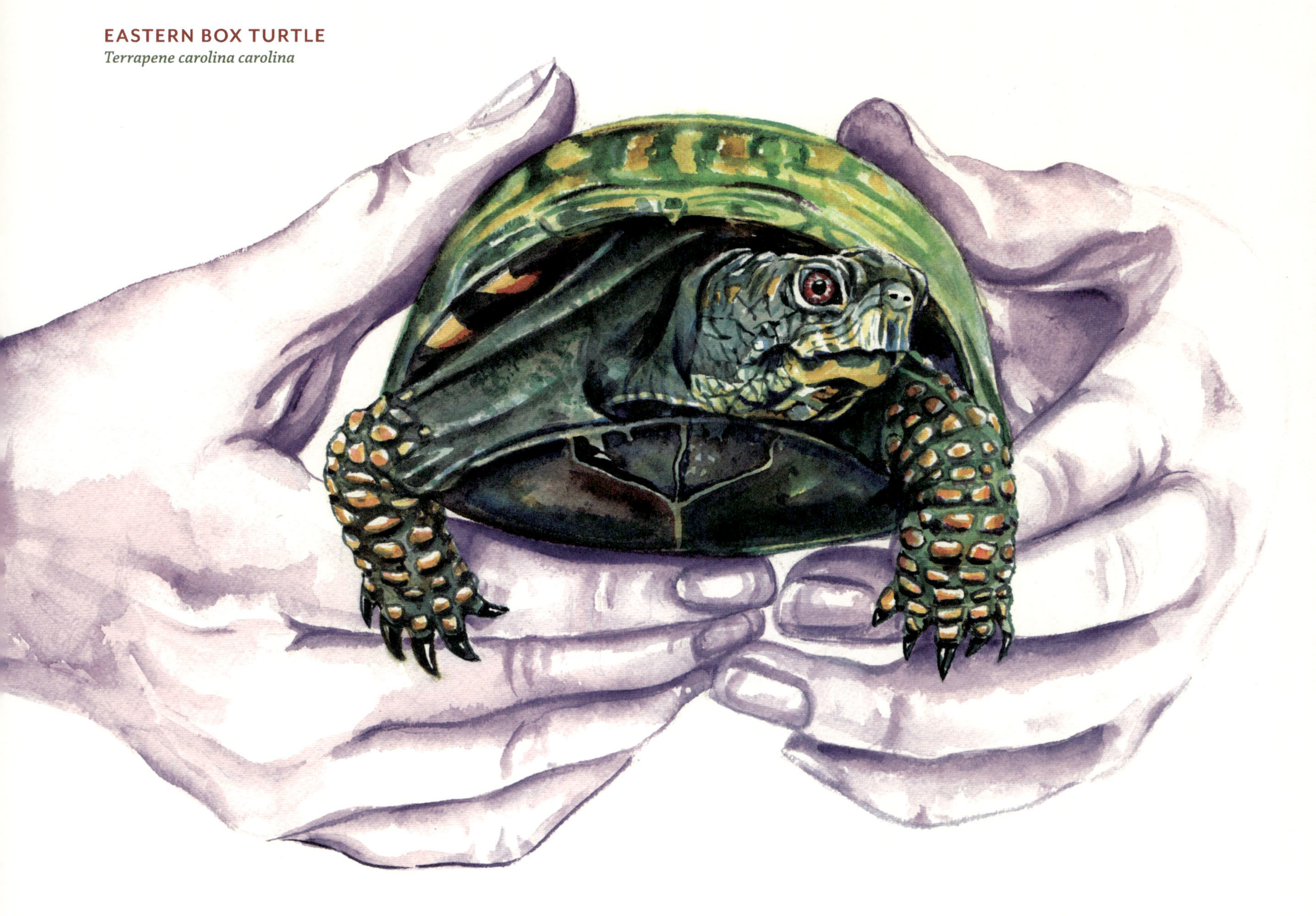

EASTERN BOX TURTLE
Terrapene carolina carolina

Helping Hands

The Wildlife Center of Virginia is just down the road from where we're staying, so I stop by to learn about what they do there. This nonprofit hospital for native wildlife treats all species indigenous to the region, from black bears and bald eagles to Virginia quail and red foxes. They have an operating room, an X-ray machine, and microsurgery capabilities so that they can perform surgery on critters as small as toads!

I am introduced to three of their ambassador animals, creatures that have been rescued and rehabilitated here but that are still too disabled to survive in the wild. The first is an opossum named Marigold, who smacks her lips together loudly as she feasts on a dish of fruits and veggies. Opossums, with their penchant for roadside litter, are some of the most frequent patients at the hospital. Marigold wound up here after being hit by a car, and she can't be released back into the wild because the collision left her blind in one eye.

The next guest I meet is Ruby the red-tailed hawk, who was also hit by a car and is missing an eye. Birds like hawks are often hit when they swoop down to catch rodents, which also like to nibble on food waste tossed from vehicles.

My favorite new friend is Sheldon the eastern box turtle, who was found in a state park with a damaged leg and a shell painted with a rainbow of nail polish. Kids will sometimes paint turtle shells, and while it may seem innocent enough, these colors prevent turtles from camouflaging themselves and can lead to injuries or death. Sheldon, who was likely kept as a pet and then released, remains at the animal hospital: with a territory of less than one square mile, box turtles can become confused and at risk if set free in a new location. The average life span of a box turtle in the wild is between twenty-five and thirty-five years, but in captivity they can live much longer—reportedly up to one hundred years old. It strikes me that Sheldon might outlast the veterinarians who have helped her.

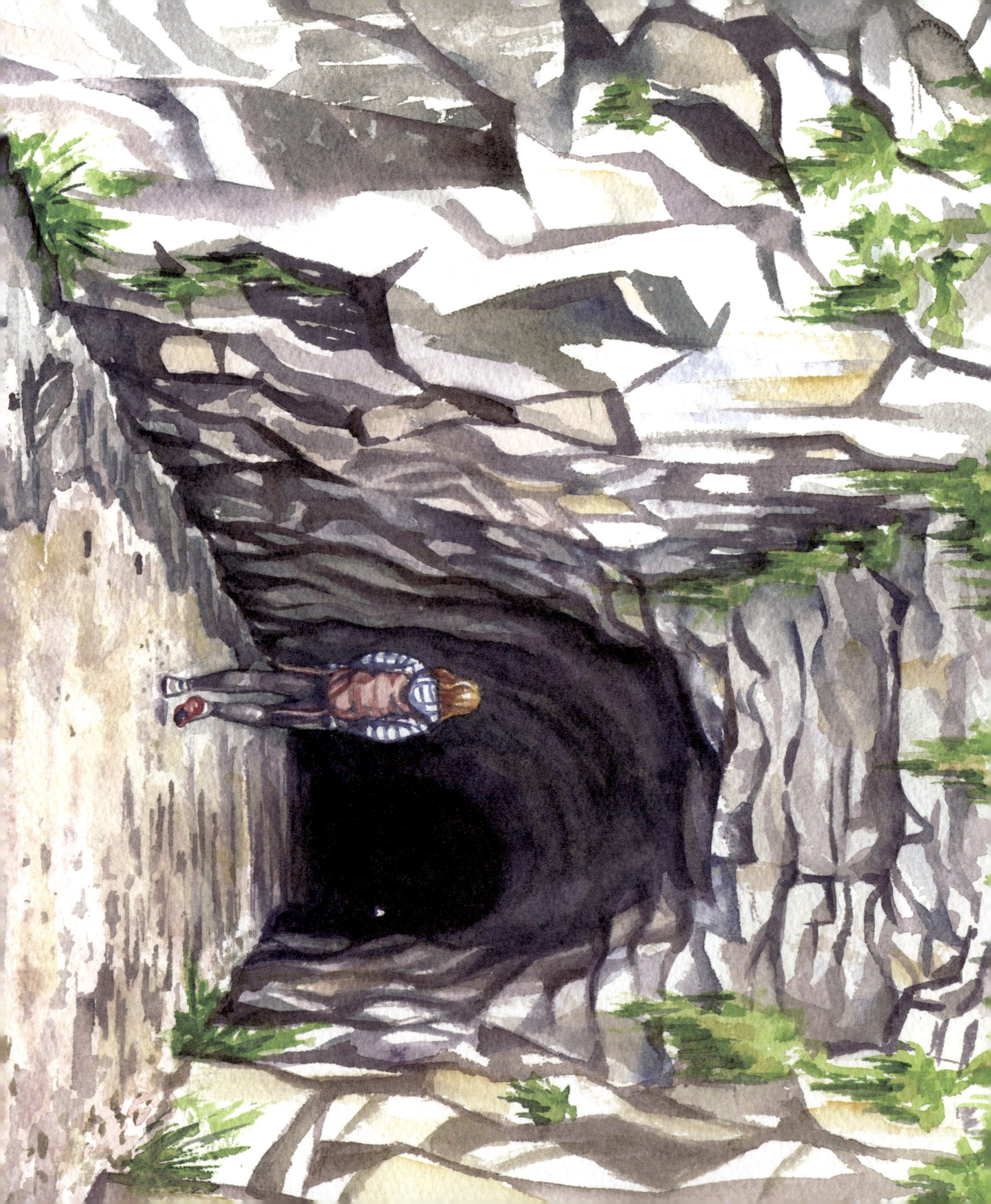

Cave Creatures

The darkness closes in on all sides and each step forward feels hesitant and unbalanced. The only sounds are a steady chorus of *drip-drip-splat*s and the faint, distorted voices of two walkers far in front of us. We're making our way through a one-mile underground tube that was blasted in the 1850s so that trains could beeline straight through the mountain instead of snaking around it. There are no lights inside, and we've decided to leave our headlamps off. We creep forward, only a pin-sized dot of light at the other end ensuring that we continue straight instead of walking into a wall.

On the return trip, we turn on our headlamps, illuminating deep ditches filled with water on either side of the tunnel and slimy, cave-like rock walls. My eye detects a flash of bright orange. It's a long-tailed salamander, curled up on a rock ledge and staring right back at me. Excited to find life in this human-made environment, I begin scanning the ditches and walls. Eventually, I notice the short, fast movements of a reddish-tan crayfish scooting backward through the ditch. By the time we finally pop out at the entrance, I've counted five crayfish, seven long-tailed salamanders, and one banded fishing spider.

LONG-TAILED SALAMANDER
Eurycea longicauda

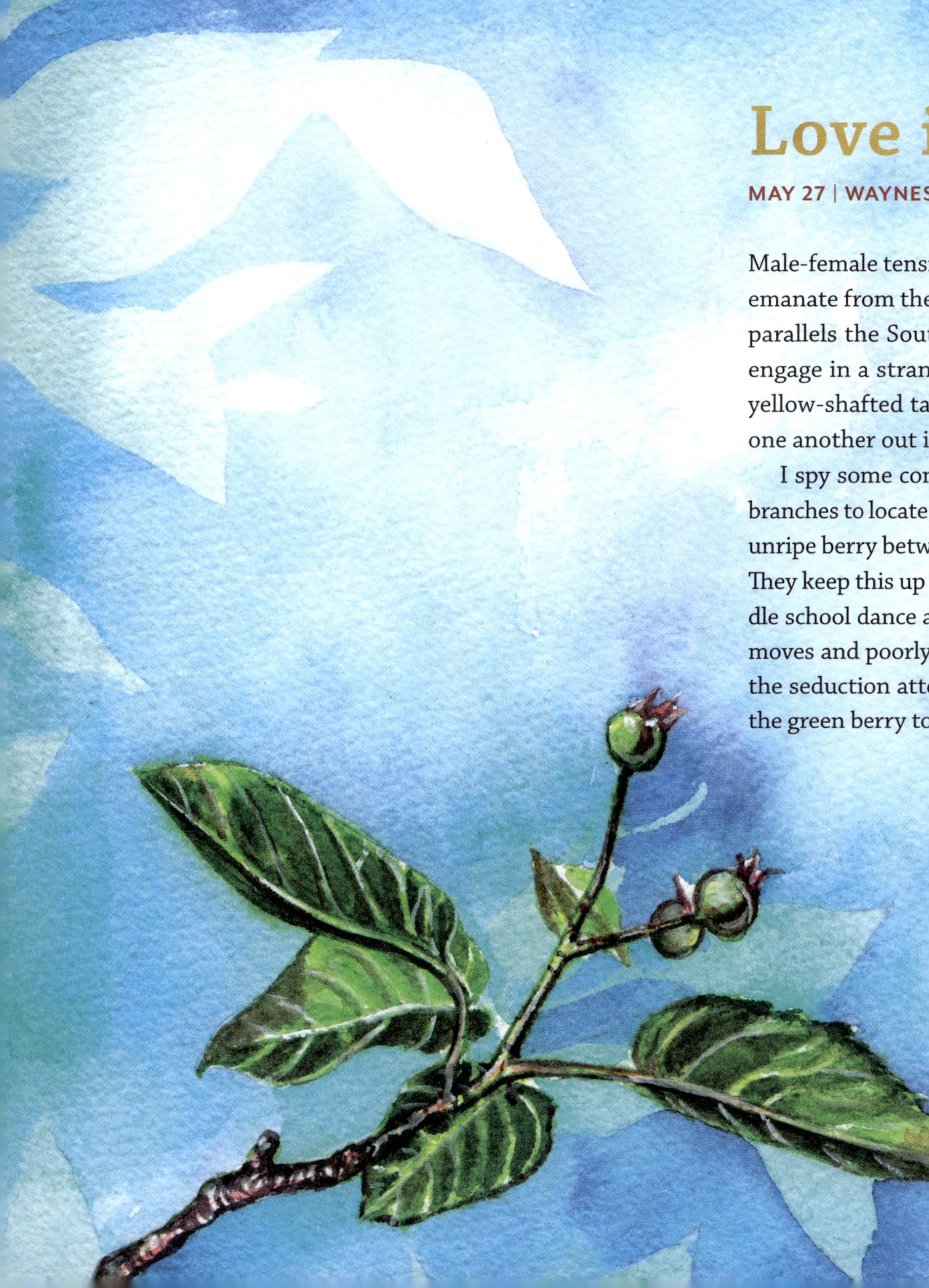

Love in the Treetops

Male-female tensions are high in the bird world today, and elaborate songs emanate from the now fully filled-out treetops along the walking trail that parallels the South River. I watch as a male and female northern flicker engage in a strange courtship dance, hopping back and forth with their yellow-shafted tails fanned out. They look like *luchadores* trying to psych one another out in the wrestling ring.

I spy some commotion in a serviceberry tree and peer up through the branches to locate its source. A male and female cedar waxwing are passing an unripe berry between their beaks while also moving from branch to branch. They keep this up for some time, and I feel like a teacher chaperoning a middle school dance as I try my best not to make fun of all the awkward dance moves and poorly executed flirting. But apparently the female doesn't find the seduction attempts as embarrassing as I do—she eventually swallows the green berry to signal that she's pleased with the gift.

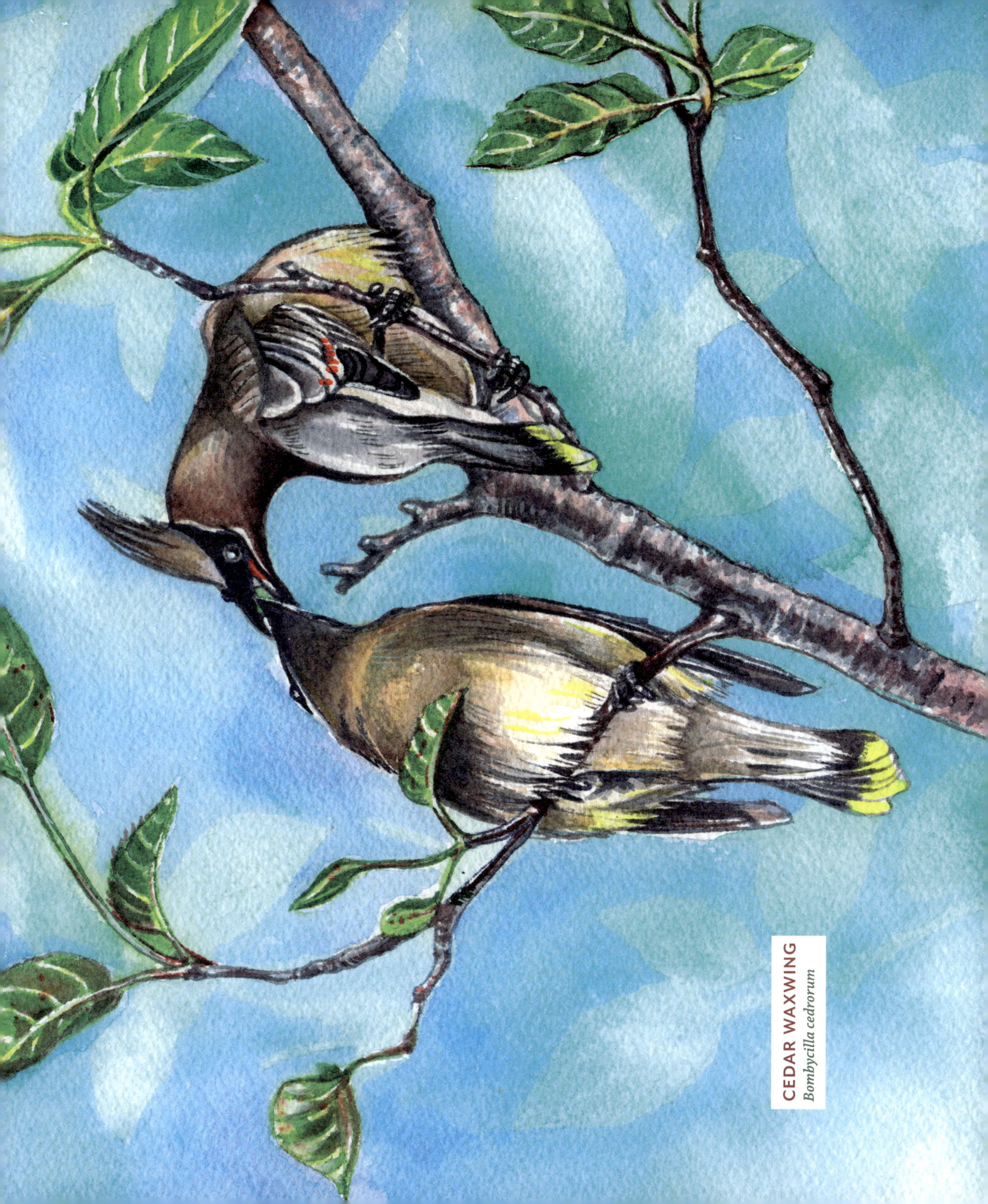

CEDAR WAXWING
Bombycilla cedrorum

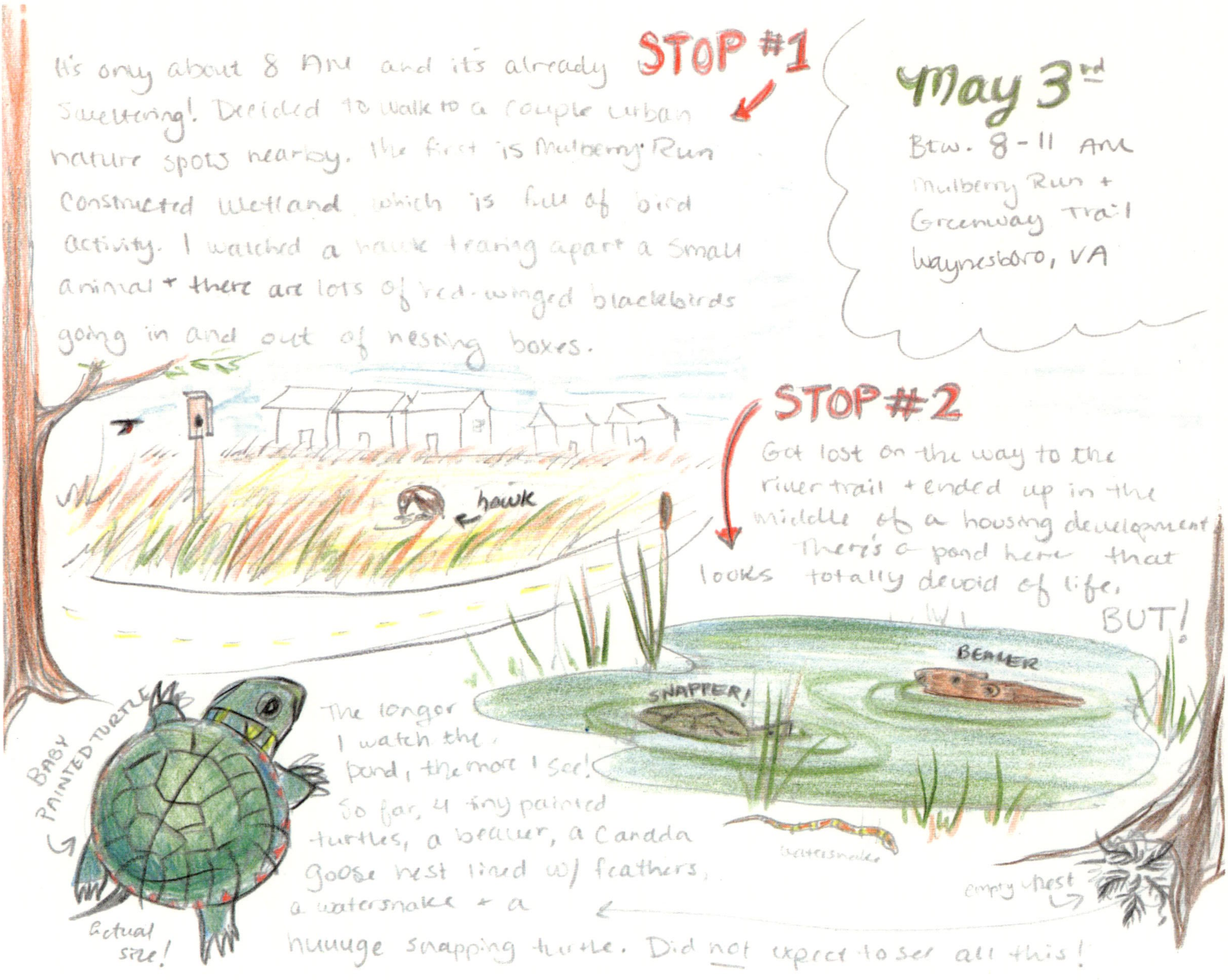

It's only about 8 AM and it's already
sweltering! Decided to walk to a couple urban
nature spots nearby. The first is Mulberry Run
Constructed Wetland which is full of bird
activity. I watched a hawk tearing apart a small
animal + there are lots of red-winged blackbirds
going in and out of nesting boxes.

STOP #1

May 3rd

Btw. 8 - 11 AM
Mulberry Run +
Greenway Trail
Waynesboro, VA

STOP #2

Got lost on the way to the
river trail + ended up in the
middle of a housing development.
There's a pond here that
looks totally devoid of life,
BUT!

The longer
I watch the
pond, the more I see!
So far, 4 tiny painted
turtles, a beaver, a Canada
Goose nest lined w/ feathers,
a watersnake + a
huuuge snapping turtle. Did not expect to see all this!

NATURE JOURNAL IDEAS

WATERFALL WATCH

A multitude of plants and critters inhabit the splash zone around waterfalls. Find a waterfall that you can safely walk on either side of. Bring a hand lens so that you can examine the mosses, liverworts, and other plants that you find growing on the soggy rock ledges. Look in crevices for frogs, salamanders, and birds that may be taking shelter in these nooks and crannies. Jot down notes about or sketch what you observe.

TOWN TRAVERSE

Map out a pedestrian-friendly route to a destination in your town that you would normally drive to. Treat this walk like a nature hike: carry a backpack with snacks and water, wear sturdy shoes, and prepare for the elements by bringing raingear and sun protection. How does approaching an urban outing with the mindset of an outdoor adventurer change the way it feels to walk through your town? What birds, trees, and flowers do you notice along the way that you might have otherwise overlooked?

STREAM SNORKEL

Snorkeling doesn't need to be something that you only do in the ocean. Find a relatively calm, clean body of water. This could be a stream, lake, or pond. When you enter the water, try not to kick up too much sediment because that will cloud your view. What fish, crustaceans, amphibians, and underwater plants do you see? How does it feel to experience this watery landscape from below? If you don't have access to safe waters for swimming, take a stream walk instead. Wear shoes that have good traction and can get wet, use a walking stick or trekking poles to keep you steady, and see how it feels to use the watercourse as your path.

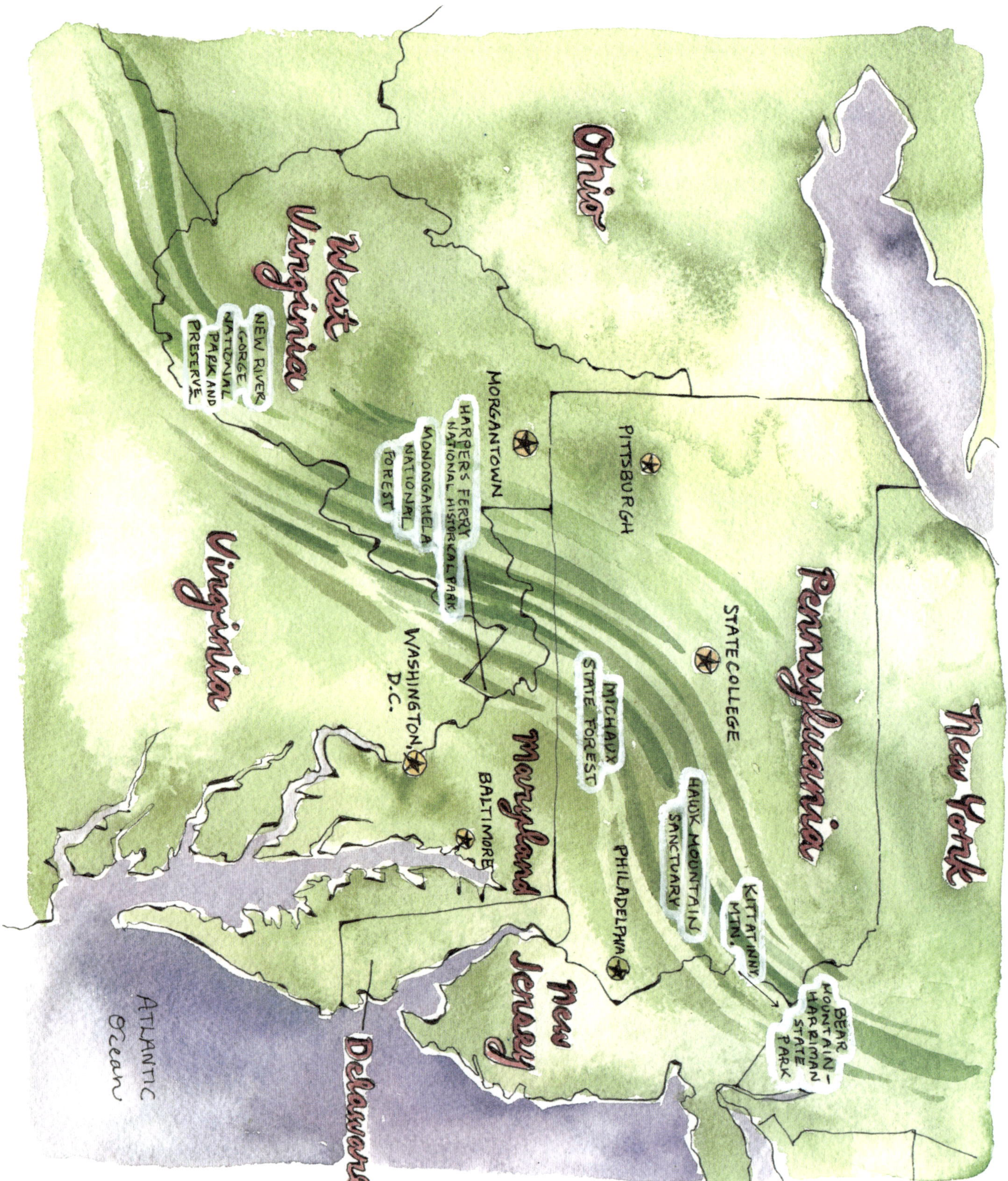

Ohio
West Virginia
Virginia
Pennsylvania
New York
Maryland
New Jersey
Delaware
NEW RIVER GORGE NATIONAL PARK AND PRESERVE
MORGANTOWN
HARPERS FERRY NATIONAL HISTORICAL PARK
MONONGAHELA NATIONAL FOREST
PITTSBURGH
WASHINGTON, D.C.
BALTIMORE
STATE COLLEGE
MICHAUX STATE FOREST
PHILADELPHIA
HAWK MOUNTAIN SANCTUARY
KITTATINNY MTN.
BEAR MOUNTAIN-HARRIMAN STATE PARK
ATLANTIC Ocean

central APPALACHIANS

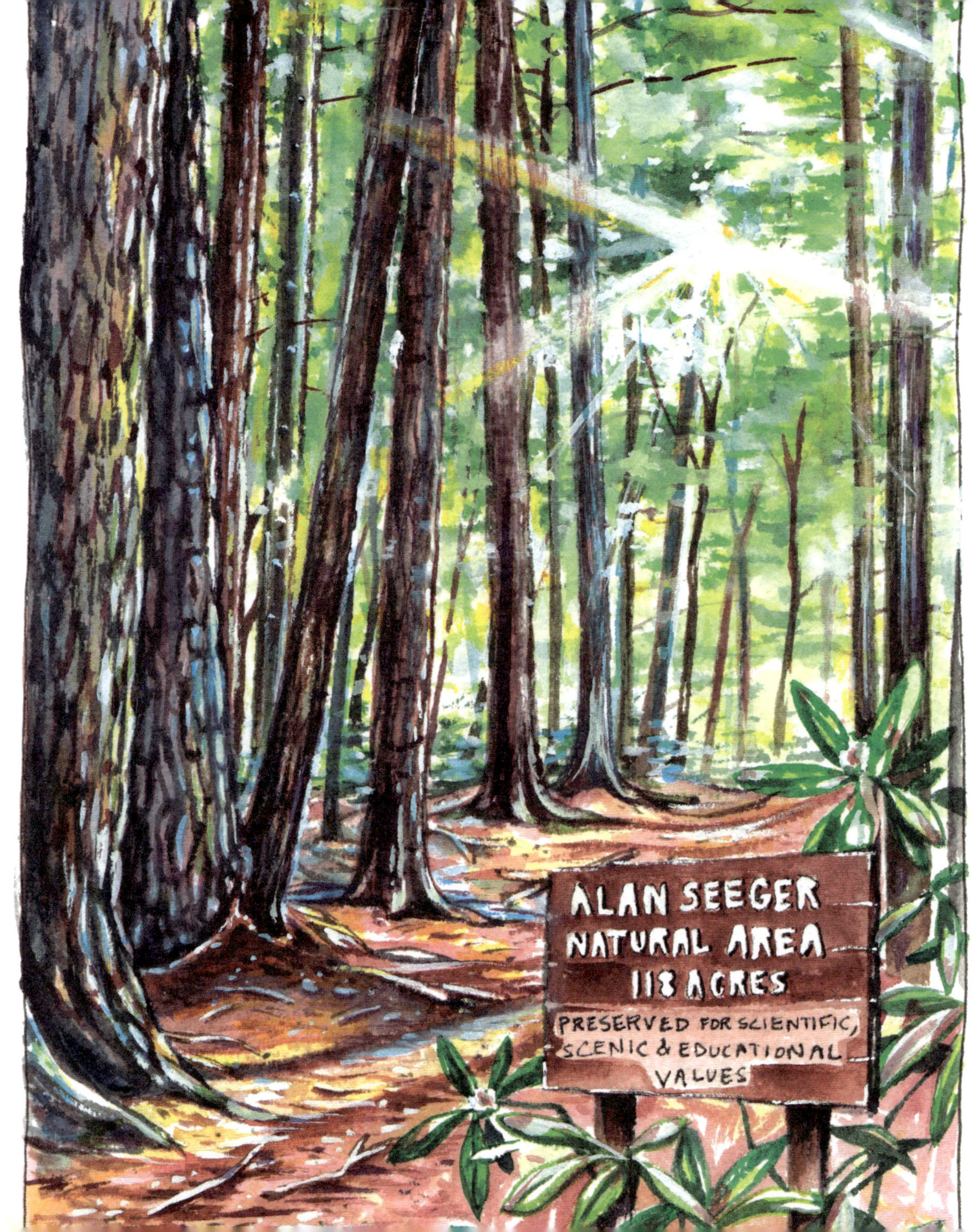

ALAN SEEGER
NATURAL AREA
118 ACRES
PRESERVED FOR SCIENTIFIC,
SCENIC & EDUCATIONAL
VALUES

REDISCOVERING THE CENTRAL APPALACHIANS

JUNE 1

We aren't sure what to expect as we cross into the more developed Ridge-and-Valley Appalachians of Pennsylvania. As the name implies, long ridges and parallel long valleys distinguish this portion of the mountains from the other subranges. In the hundreds of millions of years following the mountain-building event that created the Ridge and Valley geologic province, the softer limestone and shale eroded away to create the valleys, leaving the more durable sandstones in place to form the ridges.

This month, we'll base our explorations in south-central Pennsylvania, with some additional adventures to eastern West Virginia and western Maryland. This region is somewhat familiar to me already since I grew up just a few hours from here, but I have a feeling that my daily observations will reveal just how much more there is to learn.

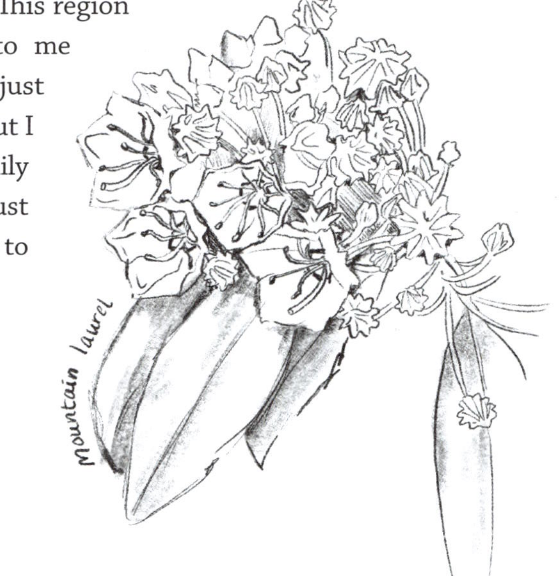

Alan Seeger Natural Area, Rothrock State Forest, PA

Swamp Dwellers

JUNE 3 | MICHAUX STATE FOREST, PA

A refreshing breeze tickles my skin as I hike along the shady, pine-needle-blanketed path that winds around a manmade lake. I enter a marshy area, and a sudden splash piques my curiosity. Moments later, a metallic, three-inch-long pickerel frog hops my way before hiding among the spongy sphagnum moss and swamp beacon fungi. An amphibious mushroom that looks like a match-stick stuck in the mud, swamp beacon fungus has the unique ability to grow—and to decompose forest floor debris—in several inches of standing water. Just think: to this little pickerel frog, each swamp beacon fungus must look as tall as a tree!

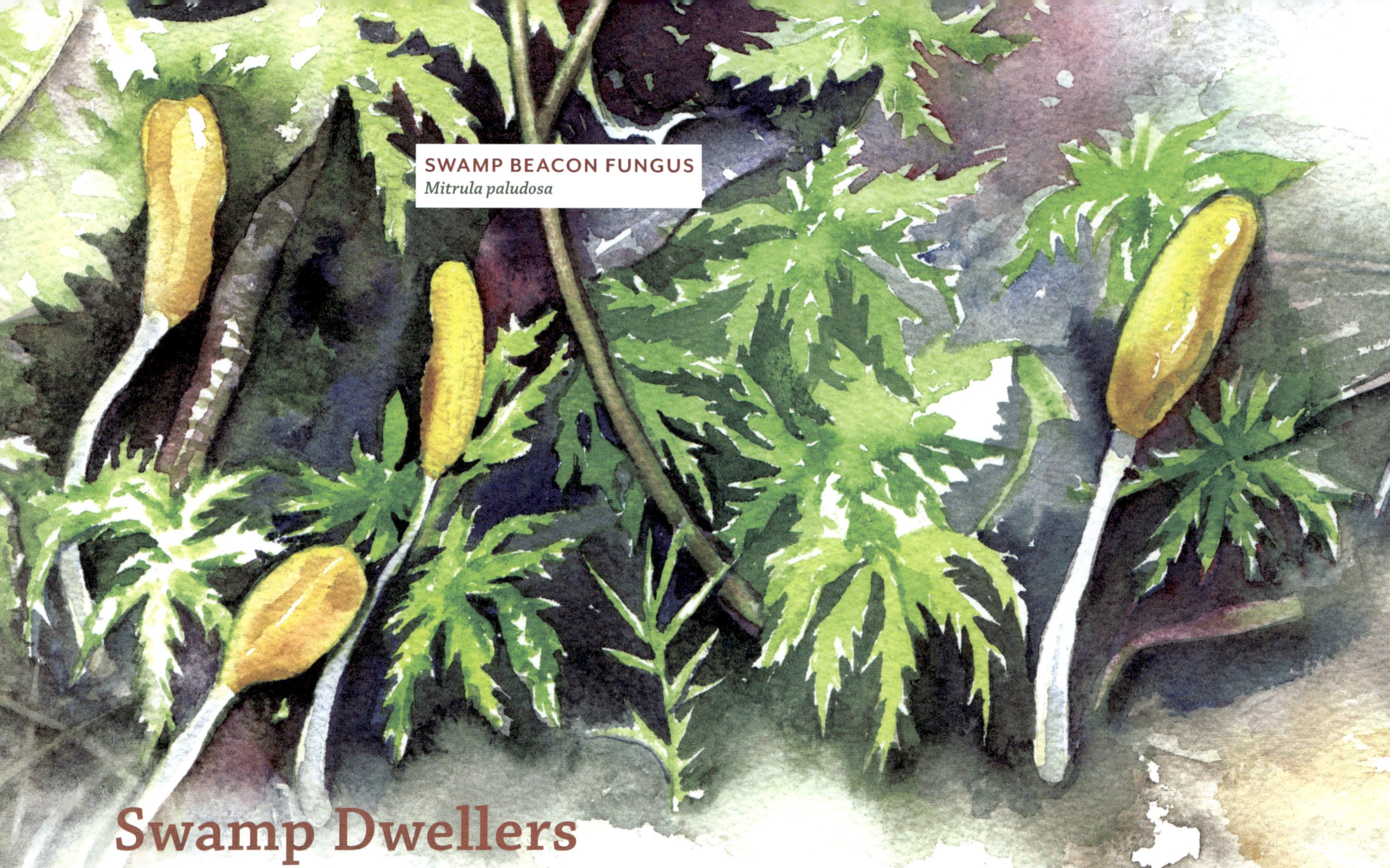

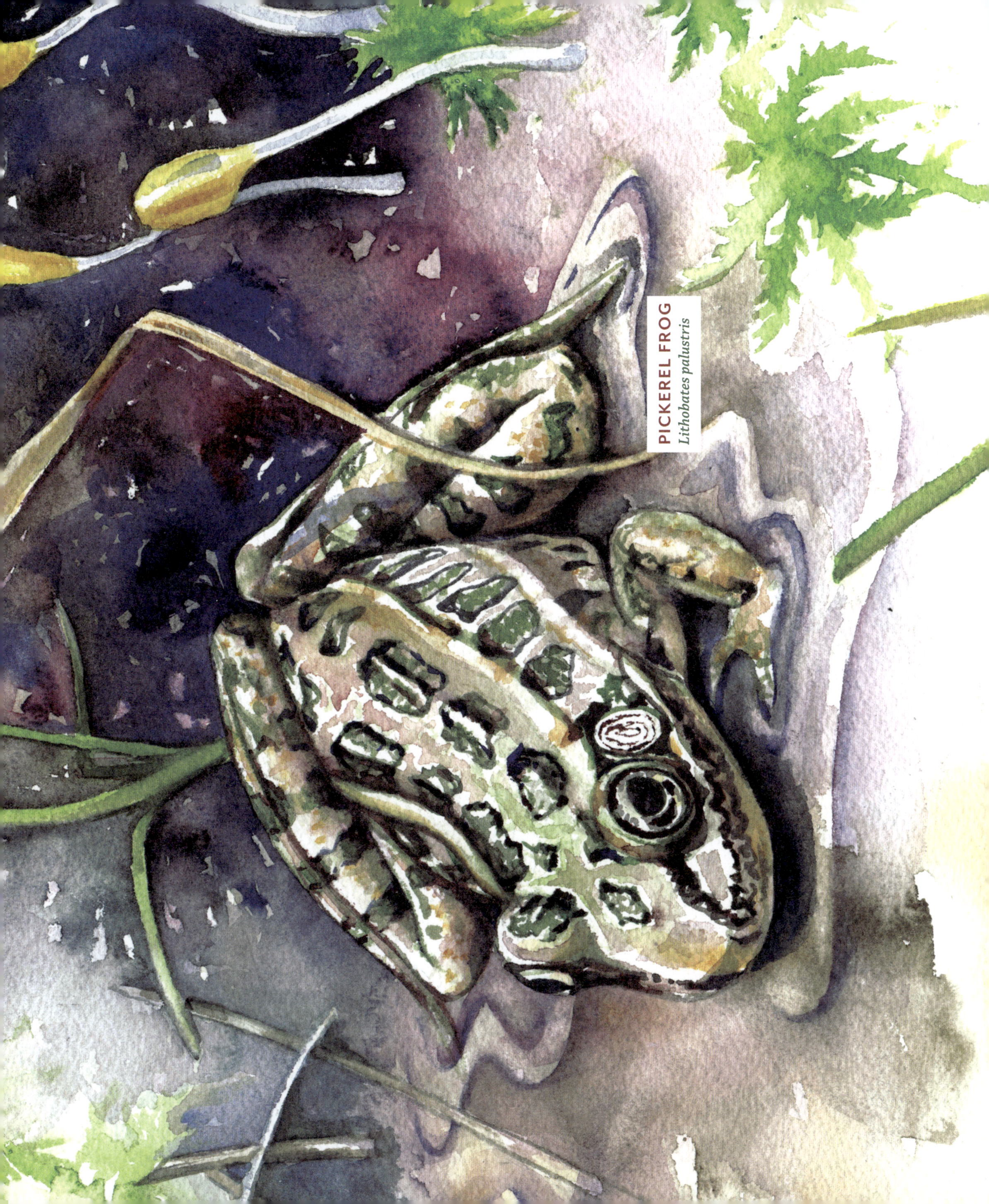
PICKEREL FROG
Lithobates palustris

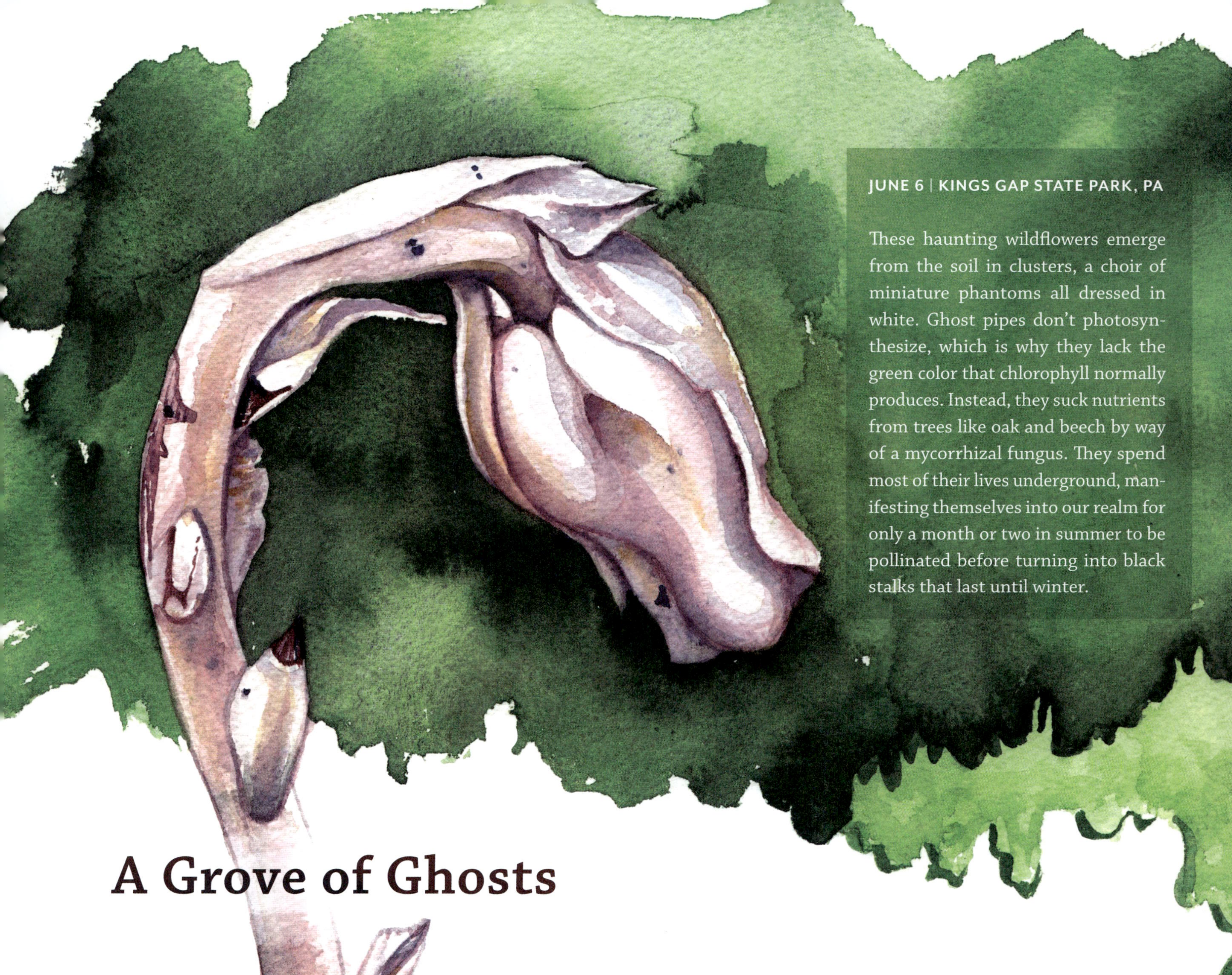

A Grove of Ghosts

These haunting wildflowers emerge from the soil in clusters, a choir of miniature phantoms all dressed in white. Ghost pipes don't photosynthesize, which is why they lack the green color that chlorophyll normally produces. Instead, they suck nutrients from trees like oak and beech by way of a mycorrhizal fungus. They spend most of their lives underground, manifesting themselves into our realm for only a month or two in summer to be pollinated before turning into black stalks that last until winter.

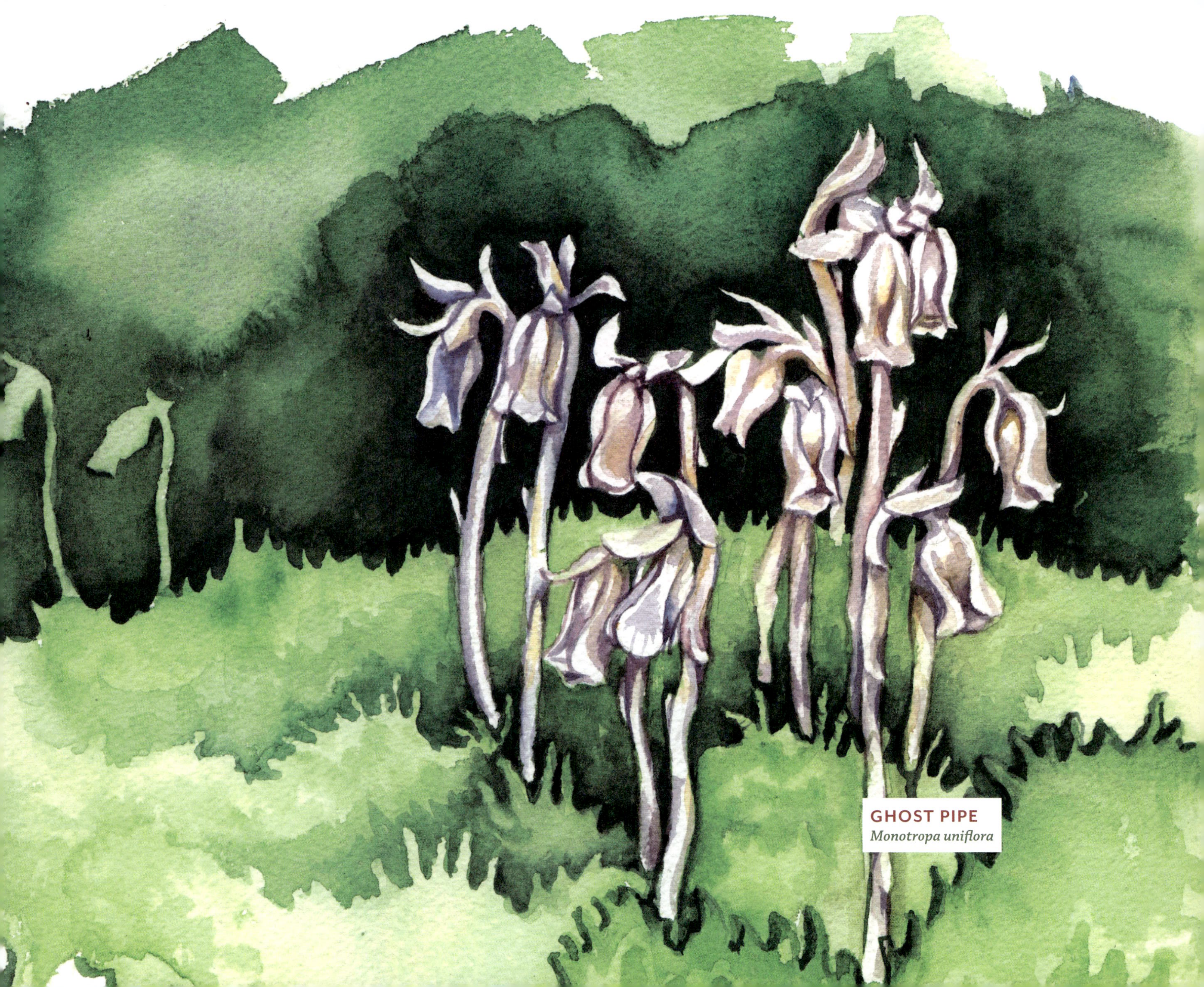

GHOST PIPE
Monotropa uniflora

Keeper of the Waters

JUNE 8 | KINGS GAP STATE PARK, PA

I'm about to trot across a wooden footbridge when I notice a giant spider standing directly in my path. With a diameter of about four inches and thornlike spikes covering its legs, the spider makes for a fearsome guard. I assess it from a distance and then move a little closer as my courage builds. Mentally, I trace its markings, taking in the surprisingly beautiful star-shaped pattern that adorns its cephalothorax (head).

But why all those spiky hairs? As it turns out, the hairs actually repel water, allowing the spider multiple modes of transportation: it can skate on the water's surface or swim underneath without really getting wet, which enables it to catch small fish, tadpoles, and crayfish. On top of that, the air bubbles that become trapped in those hairs can provide enough oxygen for it to stay underwater for up to thirty minutes.

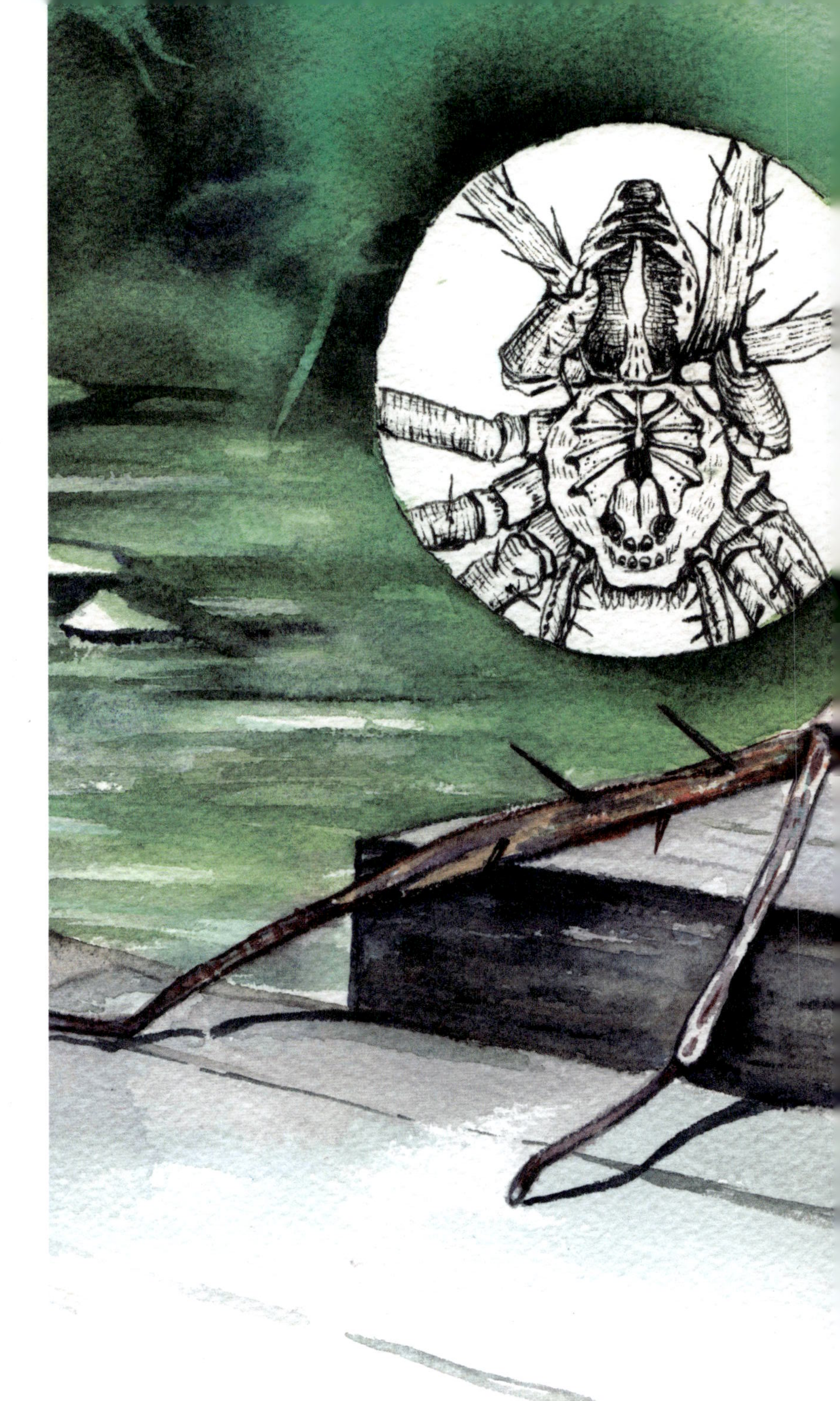

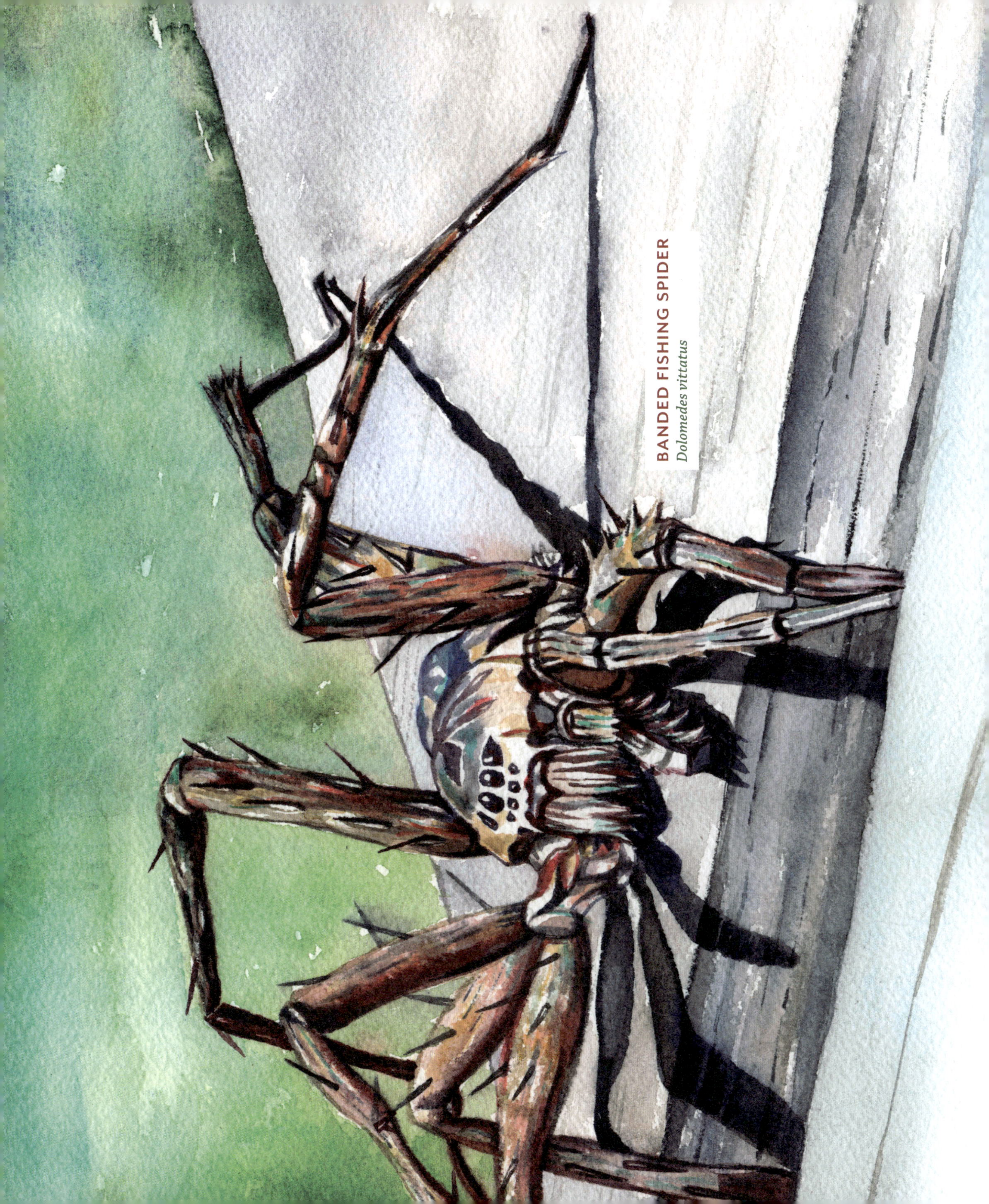

BANDED FISHING SPIDER
Dolomedes vittatus

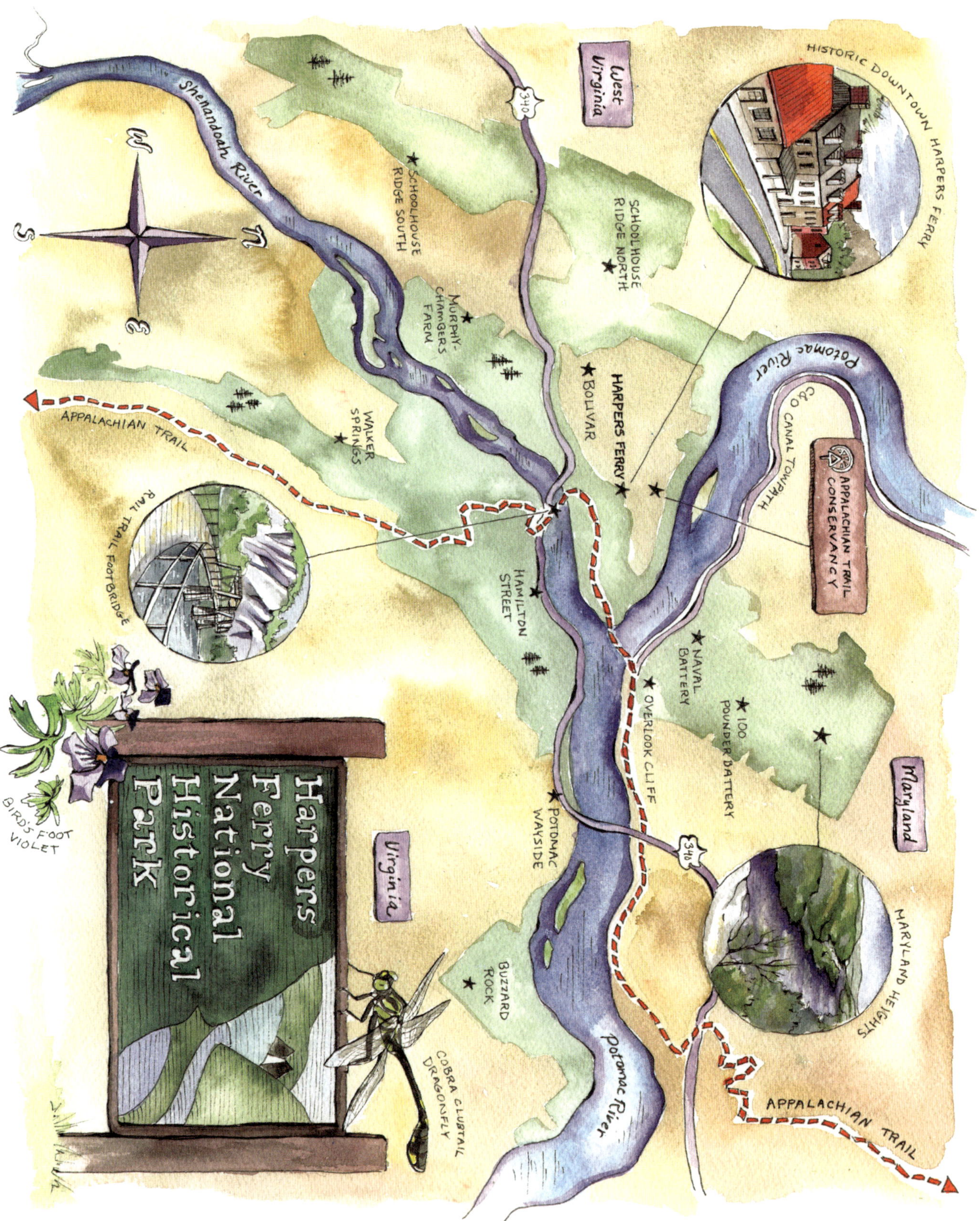

W
S
N
E
Shenandoah River
West Virginia
340
HISTORIC DOWNTOWN HARPERS FERRY
SCHOOLHOUSE RIDGE NORTH
Potomac River
Schoolhouse Ridge South
MURPHY-CHAMBERS FARM
HARPERS FERRY
BOLIVAR
APPALACHIAN TRAIL
WALKER SPRINGS
C&O CANAL TOWPATH
APPALACHIAN TRAIL CONSERVANCY
RAIL TRAIL FOOTBRIDGE
HAMILTON STREET
NAVAL BATTERY
100 POUNDER BATTERY
OVERLOOK CLIFF
Maryland
BIRD'S FOOT VIOLET
Harpers Ferry National Historical Park
Virginia
POTOMAC WAYSIDE
340
MARYLAND HEIGHTS
BUZZARD ROCK
Potomac River
COBRA CLUBTAIL DRAGONFLY
APPALACHIAN TRAIL

HARPERS FERRY NATIONAL HISTORICAL PARK

Harpers Ferry was designated a national historical park due to its significance as the site of John Brown's abolitionist attempt to capture the federal armory and arsenal in 1859. Today, visitors can wander into and among homes, churches, factories, and mills that have been preserved as museums to learn about this unassuming town's big role in the Civil War.

But the explorations aren't limited to history lessons. The 4,000-acre park is uniquely positioned at the intersection of three states (West Virginia, Virginia, and Maryland) and the confluence of two rivers (the Shenandoah and the Potomac). If you take your time to hike the trails or take a raft out on the water, you'll find a multitude of interesting critters. In fact, the park has documented 174 bird species and more than 140 insect species. Keep your eyes peeled for the evasive, emerald-hued green heron or the vibrant cobra clubtail dragonfly lurking among the wispy grasses at the water's edge.

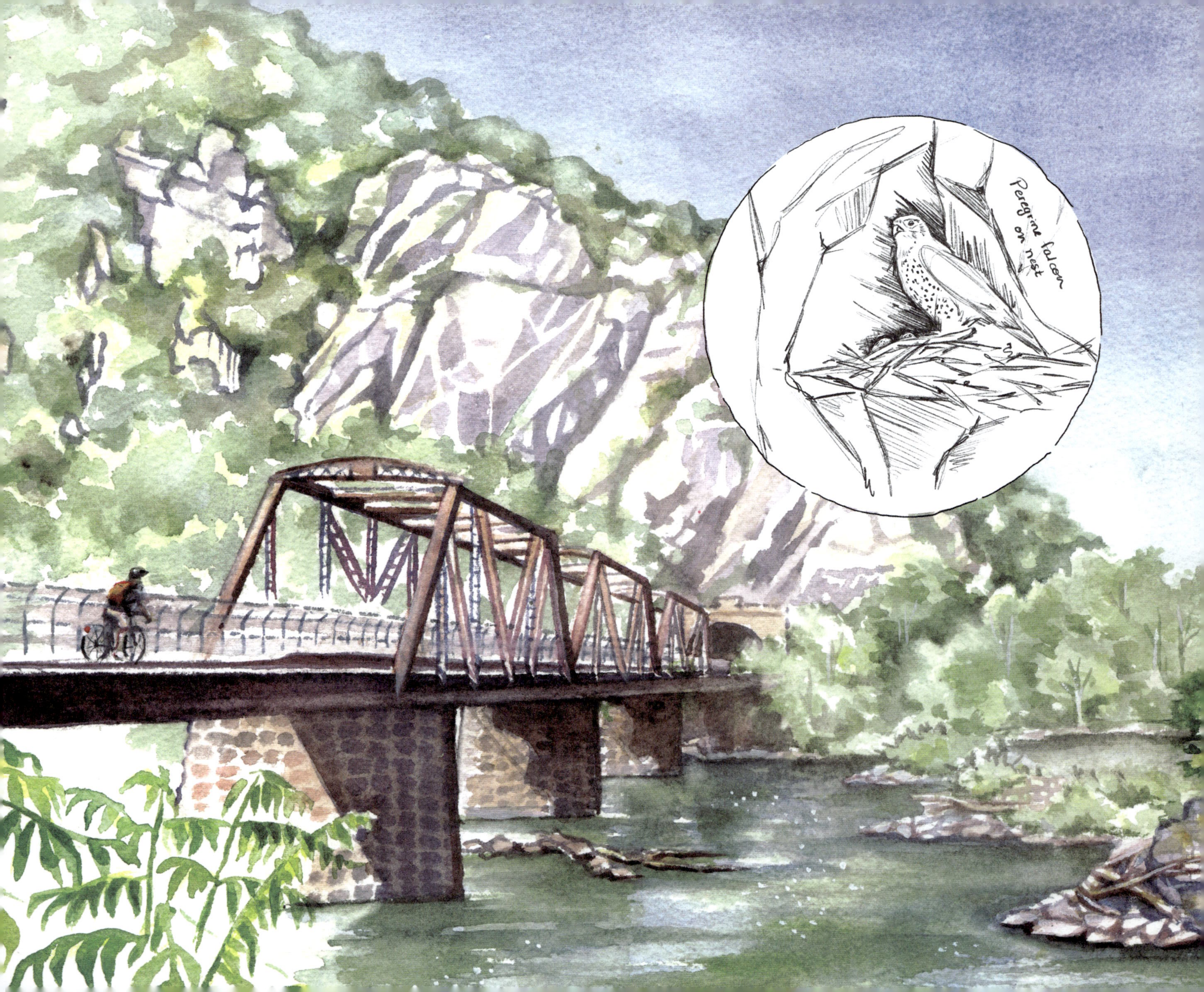
Peregrine falcon on nest

Hidden Harpers Ferry

JUNE 18 | HARPERS FERRY NATIONAL HISTORICAL PARK, WV

I cross the footbridge over the wide, rocky Potomac River and watch people on inner tubes riding the currents below. One group has formed a human chain by holding hands, reminding me of the more than twenty painted turtles I saw a few days ago, all lined up on a log to bask in the sun. As I climb the steep trail to Maryland Heights, I come upon several hikers pointing and looking into the woods beside the trail. A copperhead is weaving its way over a fallen log!

Up on the ridge, I'm greeted by a stunning vista of the town below and the two rivers as they sweep off into the Blue Ridge Mountains. I take a different trail down and along the way am treated to a slime mold I've never seen before: red raspberry slime, spread across a fallen log like a big smear of juicy jam. Then right before I finish up the loop hike, I detect the slightest movement on the ground and find a baby praying mantis that's the length of my pinky fingernail. It looks exactly like an adult, just a miniature version.

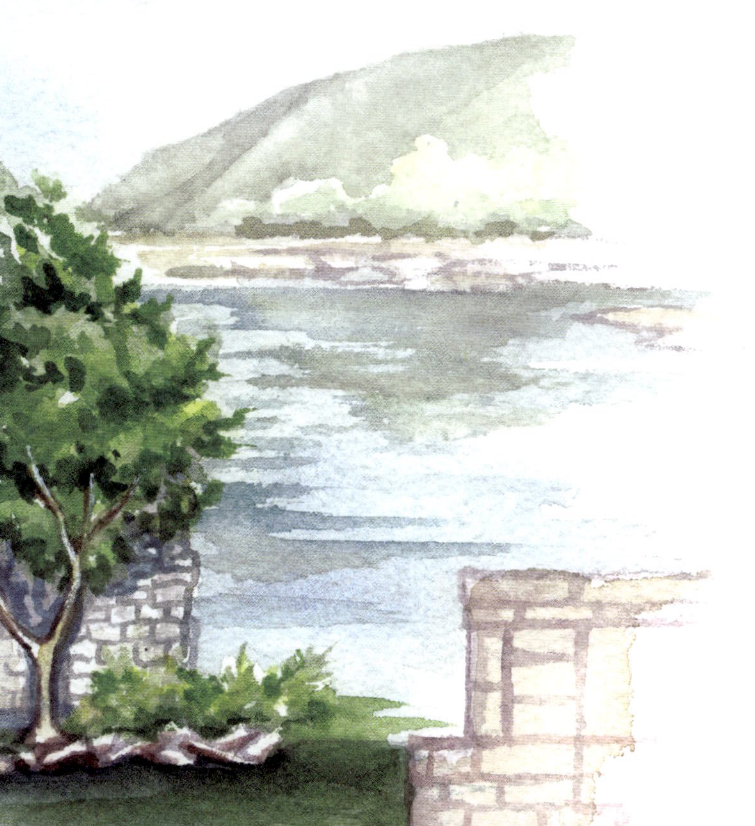

Peregrine falcons have active nesting sites on these rocky cliffs.

A Rainbow of Slime

JUNE 4–30 | CENTRAL PA AND EASTERN WV

I've become oddly obsessed with finding colorful slime molds. I usually see them on rotting logs after a good rain. While slime molds may look similar to fungi, they are actually single-celled organisms that are categorized as a type of amoeba. Slime molds can quickly change shape, color, and even location, crawling across the surface of decomposing wood and plant matter.

Scientists are discovering new attributes all the time, like the fact that despite not having a brain, slime molds possess something like a collective memory—and they can even "learn" to crawl through a maze! Whereas before I didn't pay too much attention to rotting wood unless there were mushrooms visible, I'm now hyperaware of this entirely new world of slime molds and can't help but see them everywhere.

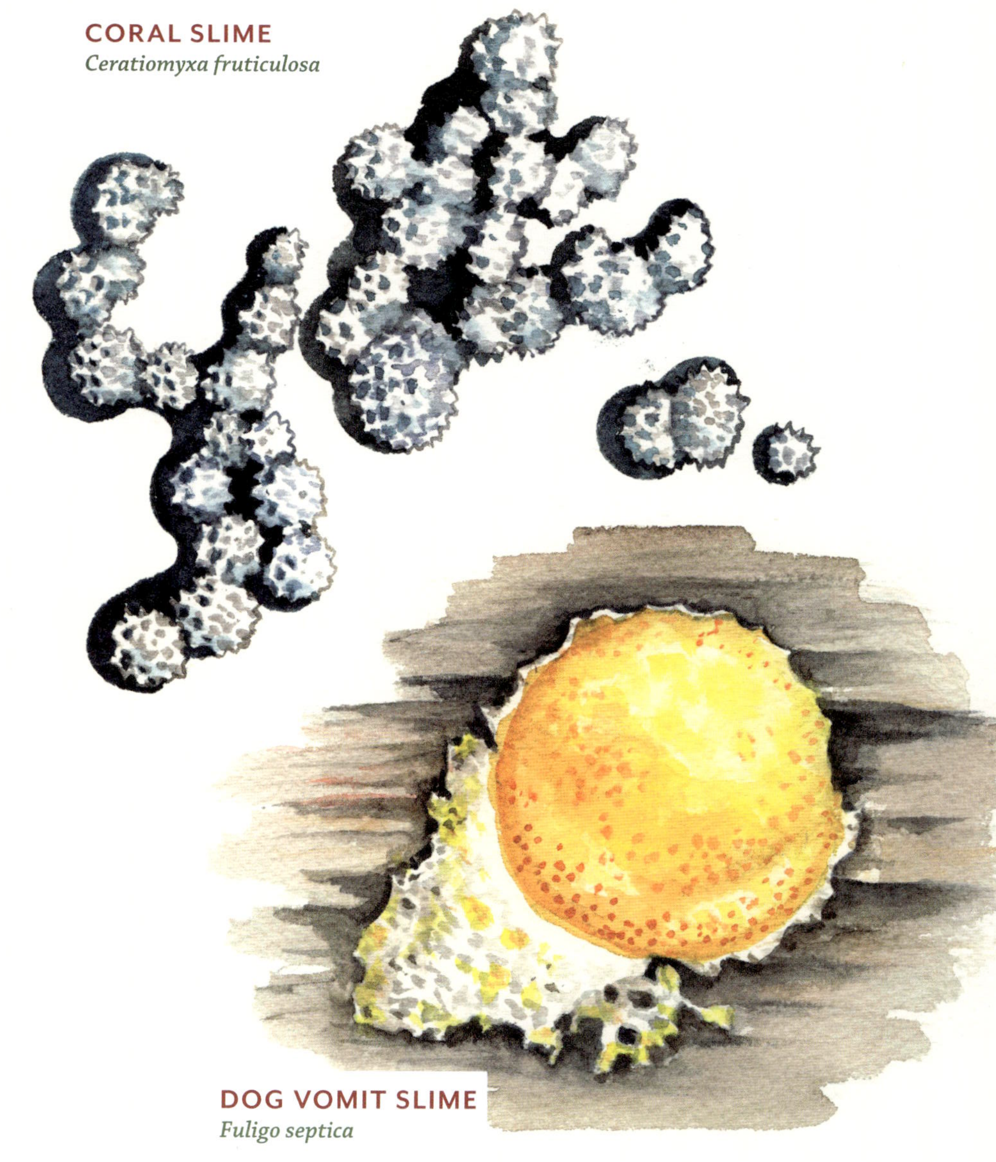

CORAL SLIME
Ceratiomyxa fruticulosa

DOG VOMIT SLIME
Fuligo septica

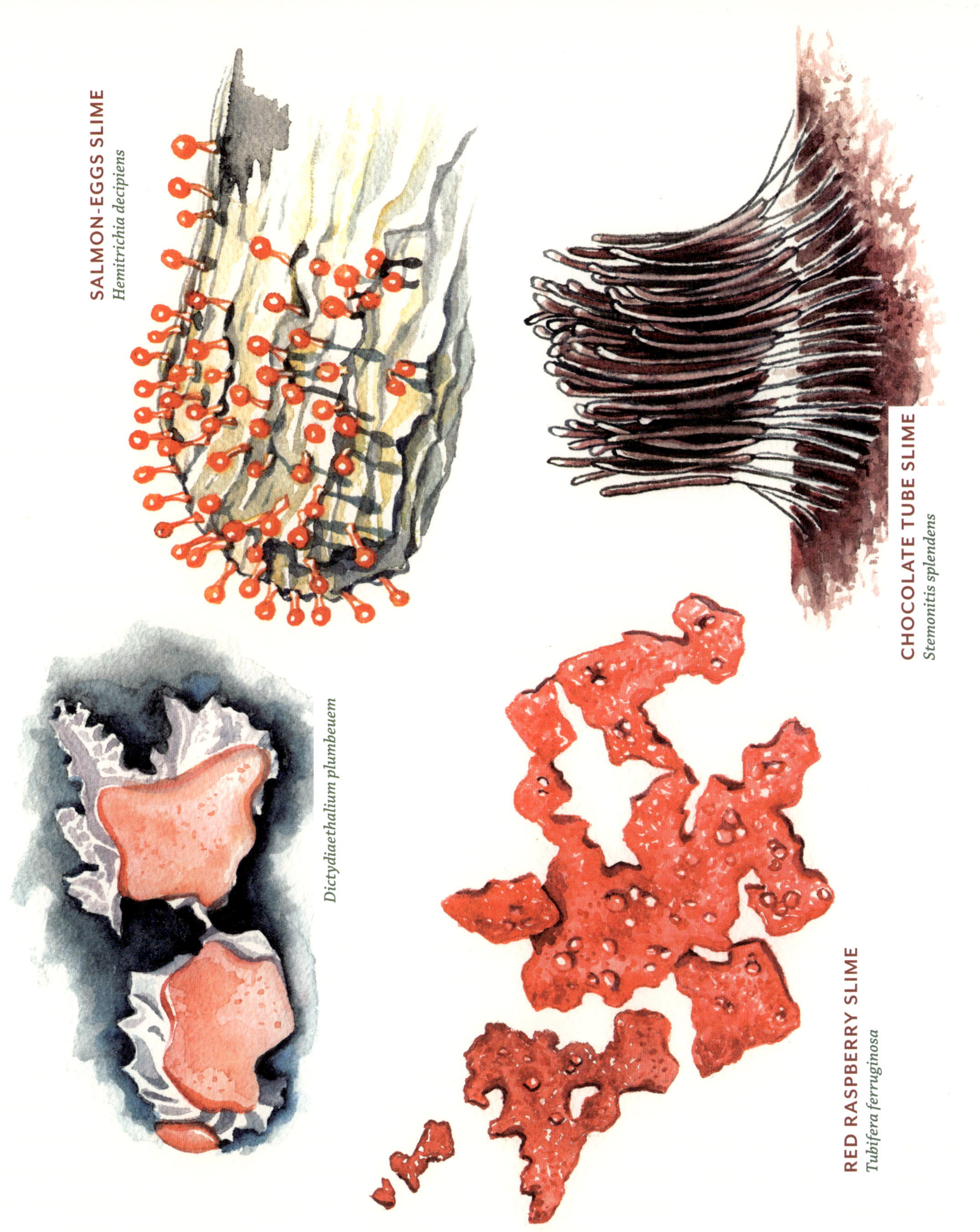

SALMON-EGGS SLIME
Hemitrichia decipiens
CHOCOLATE TUBE SLIME
Stemonitis splendens
Dictydiaethalium plumbeuem
RED RASPBERRY SLIME
Tubifera ferruginosa

Creature Discomforts

After an intensely hot and buggy uphill hike, I am relieved to arrive at Chimney Rocks in Michaux State Forest, an outcropping that will make for a nice spot to rest and reflect. I munch a few snacks, then pull out my notebook and begin to journal about how happy I am that I made the effort to do this hike today despite the heat.

The moment I put my journal away, a juvenile five-lined skink with a metallic-aqua tail scuttles across the boulder near me. But more than one reptile is enjoying this sun-soaked rock. My peripheral vision catches a meaty snake head as it emerges from behind my picnic boulder, its tongue flicking directly at me.

"Hey, you stay right there!" I command. After a pause, the black rat snake hastily slithers my way, prompting me to shatter my moment of calm and gratitude by loudly declaring, "Oh, no you don't!" I make an awkward attempt to scramble away on legs that have inconveniently turned to jelly.

This is when I notice a couple of thru-hikers resting on another rock nearby. It's obvious that they have witnessed my overreaction to this harmless black rat snake. I'm embarrassed and want to defend myself: "I wasn't actually scared! I'm brave! I hike every day!" Instead, I sheepishly head back down the trail, my snake senses fully activated. Every little rustle in the ferns has me pausing and scanning for threats.

I've seen dozens of snakes on this trip so far, almost all of them nonvenomous, and each time I am freshly startled. I just can't seem to shake that intense, split-second reaction to their presence. I've even gone so far as to try some reverse psychology on myself, treating myself to kombucha or ice cream after hikes that involve a snake sighting. So far it hasn't worked. I was also hopeful that I'd become less afraid after seeing snakes on almost every hike that I went on in Virginia, yet there is still no sign of my fear letting up. I did, however, compose an original rhyme: "If it scampers, I'm a happy camper. If it slides, stride on by."

My walks provide me with peaceful, restorative, inspiring moments. But even during those moments, there is almost always something unpleasant happening simultaneously. Maybe it's a swarm of gnats dive-bombing my face while I take in the serenity of a mountain stream at golden hour, or maybe it's the low-grade awareness of the black bears that might be lurking near my tent while I enjoy the silence of a night spent in the wilderness. Maybe it's a snake sharing a sunny picnic spot.

Such discomforts push many of us to interact with nature in a curated way. We invite it into our homes in the form of houseplants, floral bedspreads, fish tanks, pet cats, landscape paintings, and clothing with animal motifs. We visit zoos, aviaries, and botanical gardens where we can enjoy the wonders of the wild while staying clean and safe on the paved paths that meander through the displays.

Recently my dad reminded me of the time when eleven-year-old Rosalie asserted, "I enjoy nature from a safe distance." While I was content to spend weeks painting a mural of a tree on my bedroom wall, only occasionally did I seek out authentic experiences outdoors. A large factor was my paralyzing fear of snakes. After seeing a single garter snake on a path near our home, I proceeded to name the trail Snake Alley and did everything in my power to avoid it from then on. I wore knee-high rubber boots while playing outside during the hottest months

of summer, even though I was well aware that there were zero venomous snake species in our region.

How is it, then, that I now feel a deep need to seek out real, wild encounters with nature? What do they provide that I don't get from a safer, less nerve-racking trip to the zoo? I think the crucial difference lies in the fact that I feel more alive when I have immersive outdoor experiences. The threats make me hyperalert, priming me to notice more of the beauty too.

I don't know that I'll ever feel fully at peace in the woods, but I also don't think that should be my goal. Not only is this vigilance a natural evolutionary response to help keep me safe, but it also allows me to experience the wonders of nature more fully. I have to be willing to deal with the discomforts in order to access the magic. The result is raw and rewarding and always worth it.

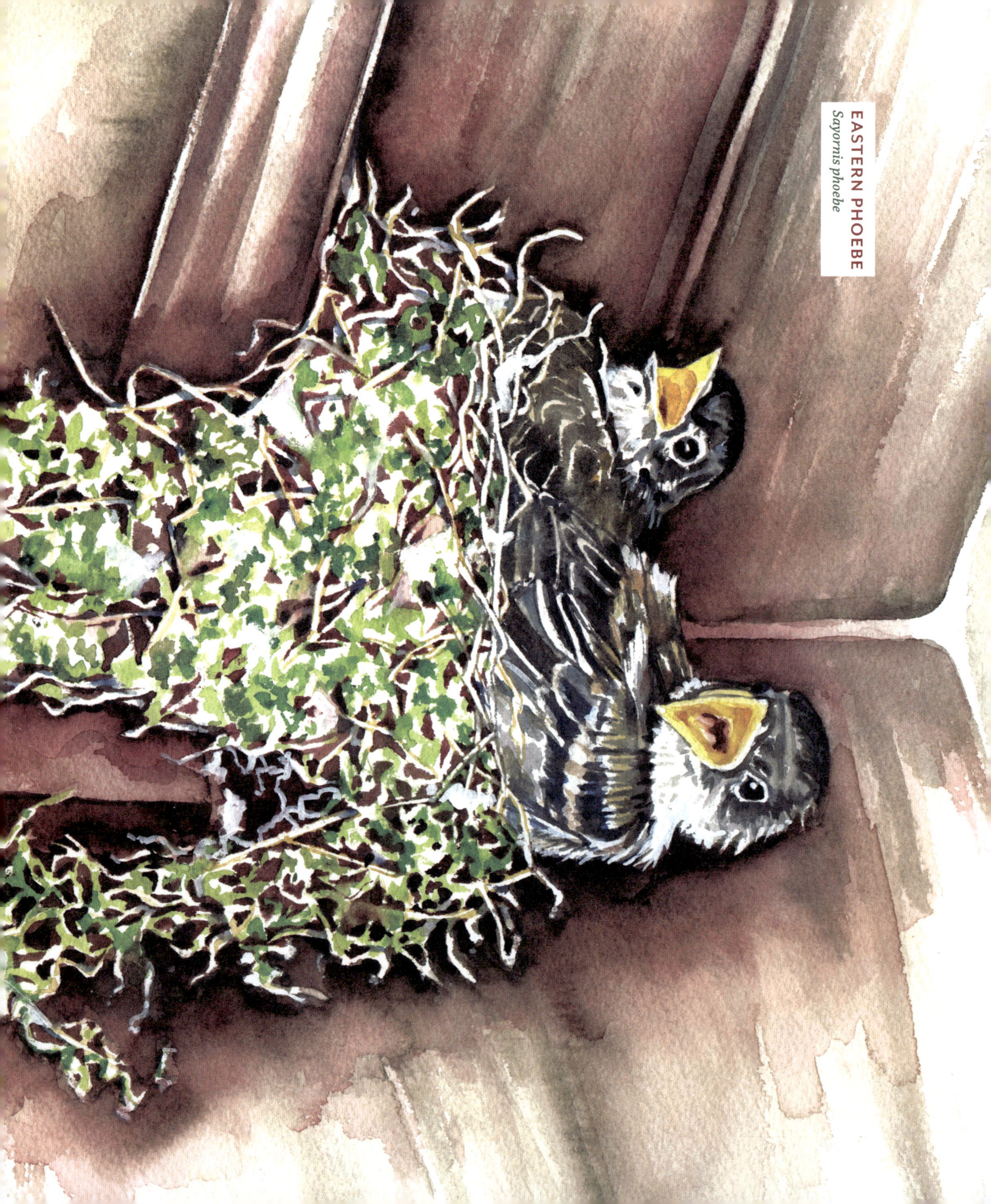

EASTERN PHOEBE
Sayornis phoebe

Oh, Babies

It's our first wedding anniversary, and Ben and I decide to celebrate by taking a kayak out on a lake. It's also Father's Day, so the park is crawling with families, many of them picnicking and swimming at the lake's sandy beach. We paddle to a quieter corner and spy on dragonflies and damselflies perched among the plants at the water's edge. Back on shore, as I'm headed into the restroom to change into dry clothes, I hear a loud cheeping emanating from a corner under the roof's overhang. I stand on tiptoe to investigate and come face-to-face with two sweet eastern phoebe babies, both shrieking from a nest fashioned from grass, moss, and mud.

I think about all the noisy toddlers who have overrun the park today. "Well, you fit right in," I say to the birds. With their mouths wide open, these wailing songbirds offer me a unique view of their clearly defined tongues, which perfectly match the yellow of their beaks. It turns out that all birds have tongues, which tend to closely resemble human ones. When these phoebes grow up, they will use their tongues to call out their quick and raspy "fee-bee"—a helpful means of identifying these otherwise nondescript brown flycatchers.

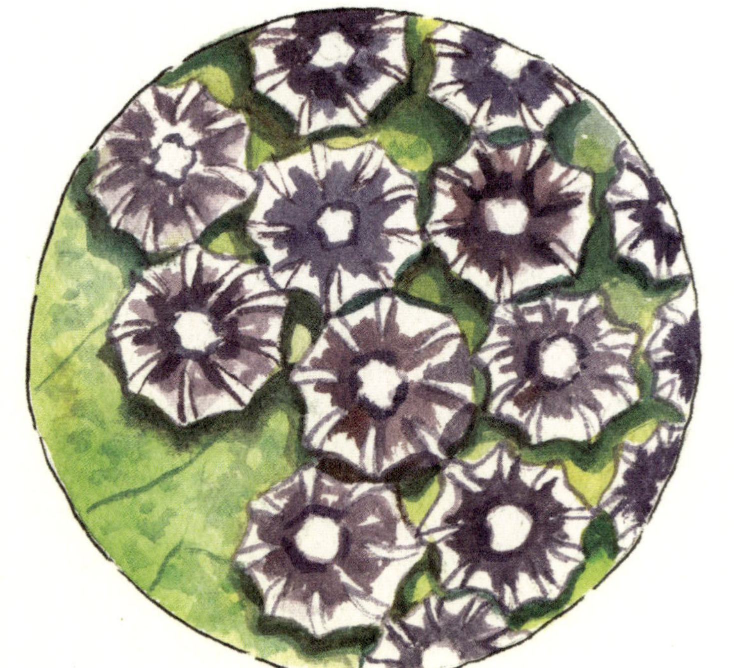

Subtle Regality

JUNE 21 | CALEDONIA STATE PARK, PA

The golden hour light drenches the pines as I take an early evening walk to see if I can find any birds, but the woods and waters are totally silent. Just as I'm about to get back in the car, I spot a gorgeous mourning cloak butterfly in the parking space next to mine. I'd seen another mourning cloak a couple of weeks ago at Kings Gap State Park, but it had startled and flown away before I was able to get a close look. This is my chance! I retrieve my camera from the trunk and, drawing on my yoga practice, tiptoe ever so slowly toward the butterfly, each movement smooth and deliberate so as not to spook it. With its wings open, this mourning cloak displays a regal combination of baby blue, maroon, and metallic gold; folded together, its wings are brown and mottled, a humble disguise that comes in handy for blending in seamlessly with the bark of oak trees. Oak sap is a favorite food during this butterfly's adult phase.

LEFT: *The eggs of the mourning cloak have an amazingly geometric shape.*

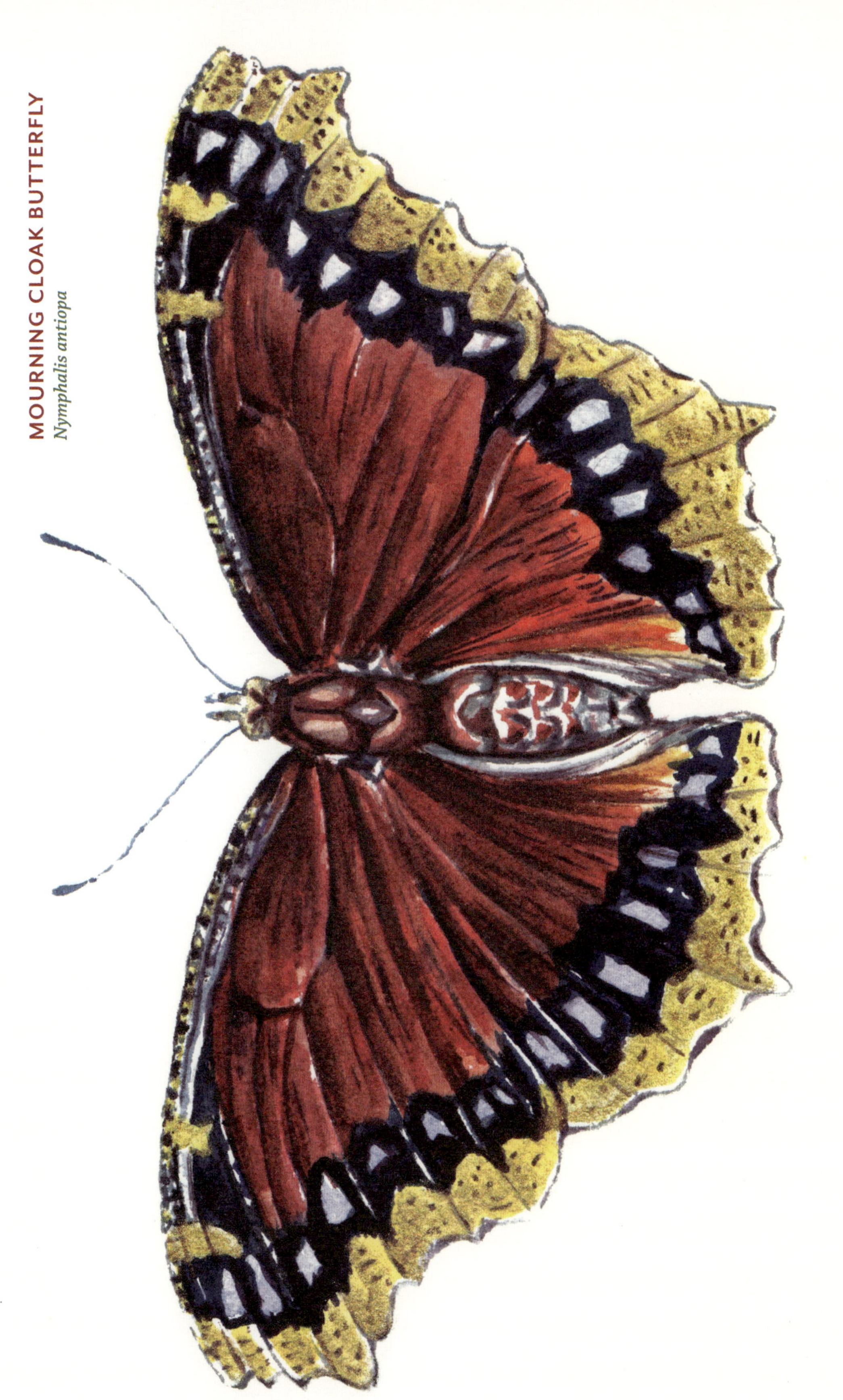

MOURNING CLOAK BUTTERFLY
Nymphalis antiopa

Moth Watch

**JUNE 25 | SHAVER'S CREEK
ENVIRONMENTAL CENTER, PA**

There is a powerful floodlight mounted outside the door of the cabin we're camping out in this weekend. Just as I'm about to go to bed, I get an idea. Since this cabin is tucked away deep in the woods with no surrounding light pollution, this is the perfect spot for some moth watching! We prepare snacks and take our seats on the front stoop, waiting for the show to begin. As midnight nears, the action picks up. Moths gravitate toward light because they use the moon and stars to orientate during flight. Tonight, we are visited by several kinds of tiger moths, some of them fluttering right into our faces.

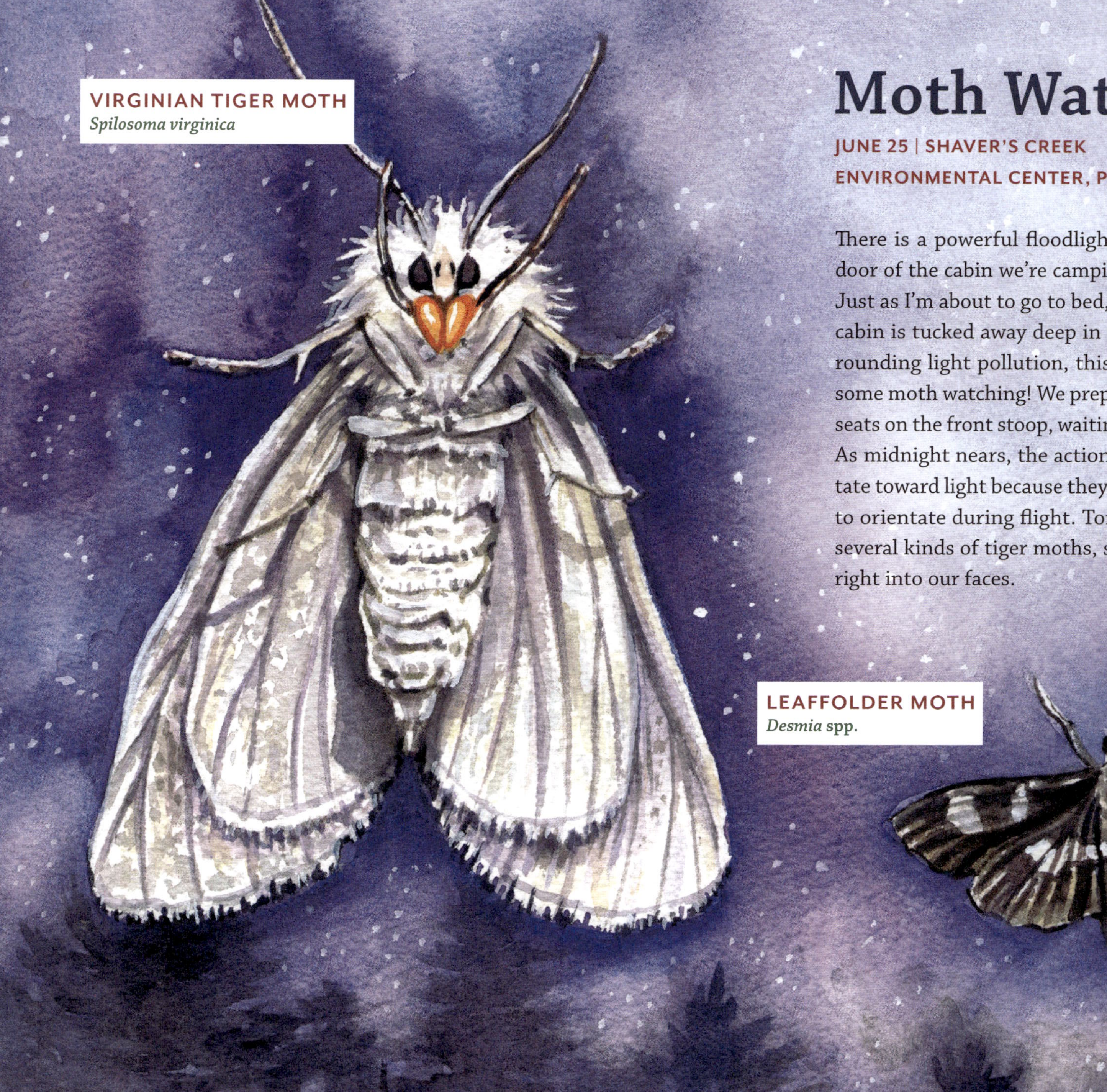

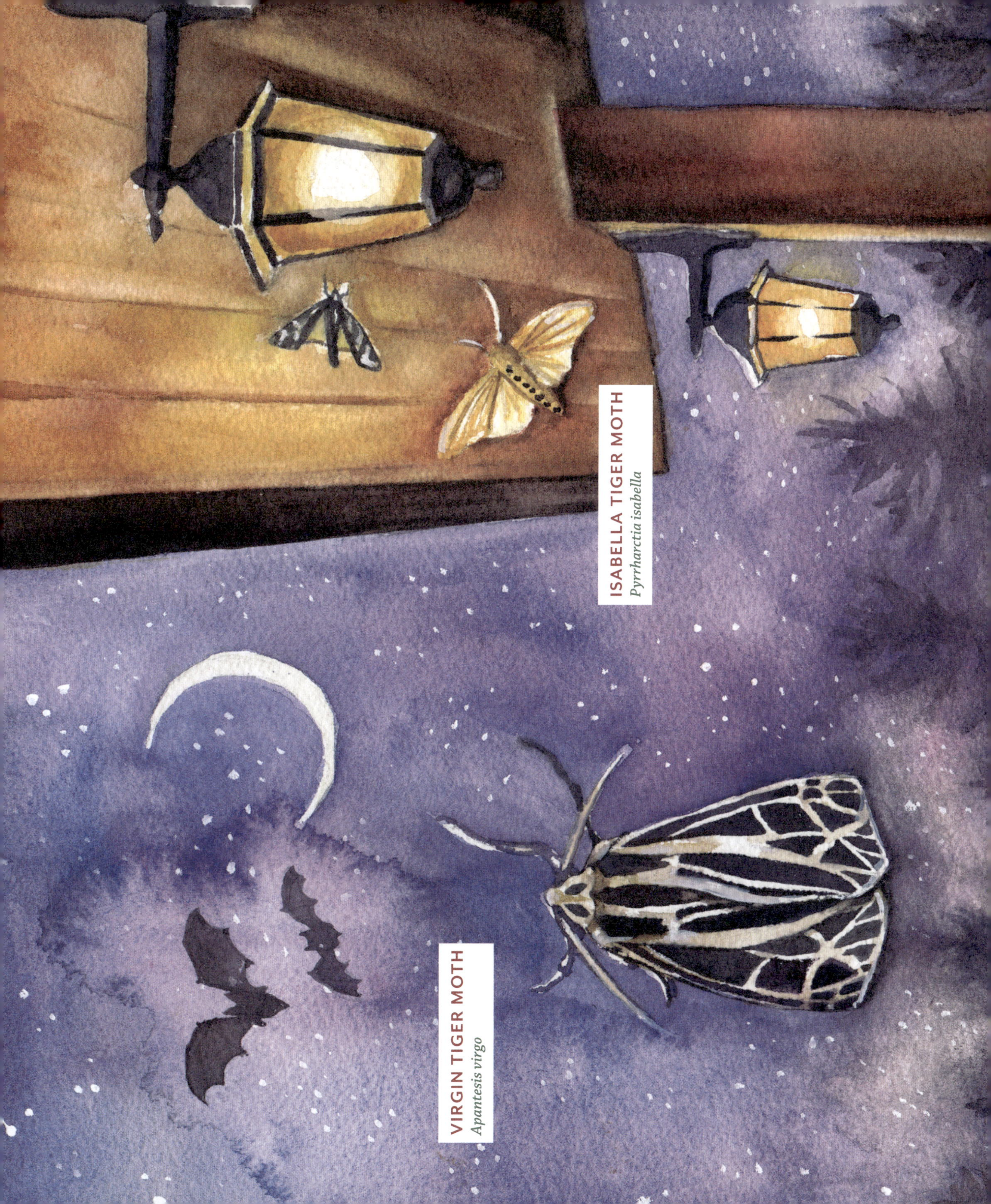

ISABELLA TIGER MOTH
Pyrrharctia isabella
VIRGIN TIGER MOTH
Apantesis virgo

Pollinator Paradise

JUNE 25 | SHAVER'S CREEK ENVIRONMENTAL CENTER, PA

I walk the loop trail around a lake, my eyes continually drawn to the fluttering movements among the plants that line the path. A few great spangled fritillary butterflies feast on the vibrant blooms of orange milkweed, or "butterfly weed," pivoting in complete circles around each flower while using their straw-like proboscises to slurp up the nectar. Every few seconds, they spread their wings and then collapse them down in one fluid motion. Butterflies open and close their wings even when they're not flying as a way to regulate their temperature.

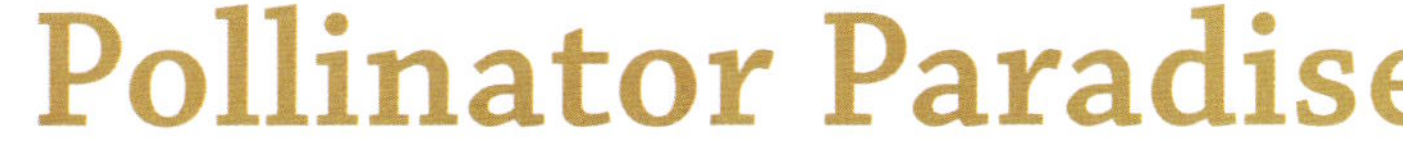

GREAT SPANGLED FRITILLARY BUTTERFLY
Speyeria cybele

BUTTERFLY WEED
Asclepias tuberosa

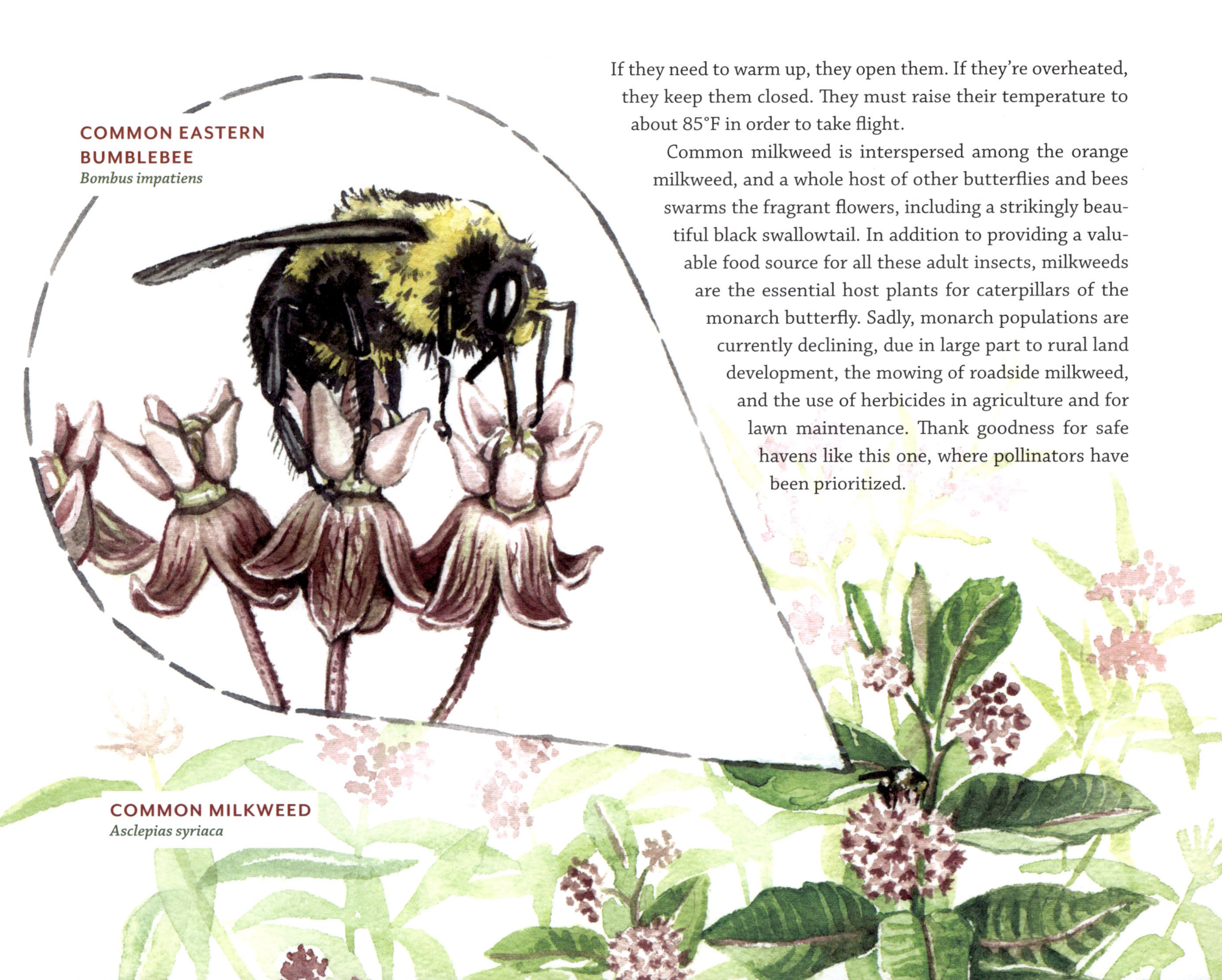

If they need to warm up, they open them. If they're overheated, they keep them closed. They must raise their temperature to about 85°F in order to take flight.

Common milkweed is interspersed among the orange milkweed, and a whole host of other butterflies and bees swarms the fragrant flowers, including a strikingly beautiful black swallowtail. In addition to providing a valuable food source for all these adult insects, milkweeds are the essential host plants for caterpillars of the monarch butterfly. Sadly, monarch populations are currently declining, due in large part to rural land development, the mowing of roadside milkweed, and the use of herbicides in agriculture and for lawn maintenance. Thank goodness for safe havens like this one, where pollinators have been prioritized.

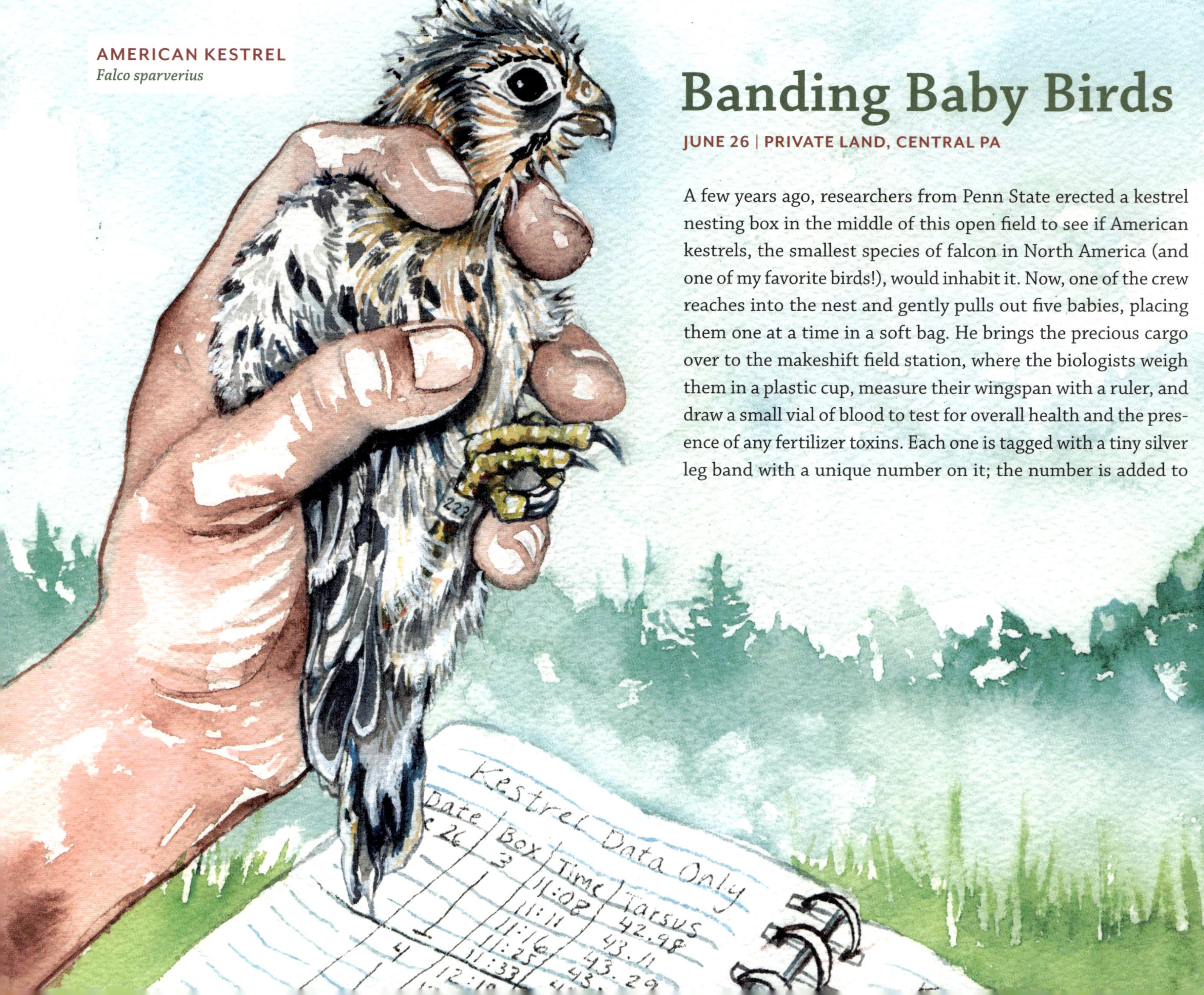

Banding Baby Birds

JUNE 26 | PRIVATE LAND, CENTRAL PA

A few years ago, researchers from Penn State erected a kestrel nesting box in the middle of this open field to see if American kestrels, the smallest species of falcon in North America (and one of my favorite birds!), would inhabit it. Now, one of the crew reaches into the nest and gently pulls out five babies, placing them one at a time in a soft bag. He brings the precious cargo over to the makeshift field station, where the biologists weigh them in a plastic cup, measure their wingspan with a ruler, and draw a small vial of blood to test for overall health and the presence of any fertilizer toxins. Each one is tagged with a tiny silver leg band with a unique number on it; the number is added to

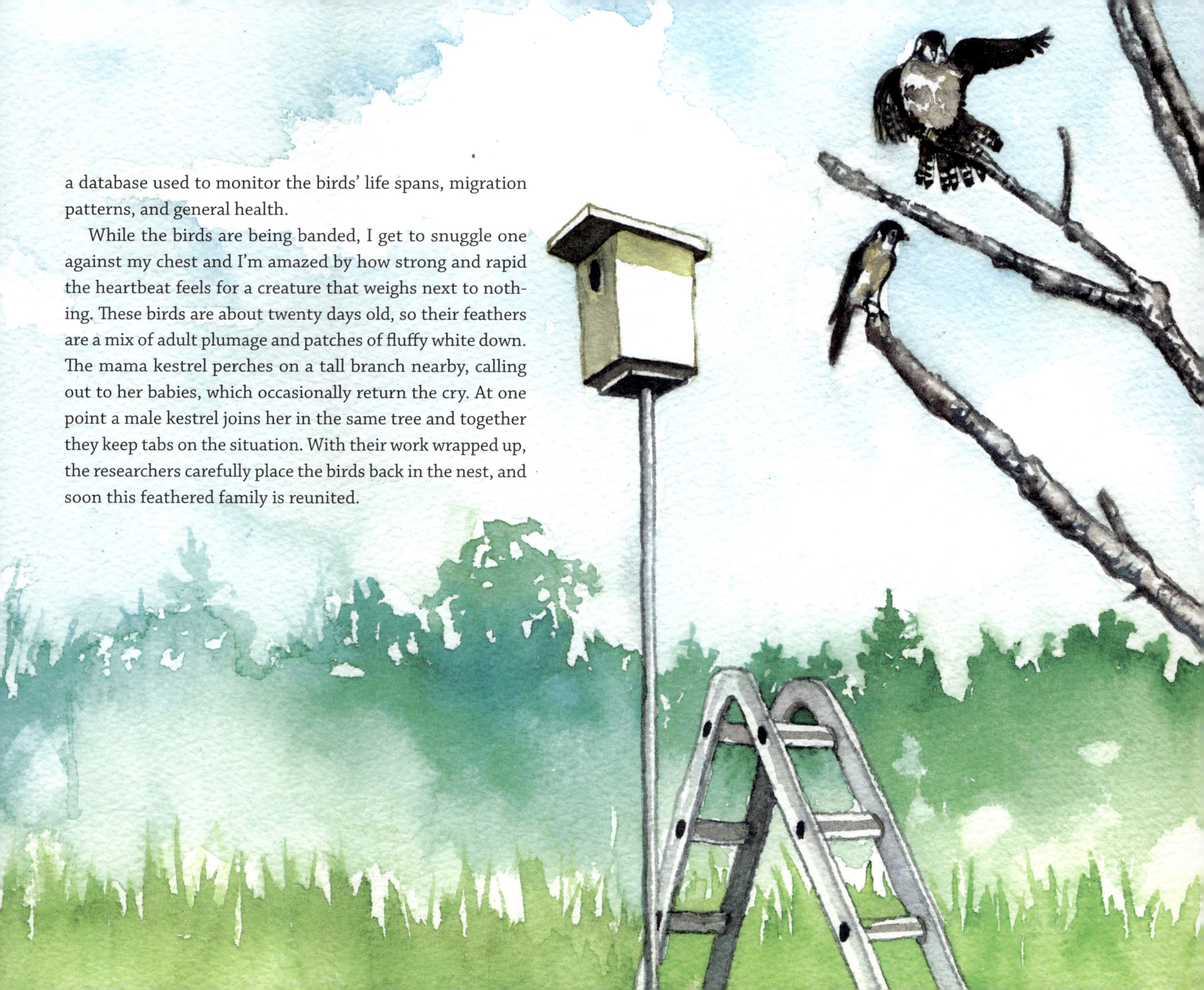

a database used to monitor the birds' life spans, migration patterns, and general health.

While the birds are being banded, I get to snuggle one against my chest and I'm amazed by how strong and rapid the heartbeat feels for a creature that weighs next to nothing. These birds are about twenty days old, so their feathers are a mix of adult plumage and patches of fluffy white down. The mama kestrel perches on a tall branch nearby, calling out to her babies, which occasionally return the cry. At one point a male kestrel joins her in the same tree and together they keep tabs on the situation. With their work wrapped up, the researchers carefully place the birds back in the nest, and soon this feathered family is reunited.

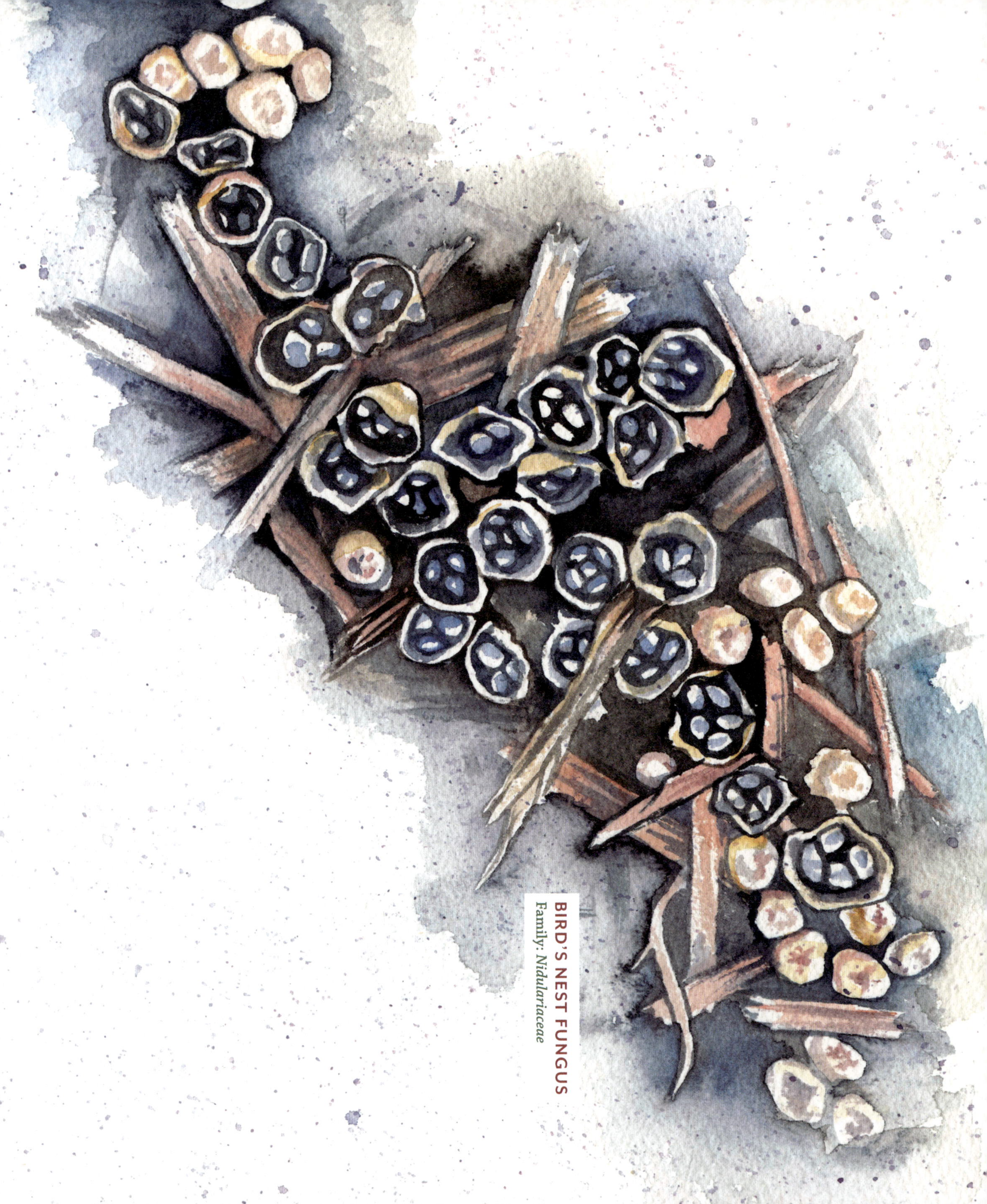

BIRD'S NEST FUNGUS
Family: Nidulariaceae

Nano Nests

I wander down a sidewalk in Chambersburg, trying to act casual as I peer into the mulched flower beds in front of each home. Whenever a car drives by, I pause my nature-snooping and pretend I'm just on a walk like a normal person.

But today I'm on a mission: I recently heard that the strange, cup-shaped bird's nest fungus can sometimes be found growing in gardening mulch. I don't have much faith that I'll actually find it, but I continue to scan the flower beds . . . until eventually my eyes land on a sea of the unusual fungi. *No way!* The smooth, dark "eggs" are actually protective sacs called peridioles, which house the mushroom's spores. When raindrops land on the sacs, the pressure of the rain springboards the peridioles up and out of the cuplike "nest." Eventually, these runaway peridioles, which can land up to four feet away, will dry out and break open to release the spores and continue the reproduction process.

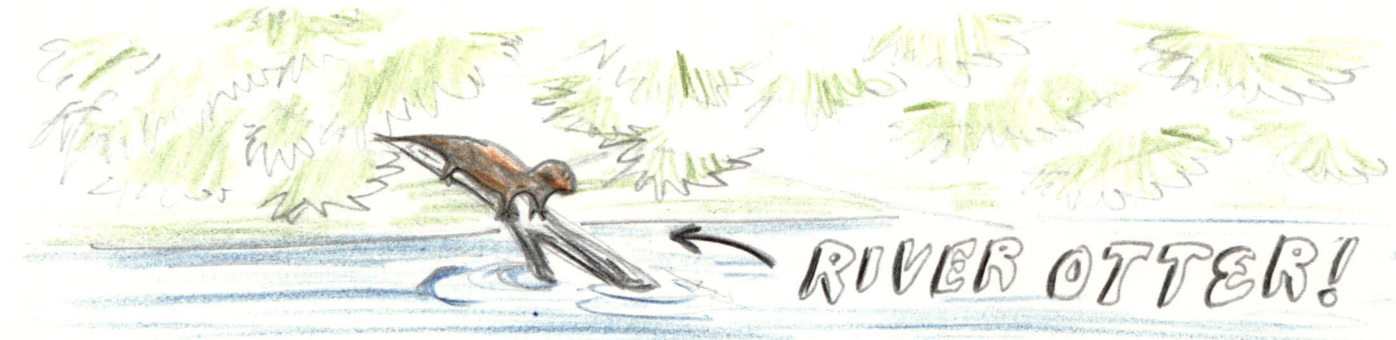

I'm sitting by the river, positioned in a good hiding spot behind a tree, just watching a few ducks float past (mallards) w/ their babies. First thing I noticed was a tiny goldfinch pretty close to me, taking a few sips at the water's edge. Definitely was proud of myself for sitting still + quietly enough that the bird felt OK taking a drink! But THEN.........

JUNE 30th
7ish PM
River that goes thru
Chambersburg, PA,
Conococheague Creek

I saw a ripple in the water coming toward me and then out popped a somewhat pointy brown snout. I thought it was a beaver and got really excited. And THEN, it swam to the opposite shore and climbed out onto a fallen log that made a kind of ramp into the bushes. And it had a NOODLE body! It's a river otter! and now it's been going in and out of the water for a while, each time looking both ways before it re-enters the water. It's definitely fishing — diving down and coming back up with its mouth full of leaves and other stuff.

NATURE JOURNAL IDEAS

MOTH PARTY

When we only nature-snoop during the day, we miss out on an entire world of nighttime activity. Your challenge is to host a moth party: If you live in a place with minimal light pollution and have a bright light outside your home, you can simply watch that light late at night and see what fluttering friends show up. Otherwise, grab a buddy, a white bedsheet, and one or two bright flashlights. Find a dark area—this is a great activity to do while camping—and hang the sheet from a tree or clothesline. Then shine your lights at the sheet and wait. Stay up until midnight or later to see members of the Saturniid family, some of the largest and most colorful nighttime flying moths.

CAPTURE AND COMPARE

On my hikes, I'm only able to see and capture a tiny sliver of all the incredible plants, animals, landscapes, and experiences that are out there. Even if someone else were to follow in my exact footsteps, their experience would be completely unique because our attention would naturally be drawn to different things. Go outside with a friend or family member and spend thirty to forty-five minutes quietly sketching and recording observations about what you see in the same spot. When the time is up, compare your notes. What did you notice that the other didn't?

SLIME STROLL

Slime molds are squishy, colorful little marvels that are super fun to look for on hikes. In the summer after you've had a bit of rain, go for a walk in the woods with the intention of finding slime molds. Keep an eye out for damp, decaying logs and wet leaves—favorite places for the bacteria and other microorganisms that comprise slime molds' diet. Take photos or sketch your findings, and once you return home, consult a slime mold field guide or the internet to get an idea of what you might have seen. If you didn't come across any slime molds on this outing, did looking closely at fallen logs reveal any other interesting discoveries?

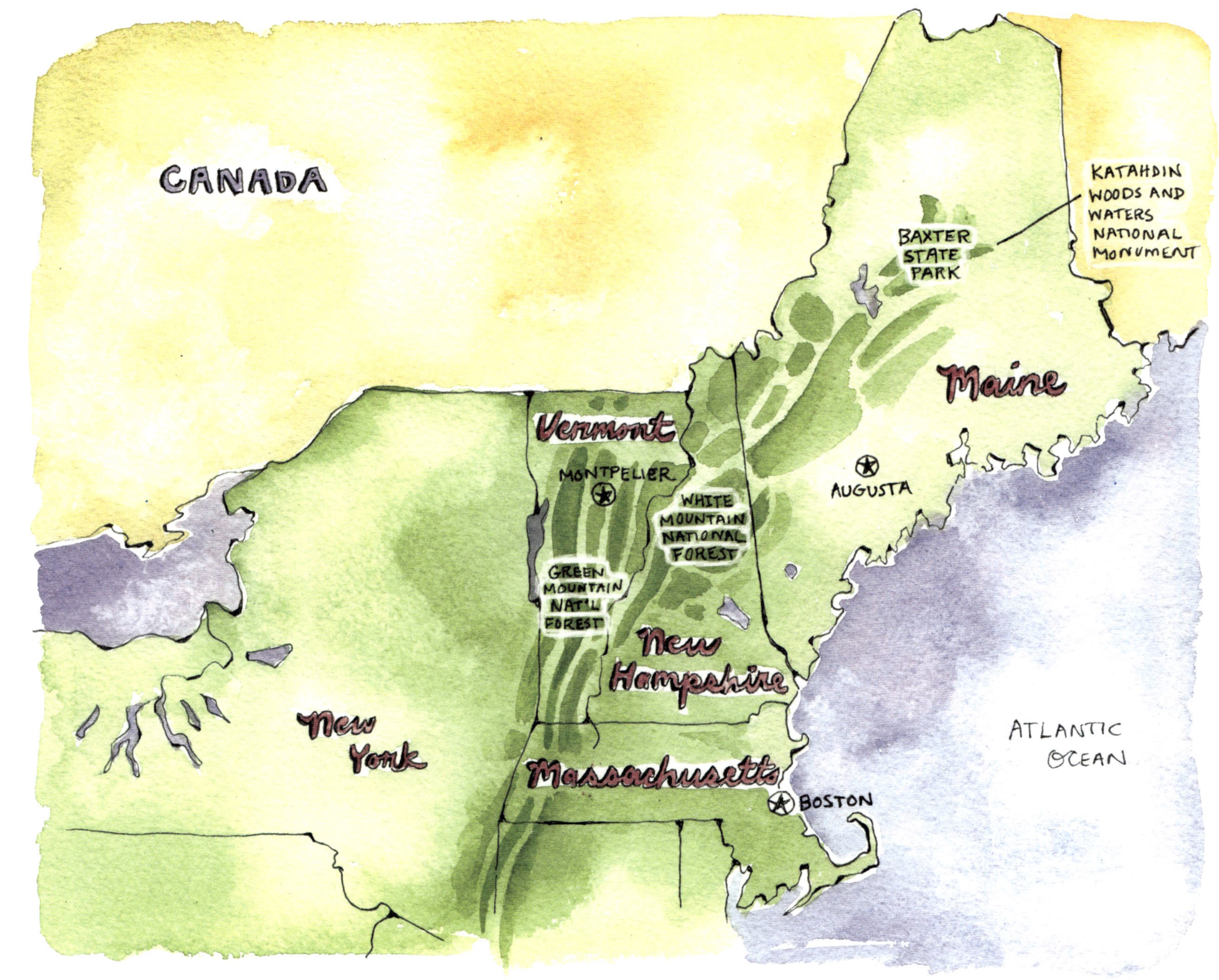

CANADA
KATAHDIN WOODS AND WATERS NATIONAL MONUMENT
BAXTER STATE PARK
Maine
Vermont
MONTPELIER
WHITE MOUNTAIN NATIONAL FOREST
AUGUSTA
GREEN MOUNTAIN NAT'L FOREST
New Hampshire
New York
Massachusetts
BOSTON
ATLANTIC OCEAN

northern APPALACHIANS

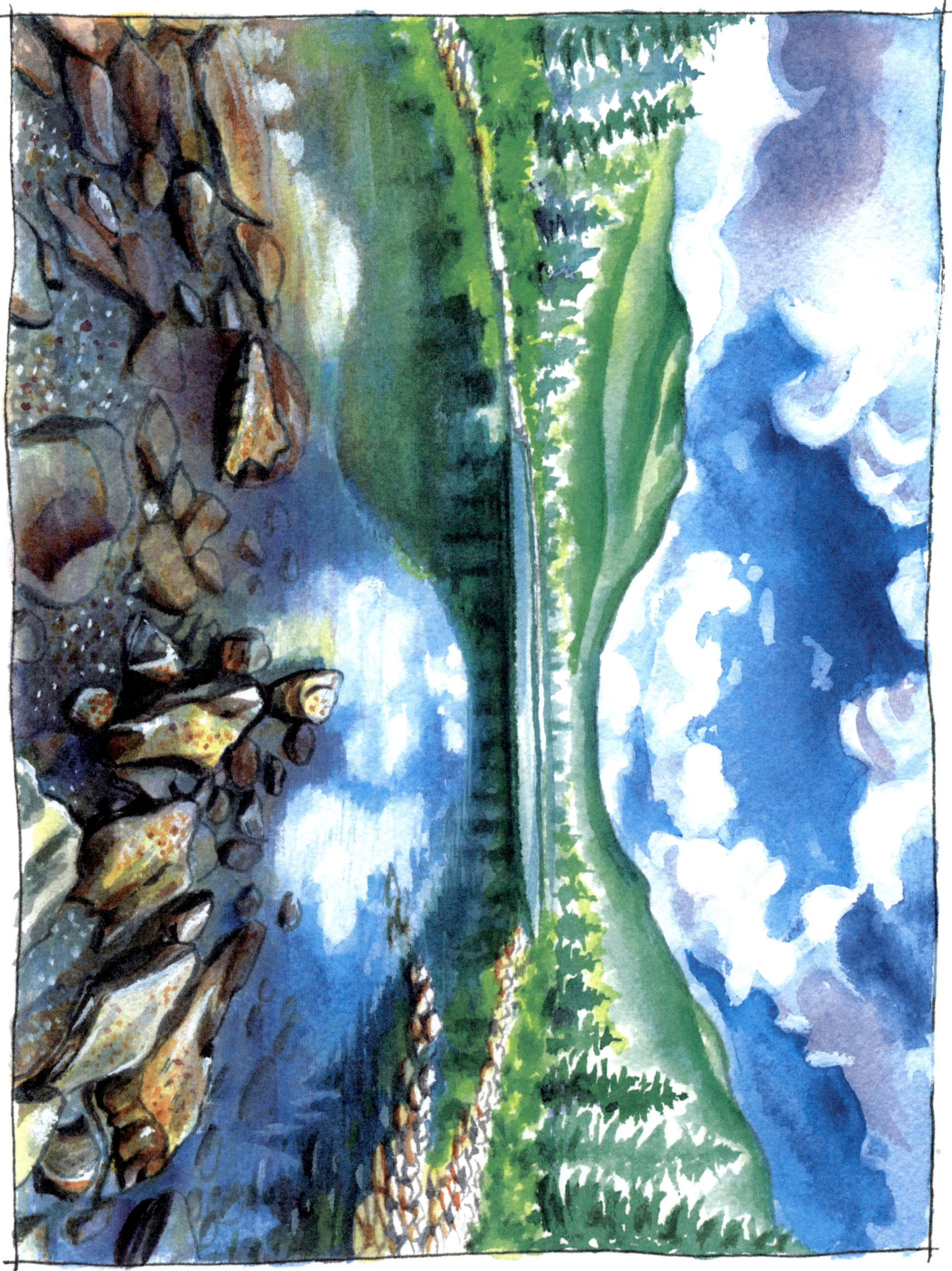

HEADING TO NEW ENGLAND

JULY 1

With a nine-hour drive ahead of us, this transition feels like the biggest one yet. And while Ben and I had a sense of what the southern and central Appalachians would be like from family vacations and other trips in the past, neither of us has ever spent any time in the White or Green Mountains. We cruise past the urban sprawl of Philadelphia, New York City, and Boston, and before long we're looking for moose at dusk while gliding along the Kancamagus Highway through the White Mountain National Forest, nothing but evergreen trees and trailheads on either side of the scenic byway.

Most people don't realize that the mountains they are so familiar with in New Hampshire, Vermont, and Maine are actually subranges of the Appalachian Mountains. When I get the opportunity to talk about my book project with locals, the reactions tend to be similar: "The Appalachian Mountains . . . *here?*"

OPPOSITE: *East Pond Trail, White Mountain National Forest, NH*

Neon Invasion

A rainstorm has just blown over, and I carefully step over and around the adorable, two-inch-long orange newts that wriggle their way across the trail every twenty feet or so. These young newts, or efts, are in their terrestrial phase, a period of one to three years during which they roam the woods, munch on insects and snails, and hibernate under rocks and logs in winter. I see them most often during and after a warm summer rain.

When eastern red-spotted newts reach their adult form (below), they lose their vibrant orange hue and return to the water to live out the rest of their lives. Amazingly, these critters can alter their life cycles to survive harsh conditions. If their pond dries up, adults can leave the water and return to their juvenile, land-dwelling form.

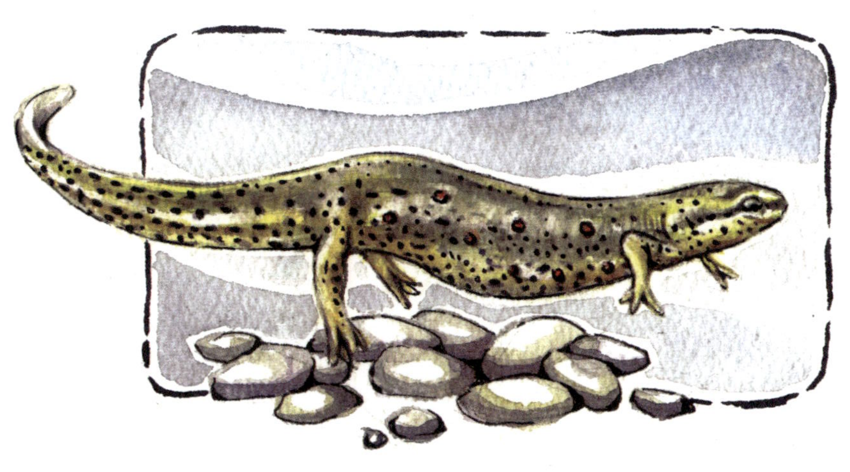

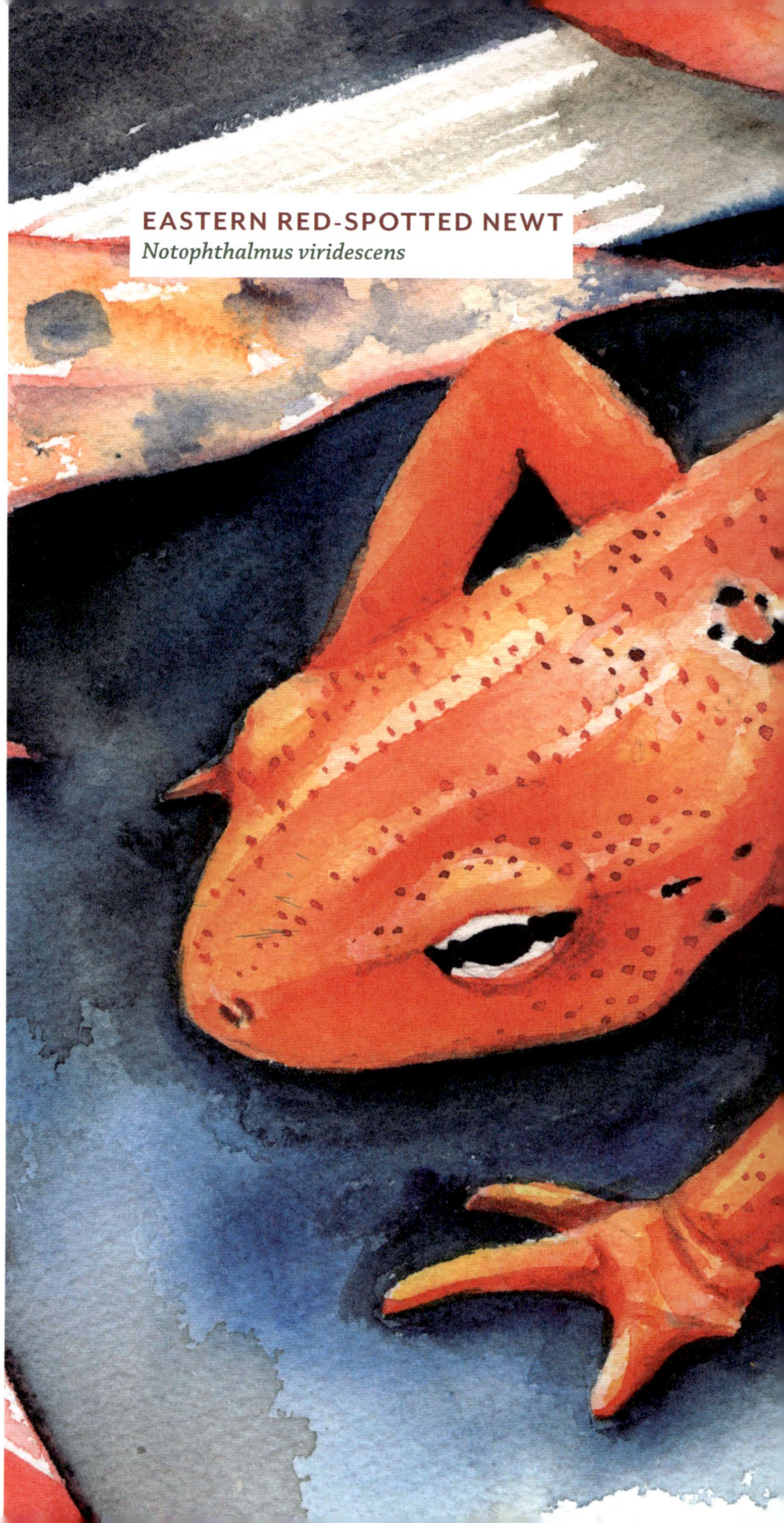

EASTERN RED-SPOTTED NEWT
Notophthalmus viridescens

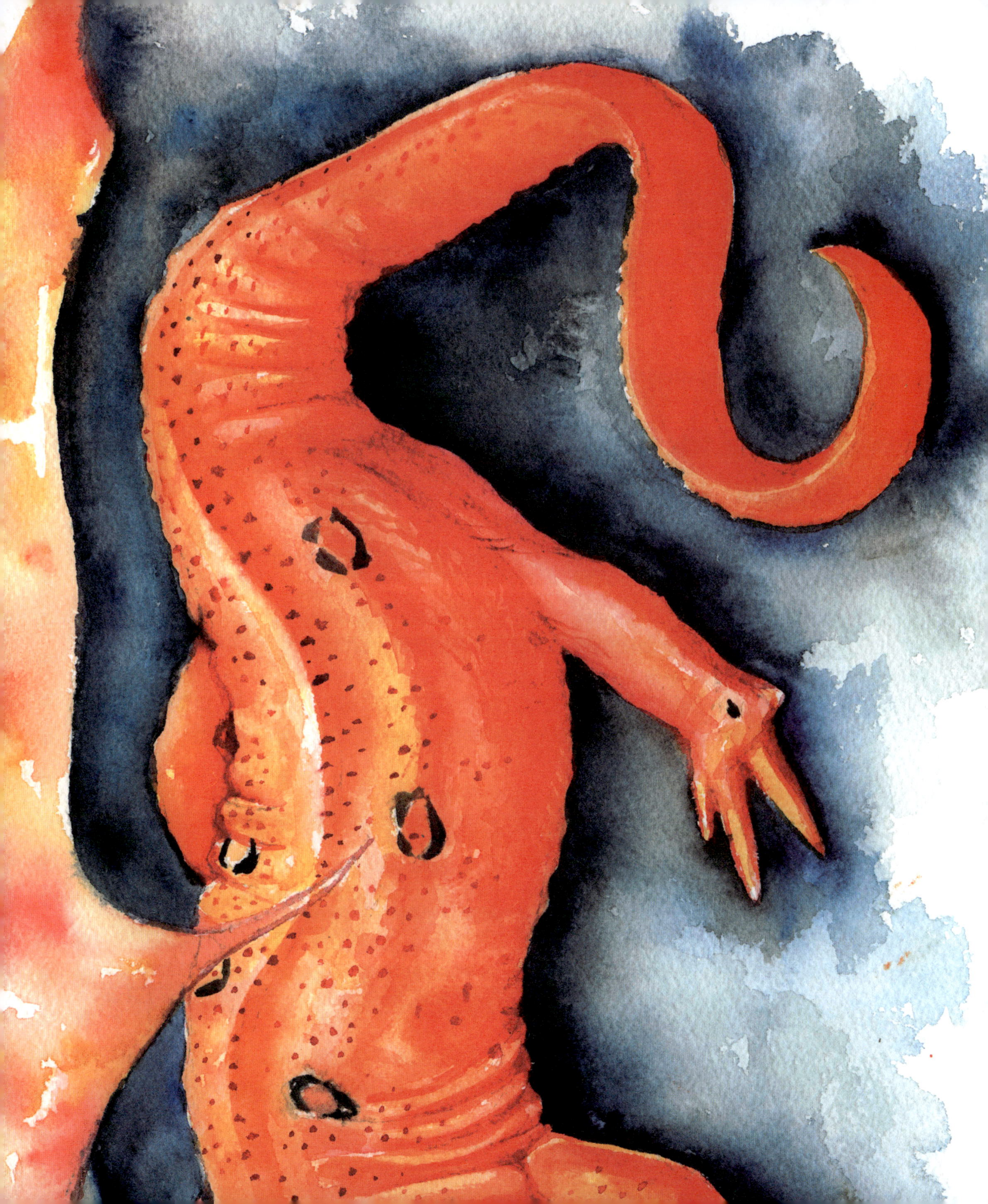

Pressing Pause

I woke up this morning feeling like my brain was going to burst. At 10:30 a.m. I am still lying in bed with the blinds drawn. Around noon I manage to pull together enough strength to pack a backpack, put on my rain jacket, and walk down the old logging road into the forest. Today feels like a waste, an unimportant blip in between better, more important days.

I shuffle slowly along the wide path. I feel drugged, because I am drugged. Three tiny white migraine pills have released their chemicals, which are sometimes effective at dulling the pain behind my right eye. But today they're not working—the pain persists—yet the medication still swaddles my brain in a thin, veil-like fog. I sit down on a mossy log, feeling dizzy and weak.

Squeezing. Pounding. Throbbing. Squeezing. Pounding. Throbbing.

A hermit thrush calls from the treetops. A mosquito bites my shoulder. A few raindrops run down my forehead. The cool wetness provides a moment of relief. If I didn't have a headache today, I'd probably be back in the cabin working. Now it's 12:30 p.m. and I haven't made progress on anything. I'm still sick, still sitting on a log—a different one now. So much for a productive Tuesday.

I resent the way this invisible illness consumes entire days, oftentimes two or three each week. But these migraine days also nudge me to value my existence beyond what I produce or create. They force me to observe and absorb, to sit in silence and notice all the tiny worlds flourishing beyond my little bubble. Even when I'm not able to paint, the black-throated green warbler still trills. The mycelium continues to root its way through the leaf litter and transform rotting wood into rich soil. The millipede busily lays hundreds of eggs underneath a damp rock.

Lately, I've felt a growing desire to learn the names of everything I see, to categorize them and become a "true" naturalist. But sometimes it's enough just to rest on a log, surrounded by a fortress of trees, and let nature hold me. Just to marinate in the comforting scents, the warm breeze, the gentle rays of sunshine peeking through the clouds. Just to receive her gifts, without trying to understand or dissect them.

On my walk back home, I stoop down to pick up a piece of bark in my path that's stained a deep green. This exquisitely tinted fungus, commonly called green stain fungus, spreads through decaying poplar, ash, aspen, and oak. In the fourteenth and fifteenth centuries, Italian woodworkers sometimes used the stained wood to create detailed green inlays. Though I often come across the fungus in the summertime, I've never found the tiny goblet-shaped mushrooms, called green elf cups, that the fungus sometimes produces. But today I feel so sick that I'm in no rush to be anywhere, so I take the time to pick up and inspect each stained piece of wood along my path. I examine my sixth stained piece and . . . Aha! Several clusters of delicate emerald-green mushrooms, each one two to five millimeters in diameter, cling to the wood.

Without painful days like this that yank me away from the demands of my to-do list, would I have found these green elf cups—or even known to look for them? Would I have ever awoken to the wonders that arise when I slow my pace? On the days when I don't have a migraine, I forget that it's even a problem for me. I snap right back into the realm of the healthy, reverting to the same get-it-done mindset, holding myself to the same high expectations. And then when I wake up in pain the next day, the disappointment strikes anew. Ah, yes, this is my other reality. If I didn't have these forced pauses, these breaths between the *go, go, go*, I bet I'd work myself to the point of burnout.

Thank you, migraine days, for teaching me how to find beauty in the pause.

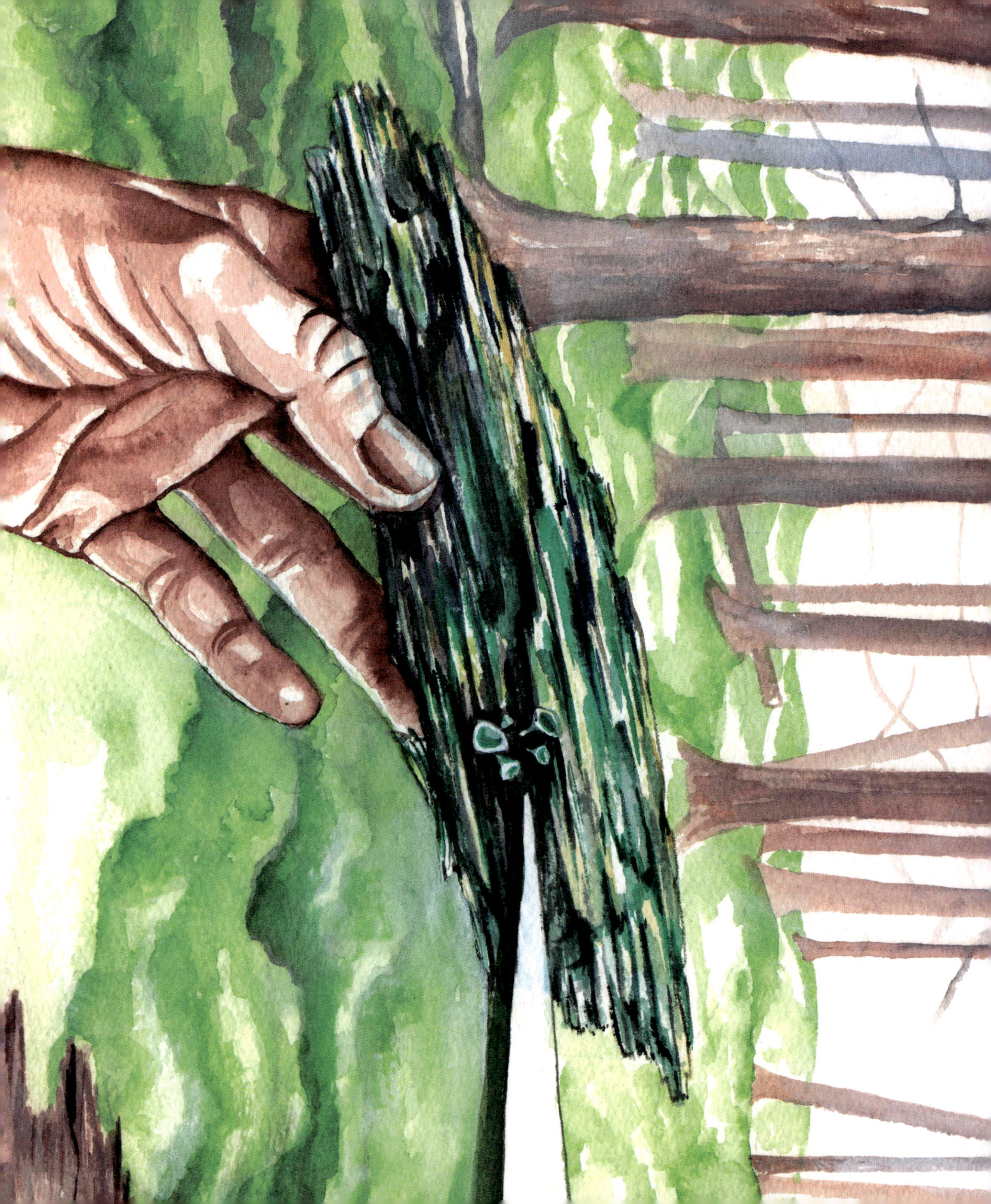

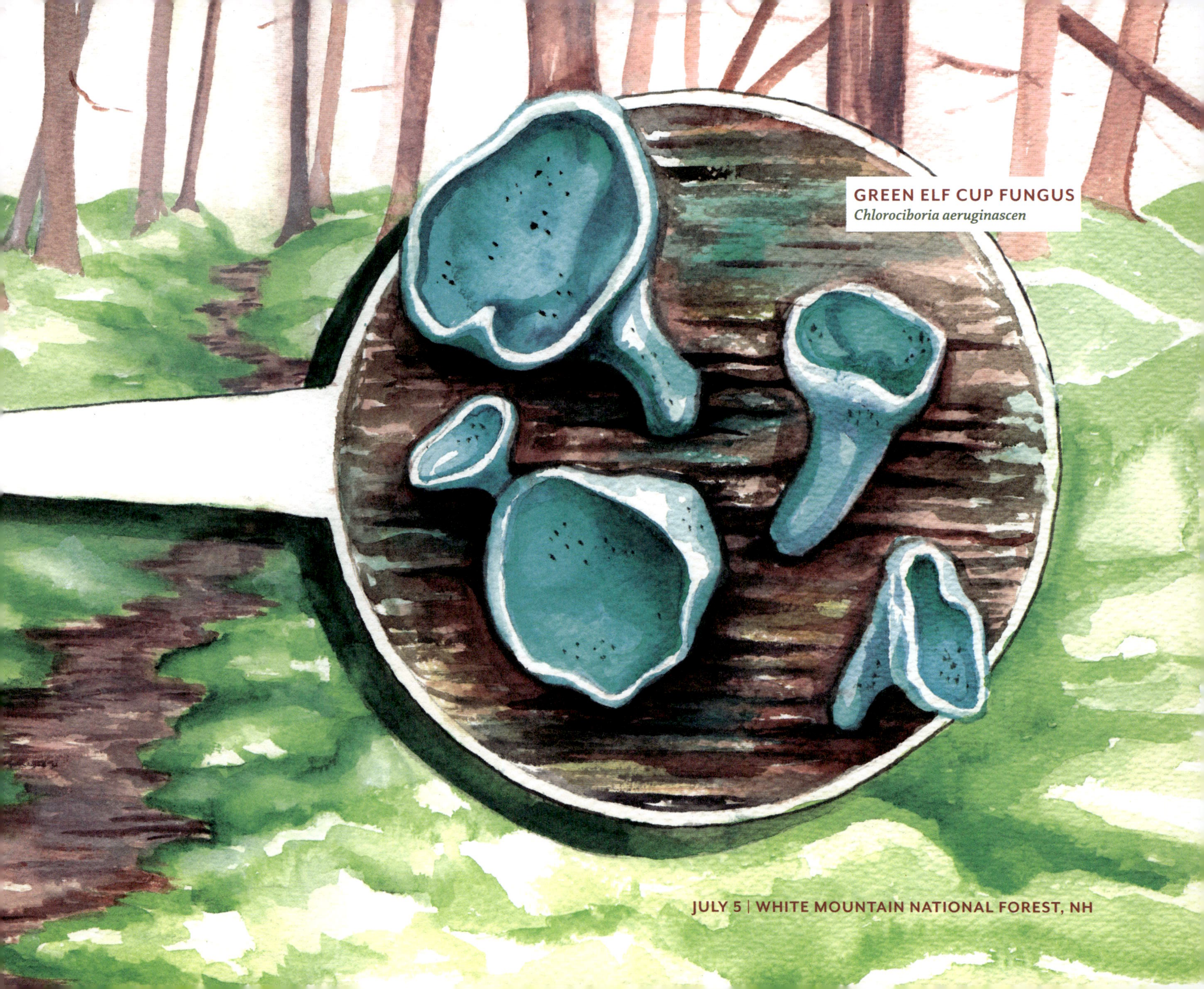
GREEN ELF CUP FUNGUS
Chlorociboria aeruginascen
JULY 5 | WHITE MOUNTAIN NATIONAL FOREST, NH

Hidden Hunters

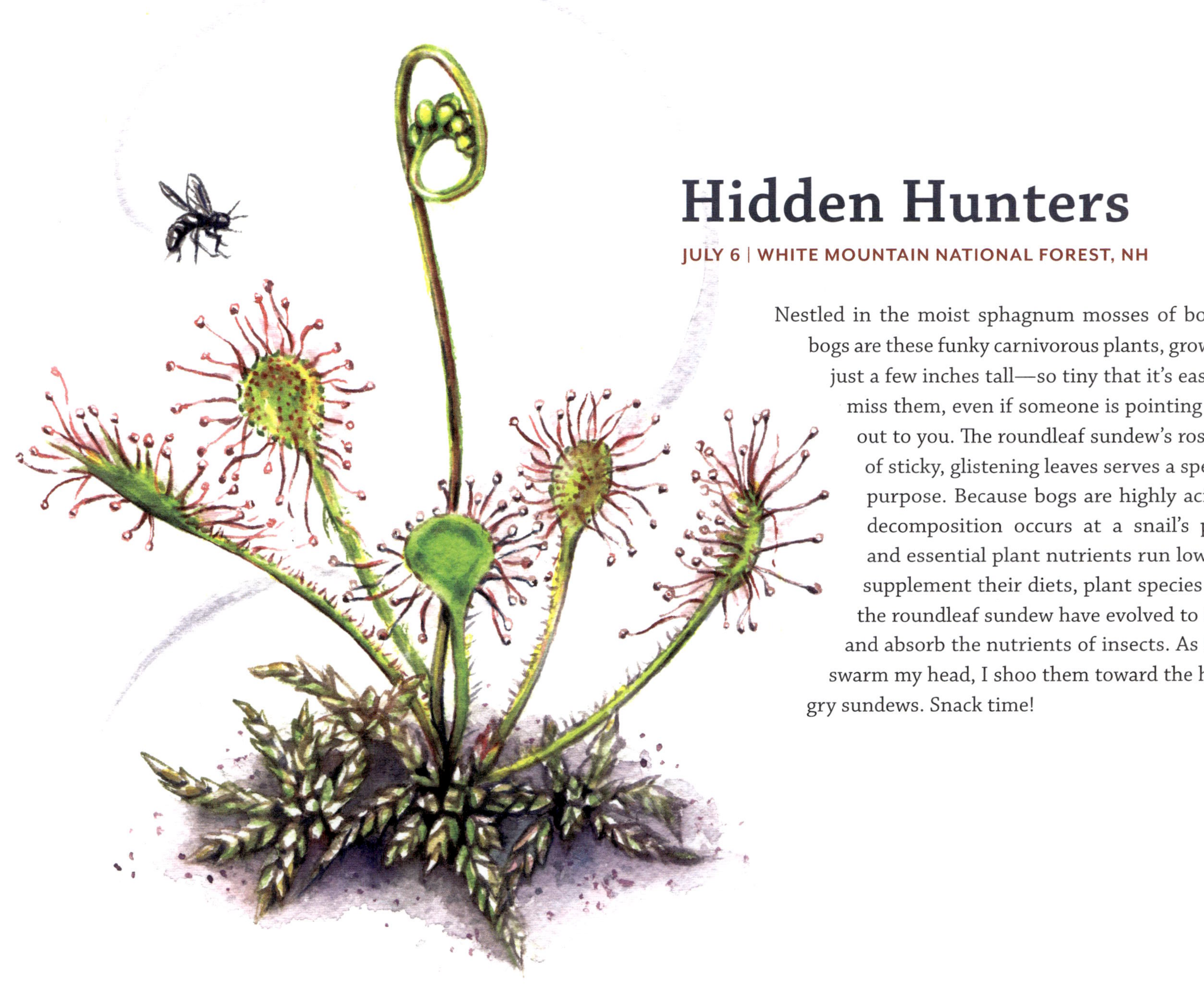

Nestled in the moist sphagnum mosses of boreal bogs are these funky carnivorous plants, growing just a few inches tall—so tiny that it's easy to miss them, even if someone is pointing one out to you. The roundleaf sundew's rosette of sticky, glistening leaves serves a special purpose. Because bogs are highly acidic, decomposition occurs at a snail's pace and essential plant nutrients run low. To supplement their diets, plant species like the roundleaf sundew have evolved to trap and absorb the nutrients of insects. As flies swarm my head, I shoo them toward the hungry sundews. Snack time!

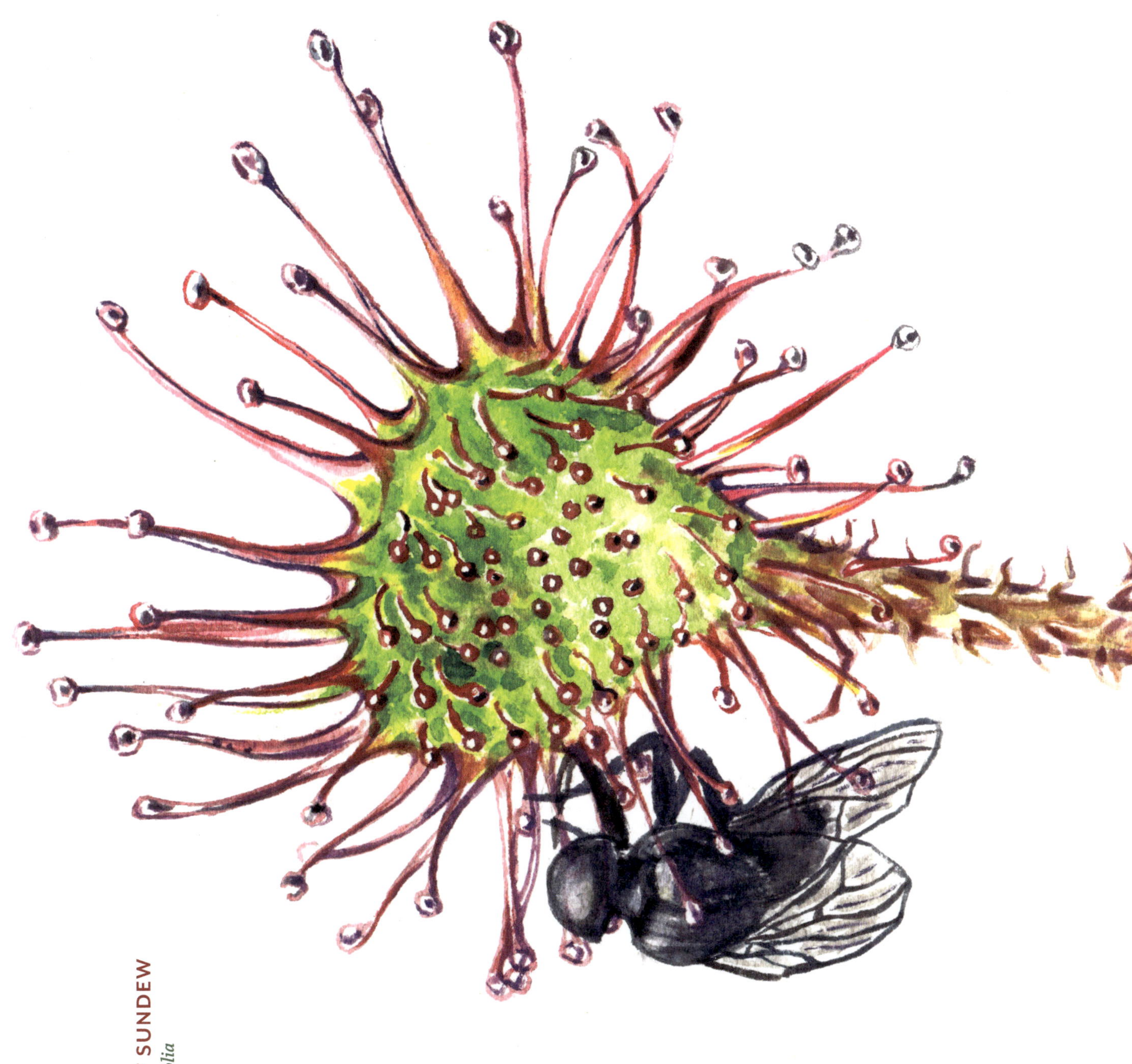

ROUNDLEAF SUNDEW
Drosera rotundifolia

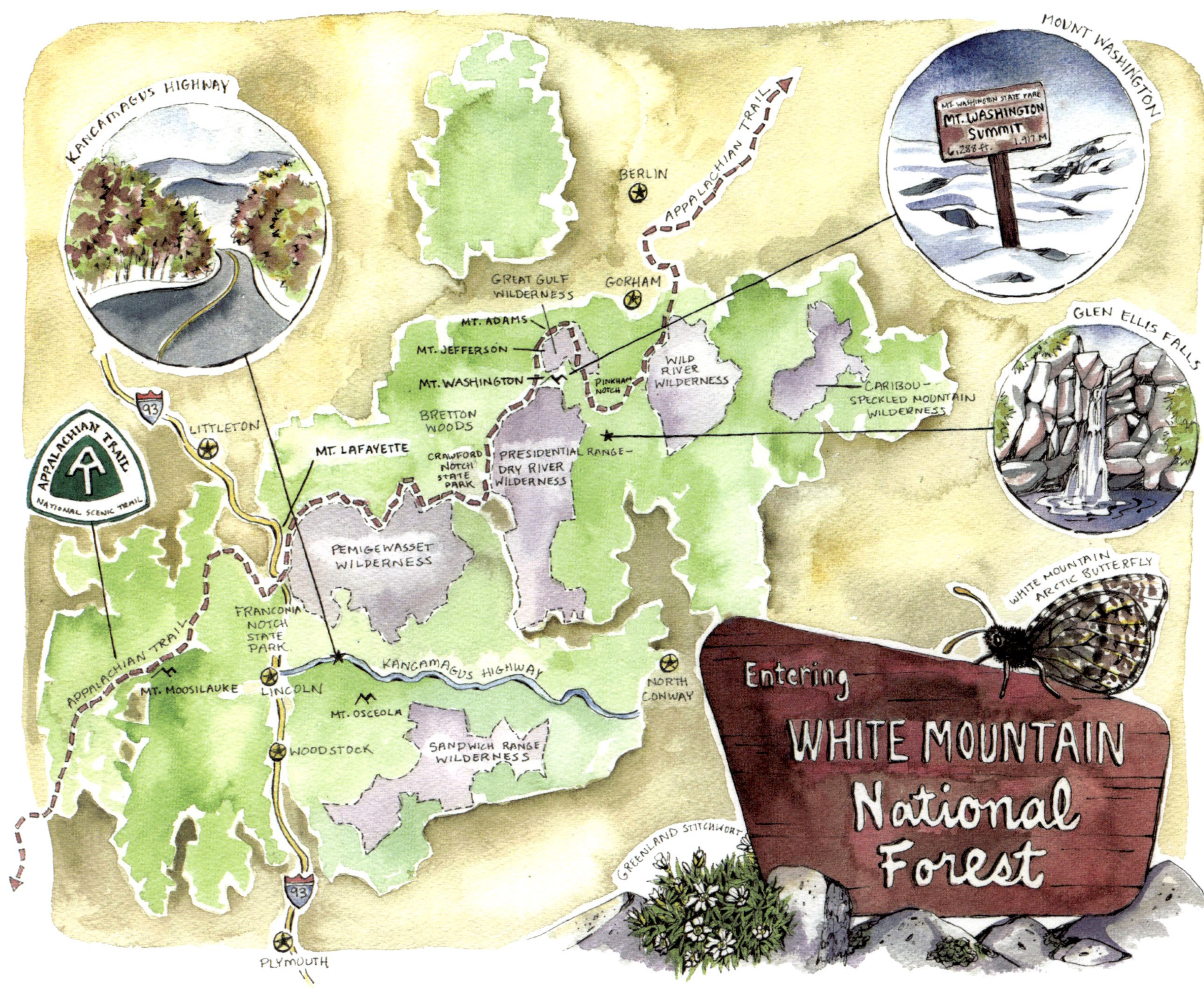

KANCAMAGUS HIGHWAY
MOUNT WASHINGTON
APPALACHIAN TRAIL
BERLIN
MT. WASHINGTON STATE PARK
MT. WASHINGTON Summit
6,288 ft.   1,917 M
GREAT GULF WILDERNESS
GORHAM
MT. ADAMS
MT. JEFFERSON
MT. WASHINGTON
PINKHAM NOTCH
WILD RIVER WILDERNESS
GLEN ELLIS FALLS
93
LITTLETON
APPALACHIAN TRAIL
NATIONAL SCENIC TRAIL
MT. LAFAYETTE
BRETTON WOODS
CRAWFORD NOTCH STATE PARK
PRESIDENTIAL RANGE — DRY RIVER WILDERNESS
CARIBOU — SPECKLED MOUNTAIN WILDERNESS
PEMIGEWASSET WILDERNESS
WHITE MOUNTAIN ARCTIC BUTTERFLY
APPALACHIAN TRAIL
FRANCONIA NOTCH STATE PARK
KANCAMAGUS HIGHWAY
NORTH CONWAY
Entering
WHITE MOUNTAIN National Forest
MT. MOOSILAUKE
LINCOLN
MT. OSCEOLA
WOODSTOCK
SANDWICH RANGE WILDERNESS
GREENLAND STITCHWORT
93
PLYMOUTH

WHITE MOUNTAIN NATIONAL FOREST

Before the White Mountain National Forest was established in 1918, much of the land had fallen victim to heavy, haphazard logging, with rampant clear-cutting contributing to erosion and widespread firestorm damage. The area has since sprung back to life, and today the national forest protects around 800,000 acres of wetlands, woods, and wilderness across New Hampshire and western Maine. The region also boasts some of New England's most challenging hiking, from the White Mountain Four Thousand Footers—forty-eight mountain peaks above 4,000 feet—to about 100 miles of the Appalachian Trail. There are ample opportunities to connect the AT with side trails to create loops for day hikes or shorter backpacking trips. The White Mountains owe their distinctive topography to a combination of magma intrusions—which left behind towering granite mountains and cliffs—and the widespread glaciation that later carved out sweeping U-shaped valleys. This range's spectacular mountain passes are referred to as "notches."

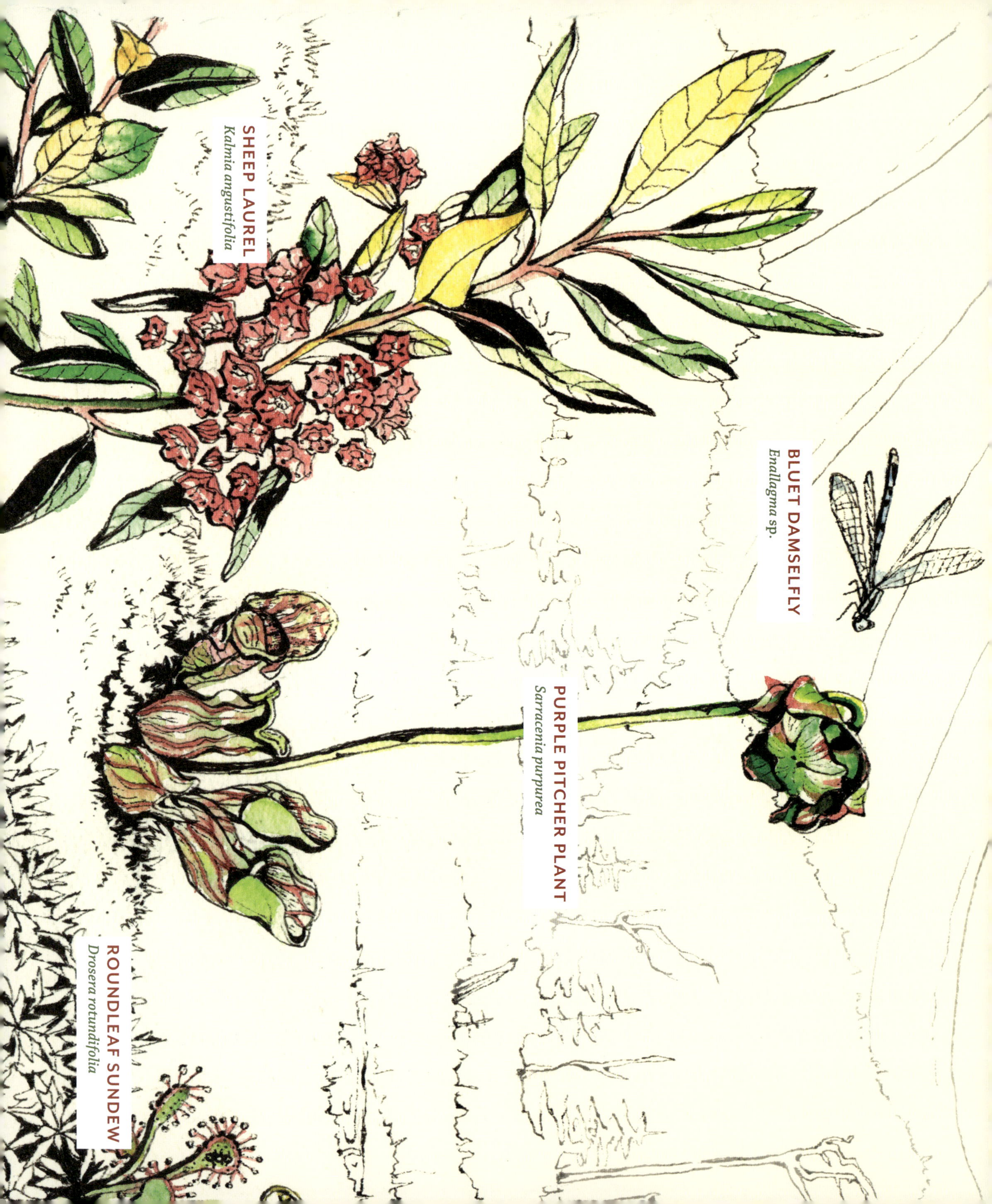

SHEEP LAUREL
Kalmia angustifolia

BLUET DAMSELFLY
Enallagma sp.

PURPLE PITCHER PLANT
Sarracenia purpurea

ROUNDLEAF SUNDEW
Drosera rotundifolia

Morning in a Mountain Bog

JULY 10 | WHITE MOUNTAIN NATIONAL FOREST, NH

It's utterly silent in the bog, the intense late-morning sun having prompted its inhabitants to seek out shade and stillness. I crouch among the laurel and green false hellebore to wait and watch, my breath quieting as I sink into invisibility. *Splat!* A green frog hops toward me. Slowly, I inch closer until we're just an arm's length apart and I'm able to see all the speckled details of the frog's eye; it looks like a shimmering gold glass marble. Behind me, something nibbles loudly on a plant—maybe a muskrat?

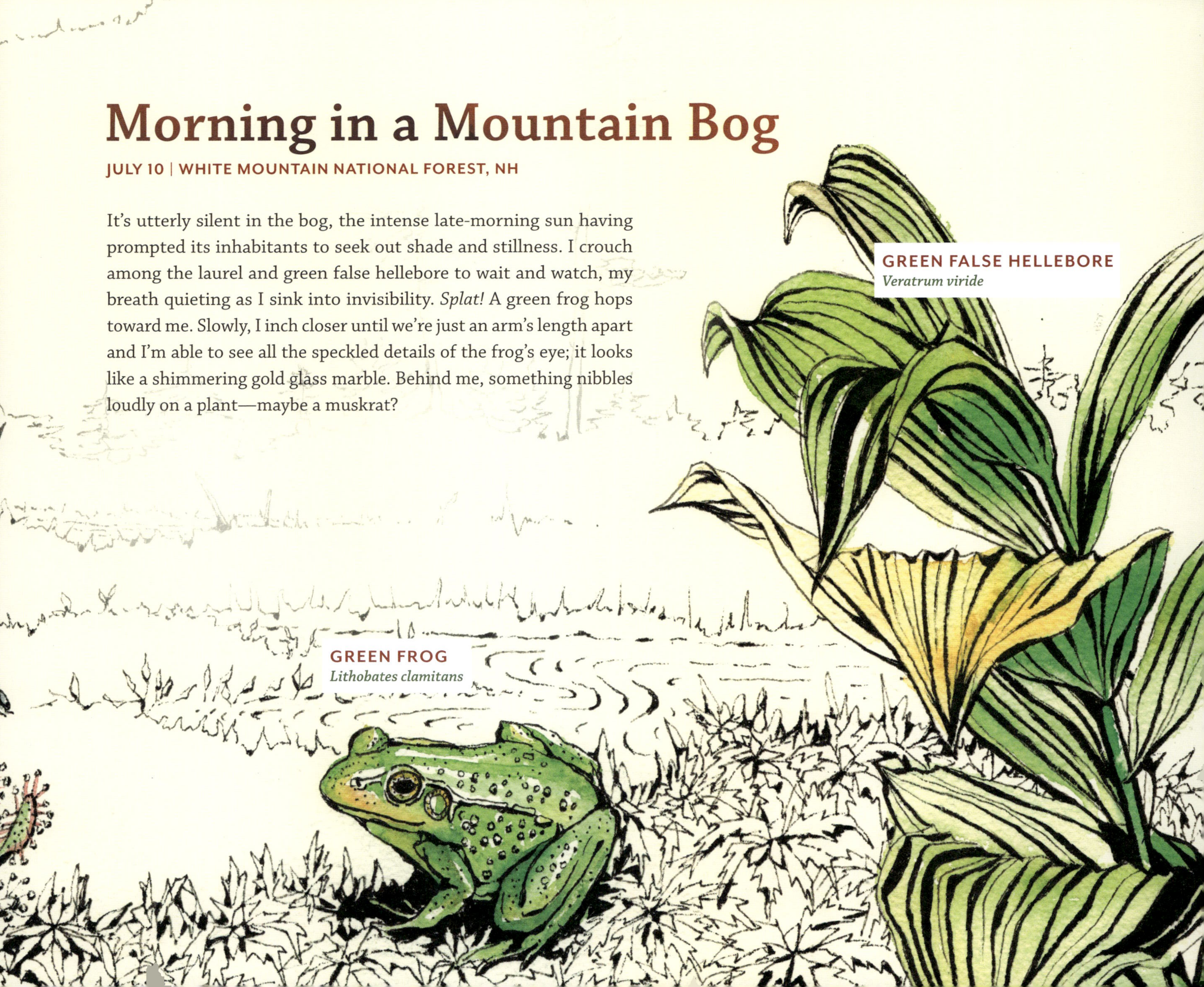

Forest Friendships

JULY 14 | TIN MOUNTAIN, NH

I step out of my car at the trailhead and wave to a small group of adults and children who are standing in the parking lot chatting. I'm here for a community nature walk and I feel the subtle nerves of showing up to an event alone. But after only a few minutes of trekking the trails with my new companions, these feelings subside, replaced with joy as I remember how easy it is to bond with strangers while tromping through the woods. Soon, even the adults in the group are buzzing with curiosity and a playful energy. After we walk into the mouth of an old cave and find giant spider egg sacs clinging to the ceiling, we pose for a group photo where we all pretend to be spiders.

Later, upon spotting a slender maritime garter snake hugging the base of a tree, we pause to delight in its pretty markings and colors. The snake seems to be shy and curious at the same time, hiding behind the tree while occasionally poking its head out to see what we're up to. This subspecies of the common garter snake hibernates beneath the frost line in order to survive New England's brutal winters, sometimes borrowing crayfish and mammal burrows to stay warm. And while most snakes give birth to litters of ten to thirty offspring, garter snakes can produce up to fifty babies at one time!

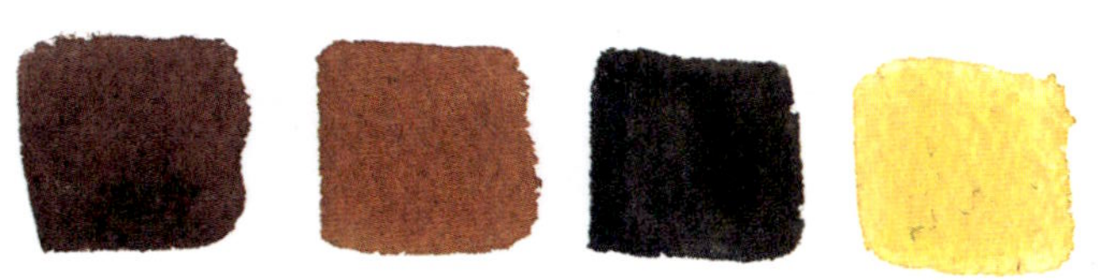

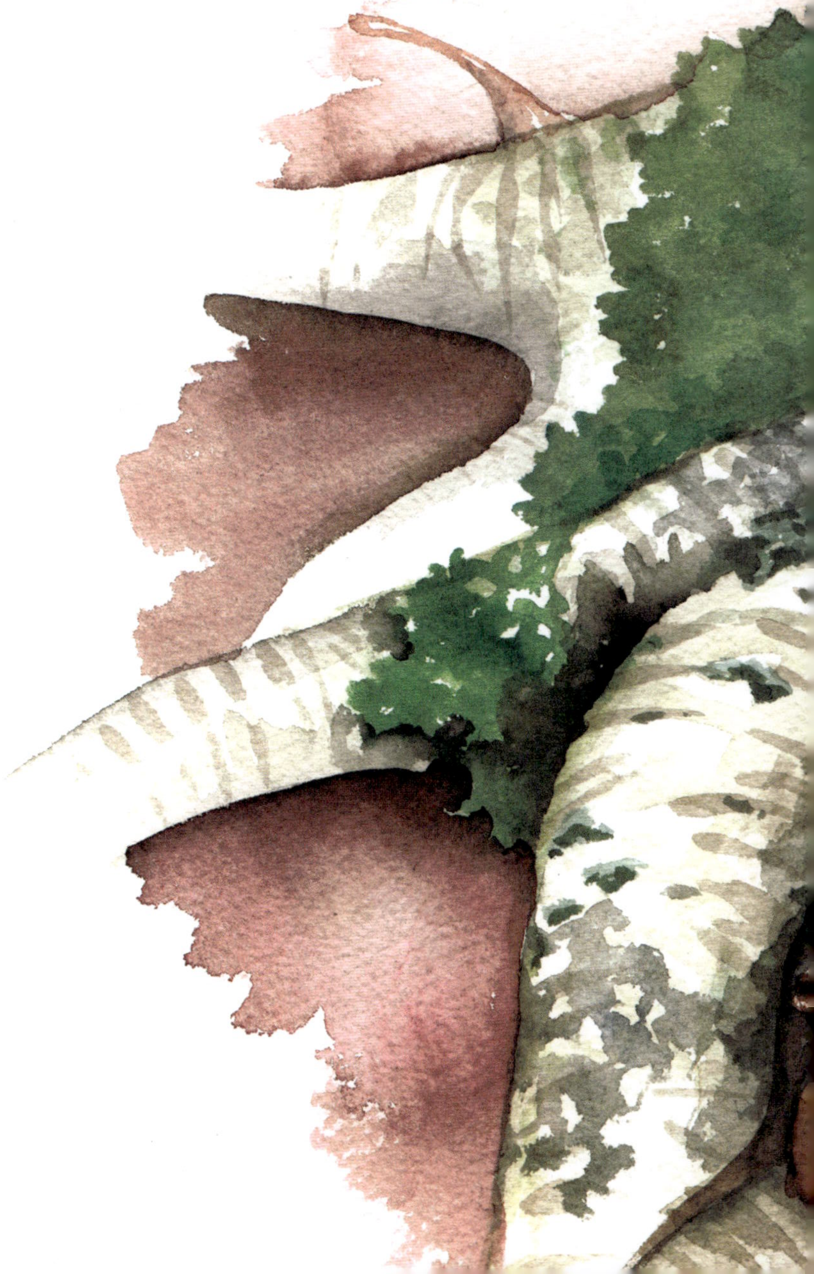

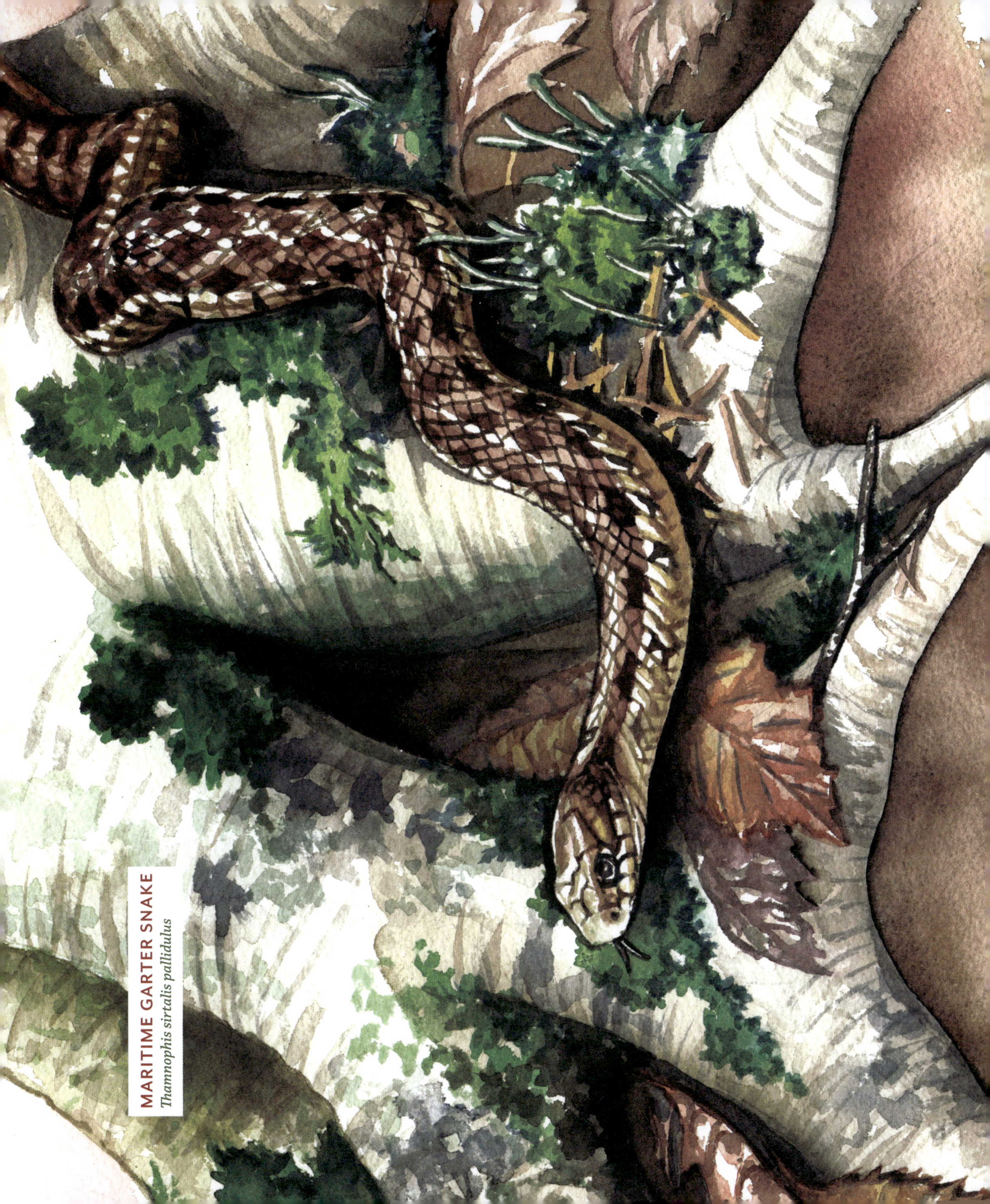

MARITIME GARTER SNAKE
Thamnophis sirtalis pallidulus

Flower Powers

When I was growing up, my family and friends called me Rosie. I didn't like the nickname, for many reasons. For one, I learned while visiting county fairs that Rosie was one of the most common female cow names. My sisters are named Clara, Ellie, and Molly—also popular cow monikers. I know that my parents didn't do this on purpose, but it still did not sit well with me.

I also didn't like that my name sounded so much like "rose." While I didn't have anything against roses in particular, I didn't understand the hype surrounding flowers in general. Every art gallery and stationery store I walked into seemed to be exploding with floral-themed paintings and products, so I wrote off flowers as decorative and uninteresting—including roses, the most romantic and overly depicted bloom of them all.

My name was not giving me the cool factor I was grasping for, so when I was ten years old I decided to take matters into my own hands. I recruited my friend Elizabeth, and together we sat on the trampoline in my backyard and brainstormed a new name for me to try out at summer camp. It had to embody the version of myself that I *wanted* to be: spunky, outgoing, bold—the total opposite of who I currently was. I settled on the name Roxy and pulled together the most clashing, colorful clothing combinations I could scrounge up in my closet for my new-and-improved self to wear at camp.

Things went pretty well—my fellow campers got to know me as Roxy—until one of my little sisters who was also at camp called me Rosie in front of a group, causing a stir of confusion. *How dare she?!* I carried Roxy back to my normal, post-camp life, and my family thought it was funny so they went along with it. But when I tried to change my name back to Rosie a few months later, I realized that it was too late. Habits had been formed, and to this day certain family members call me Roxy just as often as Rosie or Rosalie.

My appreciation for flowers eventually began to germinate, as I learned more about plants and realized that their incredible survival mechanisms rival their beauty. On the highest peaks in New Hampshire, I saw tiny *Diapensia* plants, barely two inches tall, clinging for dear life to the thin summit topsoil. The leaves of this flower lock together in incredibly dense, interwoven mats to prevent the topsoil from blowing away in some of the windiest places on Earth. Some wildflowers, like bear corn and ghost pipes, look so bizarre as to resemble a fungus; neither relies on the sun at all, instead tapping into the roots of trees to gain their nutrients. I've also come across flowers that are literally flesh-eating hunters, such as the purple pitcher plant, which traps and swallows insects to offset the lack of nutrients in the boggy areas where it grows. And this isn't even to mention the critical role of flowering plants in the ecosystem: they provide nectar for bees, seeds for birds and rodents, and leaves for the larvae of moths and butterflies.

I've even learned some interesting things about roses that have totally changed the way I view them. There's one rosebush growing on a cathedral in Germany that's believed to be over 1,000 years old. Even when the cathedral was bombed during World War II, the rosebush survived, and it still blooms every spring. Another delightful tidbit is that every species of rose is edible to humans.

At my wedding celebration, my dad requested the song "Wildflowers" by Tom Petty as a last-minute father-daughter dance. With its lyrics about freedom, love, and belonging, it felt like the perfect song—and even more so because I now see flowers as resilient, resourceful, and integral to their communities.

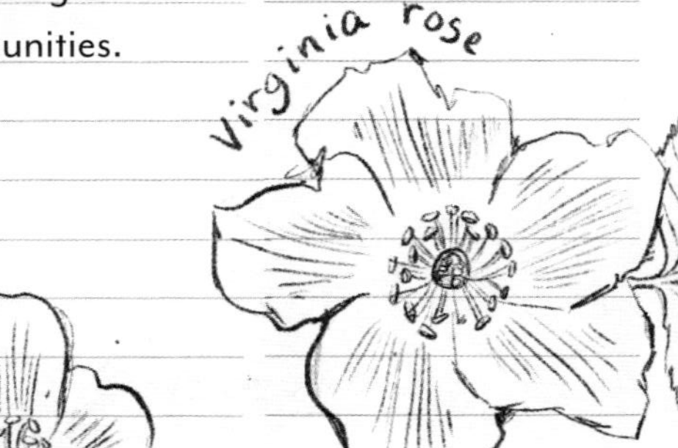
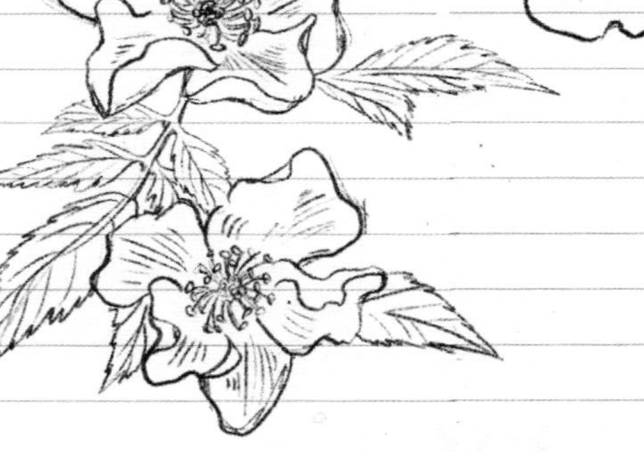

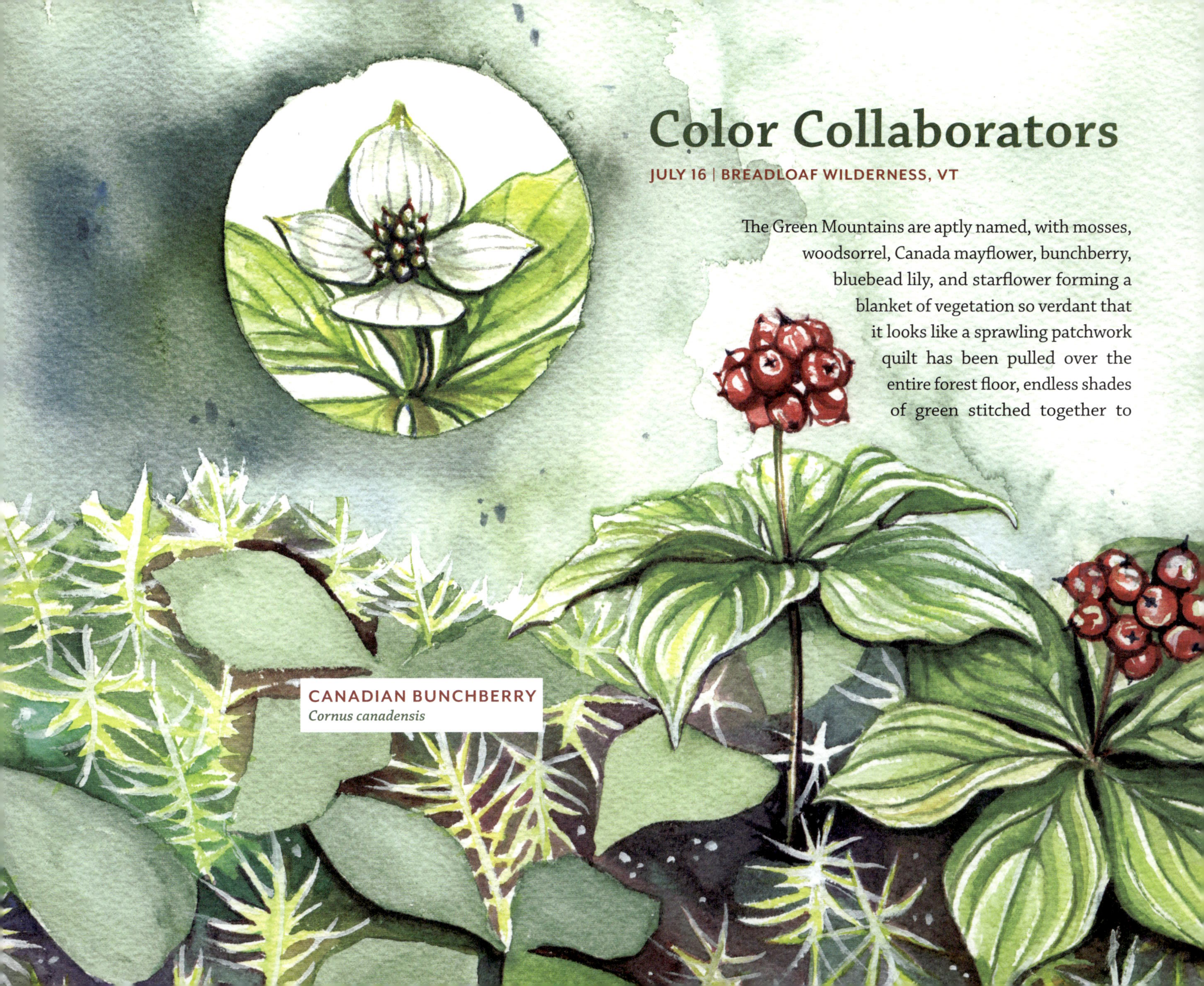

Color Collaborators

JULY 16 | BREADLOAF WILDERNESS, VT

The Green Mountains are aptly named, with mosses, woodsorrel, Canada mayflower, bunchberry, bluebead lily, and starflower forming a blanket of vegetation so verdant that it looks like a sprawling patchwork quilt has been pulled over the entire forest floor, endless shades of green stitched together to tuck in the evergreen trees for a soft slumber. At the beginning of the month, both bluebead lily and bunchberry were still flowering. But over the past two weeks, I have witnessed a gradual transformation: first, the flowers gave way to small green berries, which have now ripened into a bright cobalt blue and a deep red, respectively. I frequently see red squirrels and songbirds nibbling on the shiny bunchberries.

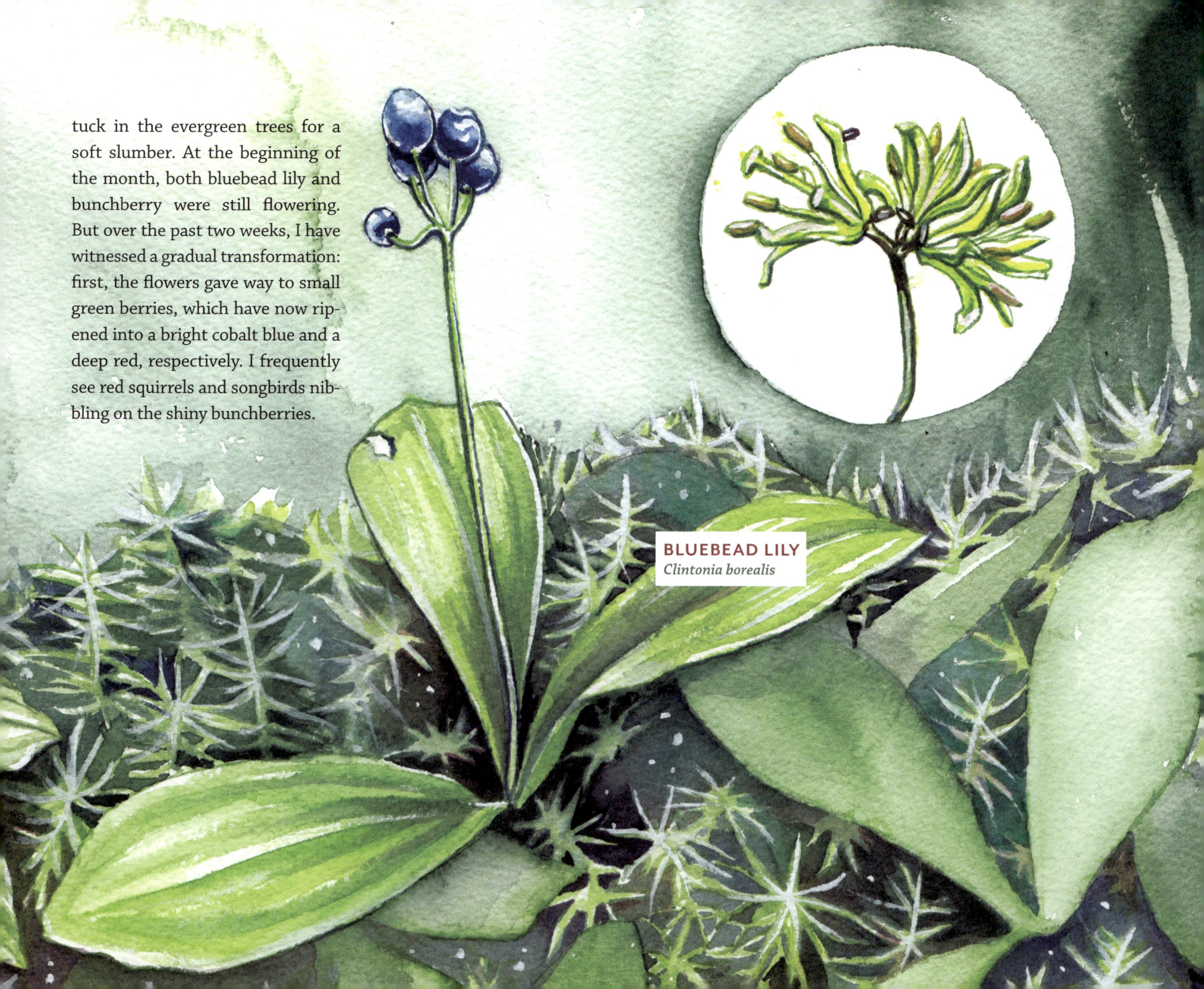

Quirky Crawlies

I sit quietly at the wetland viewing area, but I don't notice any birds or other wildlife. Although moose are frequently spotted here, there's no luck on that front for me today. But what is that bizarre green-and-white creature wiggling across the leaves beside my bench? It's unlike anything I've ever seen before: a caterpillar entirely covered in fuzzy white orbs. This hornworm caterpillar has been parasitized by *Apanteles* wasps, tiny wasps that lay their eggs inside the caterpillar's body. The eggs then hatch into maggot-like larvae that live inside the caterpillar and feed on its fat reserves. Toward the end of the parasitization process, the larvae craft the miniature white cocoons on the exterior of the caterpillar's body that now catch my eye, from which a new

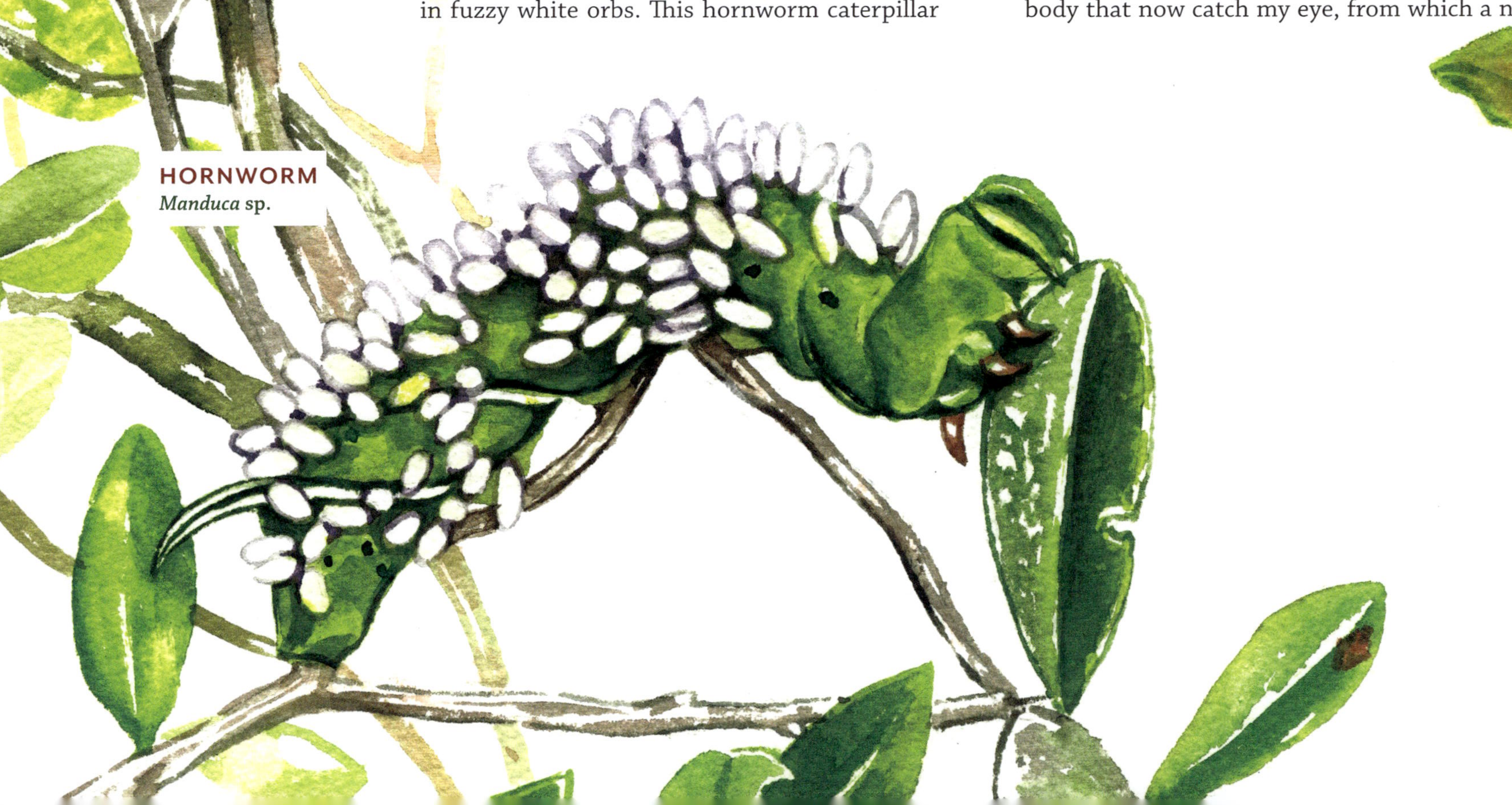

HORNWORM
Manduca sp.

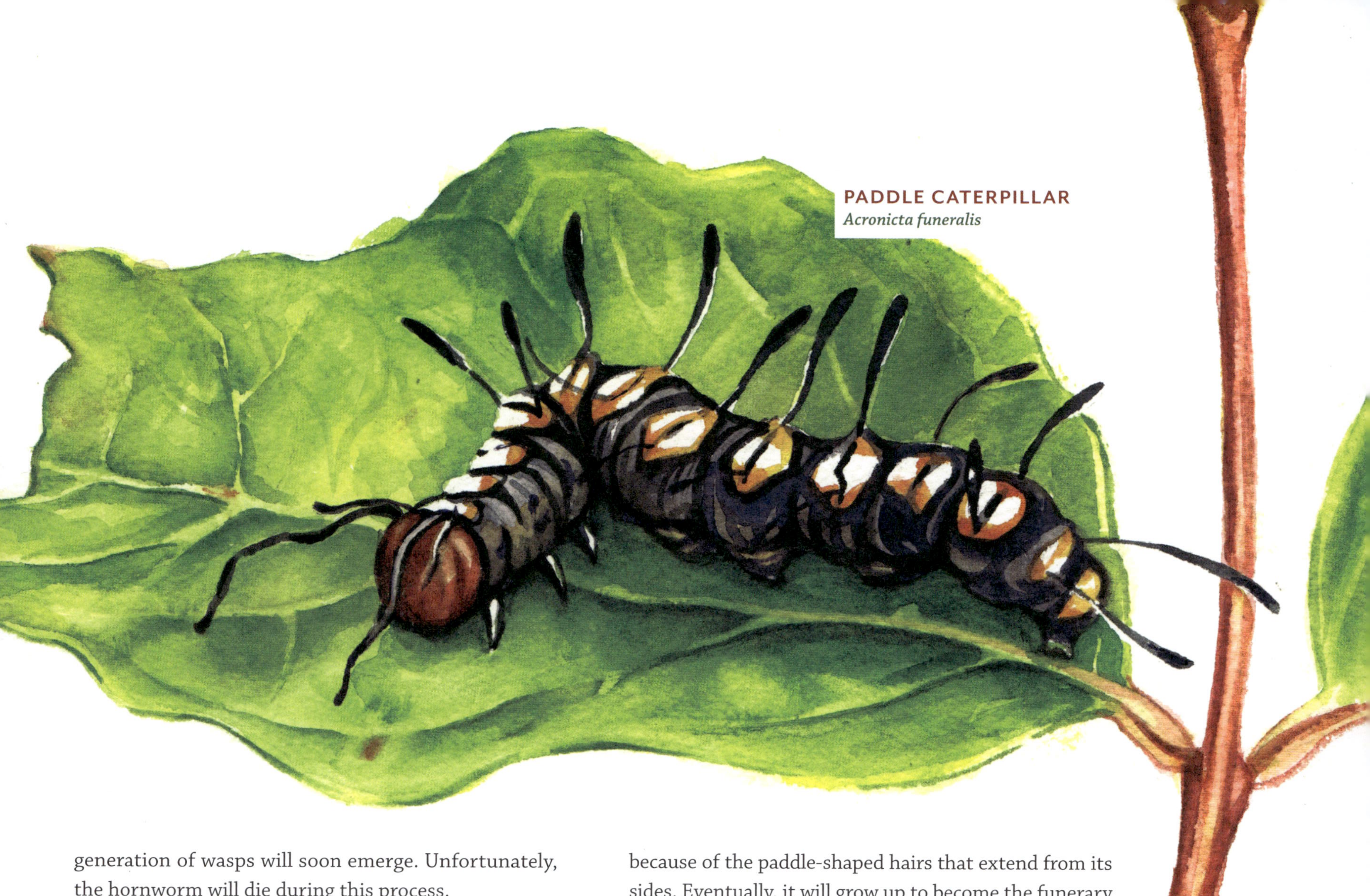

generation of wasps will soon emerge. Unfortunately, the hornworm will die during this process.

On my hike back to the trailhead, I see another wacky-looking caterpillar inching along a leaf. This gooey guy has been nicknamed the "paddle caterpillar" because of the paddle-shaped hairs that extend from its sides. Eventually, it will grow up to become the funerary dagger moth, exchanging its funky larval appearance for a comparatively drab cloak. Enjoy it while you have it, little caterpillar!

Green Mountain Melody

It's eight in the morning in the Green Mountains and I'm still hunkered down in my sleeping bag. We're backpacking a short section of the 272-mile Long Trail, the oldest long-distance hiking trail in the United States.

The cheerful chattering of birdsong slowly awakens my groggy brain. A white-throated sparrow! I learned to recognize a few of its triumphant songs recently and now I can't help but notice them wherever I go. Sometimes the song starts low and escalates in pitch, and sometimes it does the opposite, but the sparrow uses the same notes for both iterations. Occasionally I hear the same song in a much higher pitch, reminiscent of the moment toward the end of a song when the singer switches to a higher key and the listener knows something big is coming. "Oh yeah, here we go!"

But then my ear catches another unusual twist: it sounds like two white-throated sparrows are harmonizing with one another! I listen longer and hear it again. How could two birds sync their songs so seamlessly? With my head still resting on my makeshift stuff-sack pillow, I come up with an idea. I'll take some audio recordings of these birdsongs, and when we get back to our rental cabin, I'll try to play them on the fiddle. I've been taking lessons for about a year, so I can easily manage the slower, simpler melodies, right?

As I trudge along the trail later that day, I think about what it will be like to perfectly mimic the gorgeous songs of my favorite birds. If ever fate demanded that I become famous for a weird hidden talent, flawlessly imitating birdsongs on my violin would be a skill that I could get behind.

Back at the cabin, I hit play on the recordings I took during the hike to refresh my memory. Then I lift my fiddle to my chin, touch the bow to the strings, and . . . the screeches that emerge sound more like a frog pretending to be a bird. I try again and again, but each time my version is pitchy and forced and nothing like the graceful fluidity that charmed me in the forest.

For some reason I expected the birdsongs to fall somewhere within the keys and timing that I'm familiar with, but each one I try to imitate seems to chart a melodic path all its own. Thirty minutes pass fruitlessly before I remember that there are people staying in the other rental that's attached to ours and I become too self-conscious to continue. No going viral for me.

My ego recovers a little bit when I read that it sometimes takes songbirds a full year to master the art of their songs. Recordings of immature birds reveal iterations almost as awkward as the ones I was producing with my violin. These early attempts are often characterized by hesitant, disorganized sequences of notes and even singing in the wrong key.

I also learn that white-throated sparrow songs are fluid, evolving slightly as groups from different regions intermingle in their southern wintering grounds. Certain individuals latch on to new dialects because it seems that female white-throated sparrows might favor unique spins on the traditional songs when choosing a mate. And perhaps most mind-blowing of all, I discover that some birds can harmonize with *themselves*. While humans are limited to singing one note at a time, songbirds have a different anatomy that allows them to create two different sounds simultaneously. Even if I were to perfect the birds' songs on the fiddle, I'd never be able to harmonize with myself. I guess some things are best left to the birds.

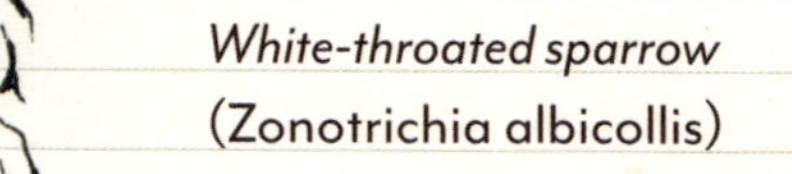

White-throated sparrow
(Zonotrichia albicollis)

Mount Washington Miniatures

JULY 23 | MOUNT WASHINGTON, NH

The wind whips right through our puffy jackets; the temperatures on top of the mountain are 25°F cooler than they were at the base. At 6,288 feet, this is the highest peak in the northeastern United States, and it boasts one of the fastest recorded wind speeds in history: a 231-mile-per-hour gust, clocked in 1934. Yet even on this inhospitable summit, I notice tiny worlds. Saxifrage wildflowers survive up here, along with many different kinds of lichen. Before we head back down, I pause to watch a white-spotted sawyer beetle moseying along a stone wall, its curved antennae longer than its entire body. This wood-boring insect is ecologically important because it infests trees that are already dying and speeds up their decomposition, thus accelerating the nutrient cycle.

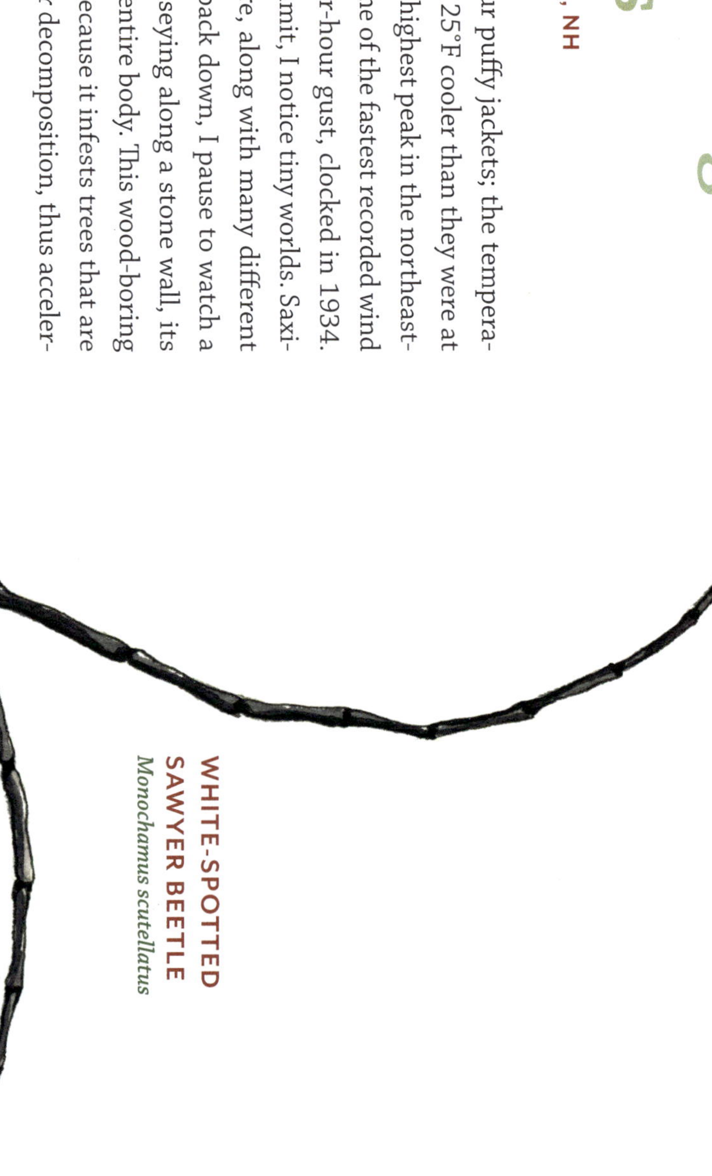

WHITE-SPOTTED SAWYER BEETLE
Monochamus scutellatus

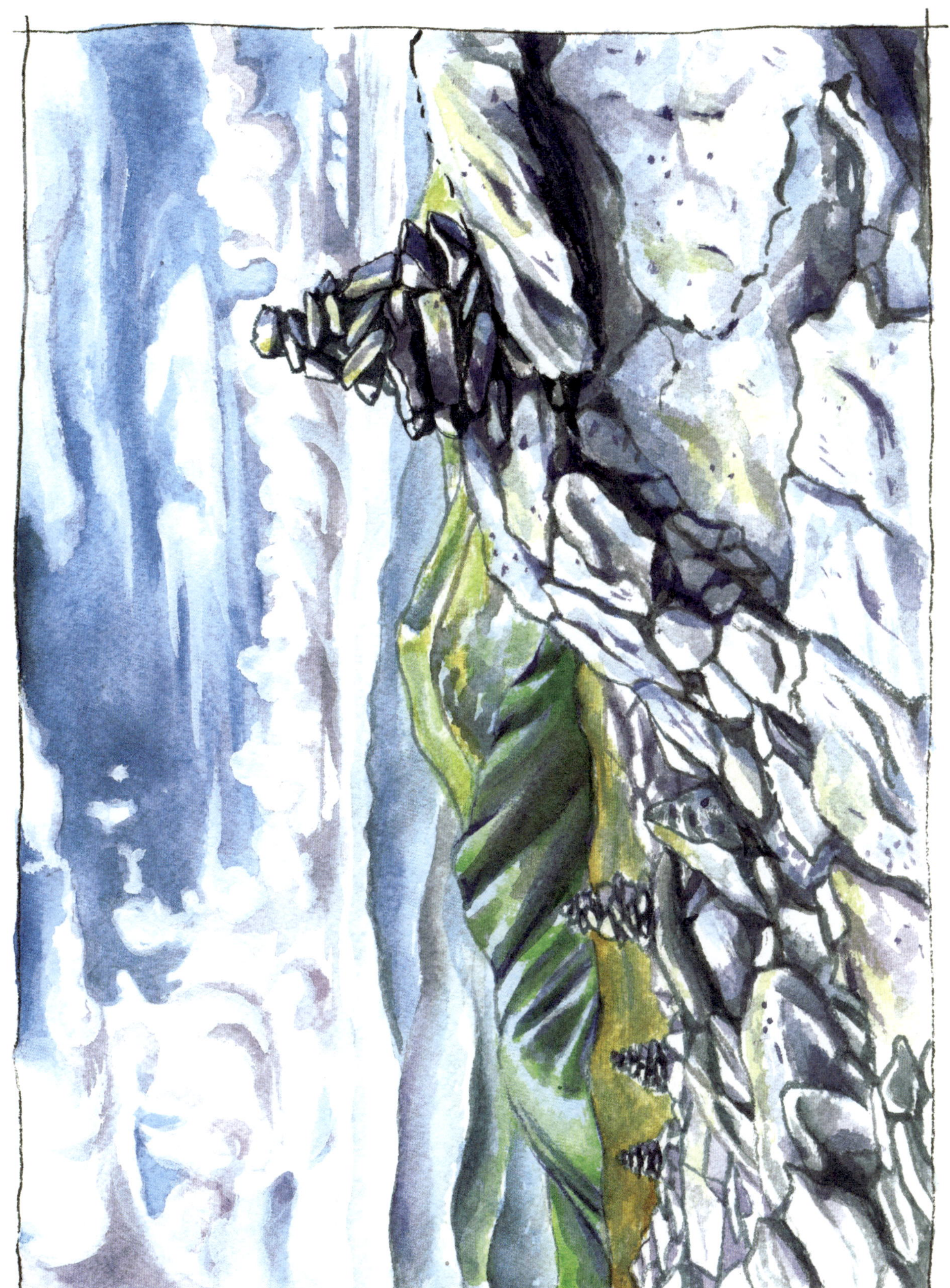

Mount Washington summit, Mount Washington State Park, NH

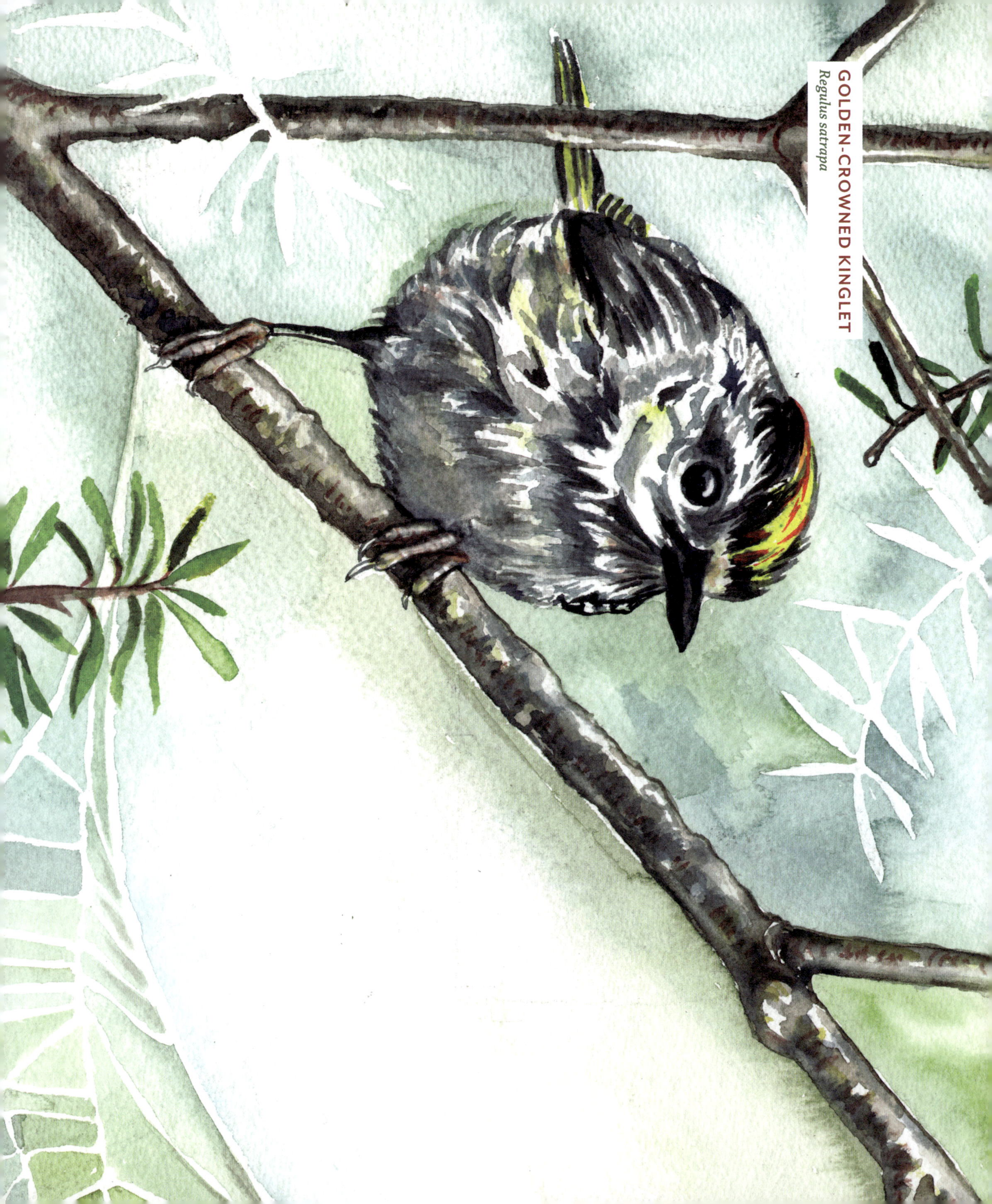

GOLDEN-CROWNED KINGLET
Regulus satrapa

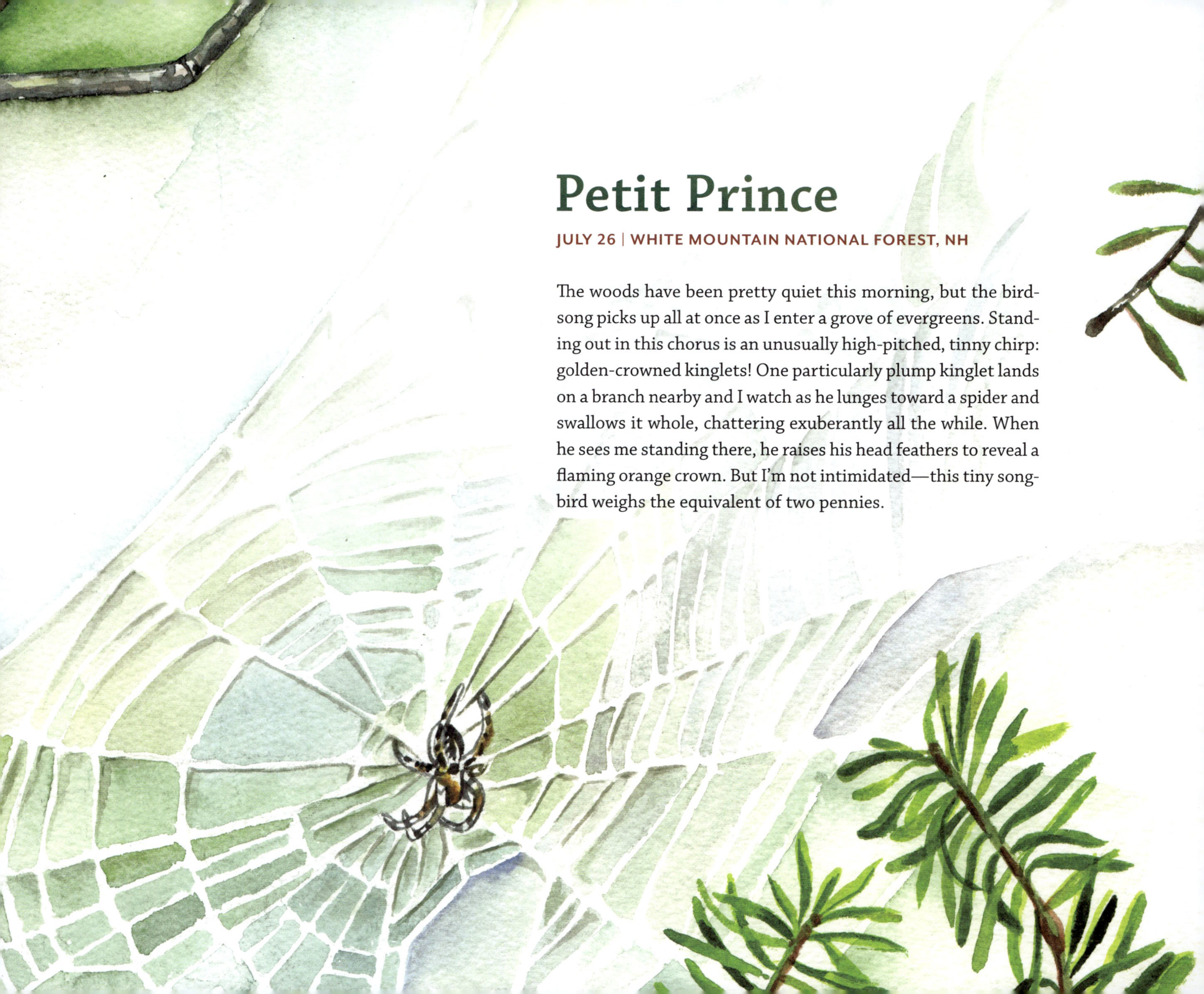

Petit Prince

The woods have been pretty quiet this morning, but the birdsong picks up all at once as I enter a grove of evergreens. Standing out in this chorus is an unusually high-pitched, tinny chirp: golden-crowned kinglets! One particularly plump kinglet lands on a branch nearby and I watch as he lunges toward a spider and swallows it whole, chattering exuberantly all the while. When he sees me standing there, he raises his head feathers to reveal a flaming orange crown. But I'm not intimidated—this tiny songbird weighs the equivalent of two pennies.

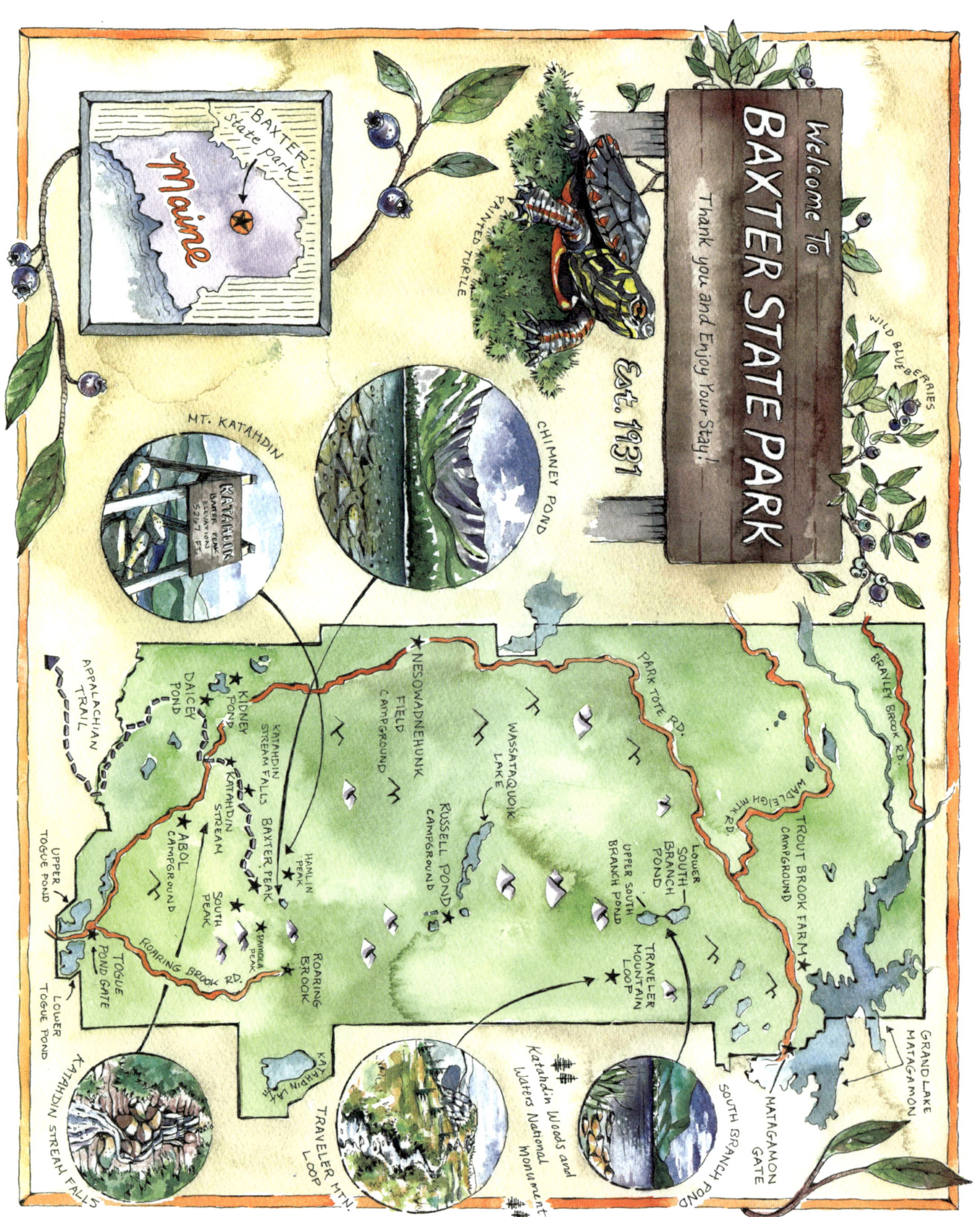

BAXTER State Park
Maine
Welcome To
BAXTER STATE PARK
Thank you and Enjoy Your Stay!
Est. 1931
PAINTED TURTLE
WILD BLUEBERRIES
MT. KATAHDIN
KATAHDIN BAXTER PEAK 5,267 FT.
CHIMNEY POND
APPALACHIAN TRAIL
DAICEY POND
KIDNEY POND
KATAHDIN STREAM FALLS
KATAHDIN STREAM
NESOWADNEHUNK FIELD CAMPGROUND
PARK TOTE RD.
BRAYLEY BROOK RD.
BAXTER PEAK
HAMLIN PEAK
WASSATAQUOIK LAKE
WADLEIGH MTN. RD.
TROUT BROOK FARM
ABOL CAMPGROUND
SOUTH PEAK
PAMOLA PEAK
RUSSELL POND CAMPGROUND
UPPER SOUTH BRANCH POND
LOWER SOUTH BRANCH POND
CAMPGROUND
UPPER TOGUE POND
TOGUE POND GATE
ROARING BROOK RD.
ROARING BROOK
TRAVELER MOUNTAIN LOOP
LOWER TOGUE POND
KATAHDIN STREAM FALLS
KATAHDIN LAKE
TRAVELER MTN. LOOP
Katahdin Woods and Waters National Monument
SOUTH BRANCH POND
MATAGAMON GATE
GRAND LAKE MATAGAMON

BAXTER STATE PARK

Baxter State Park is a preserved wilderness hidden deep in the woods of north-central Maine. While many state parks cater to family recreation with colorful playgrounds and sprawling RV campgrounds, the majority of Baxter State Park is managed as a wildlife sanctuary, with day-use and camping passes strictly regulated to ensure low visitation numbers at any given time. The protection pays off: throughout the park's 200,000 acres—an expanse that includes more than fifty lakes—countless creatures thrive, everything from larger fauna like moose, black bears, bobcats, and river otters to a wide array of smaller inhabitants. The park is also home to 5,267-foot Mount Katahdin, the highest peak in Maine and the northern terminus of the Appalachian Trail. Katahdin means "Great Mountain" and was named by the Indigenous Penobscot people.

Just outside the park, in Katahdin Woods and Waters National Monument, hikers can hop onto the International Appalachian Trail and continue their journey into Canada.

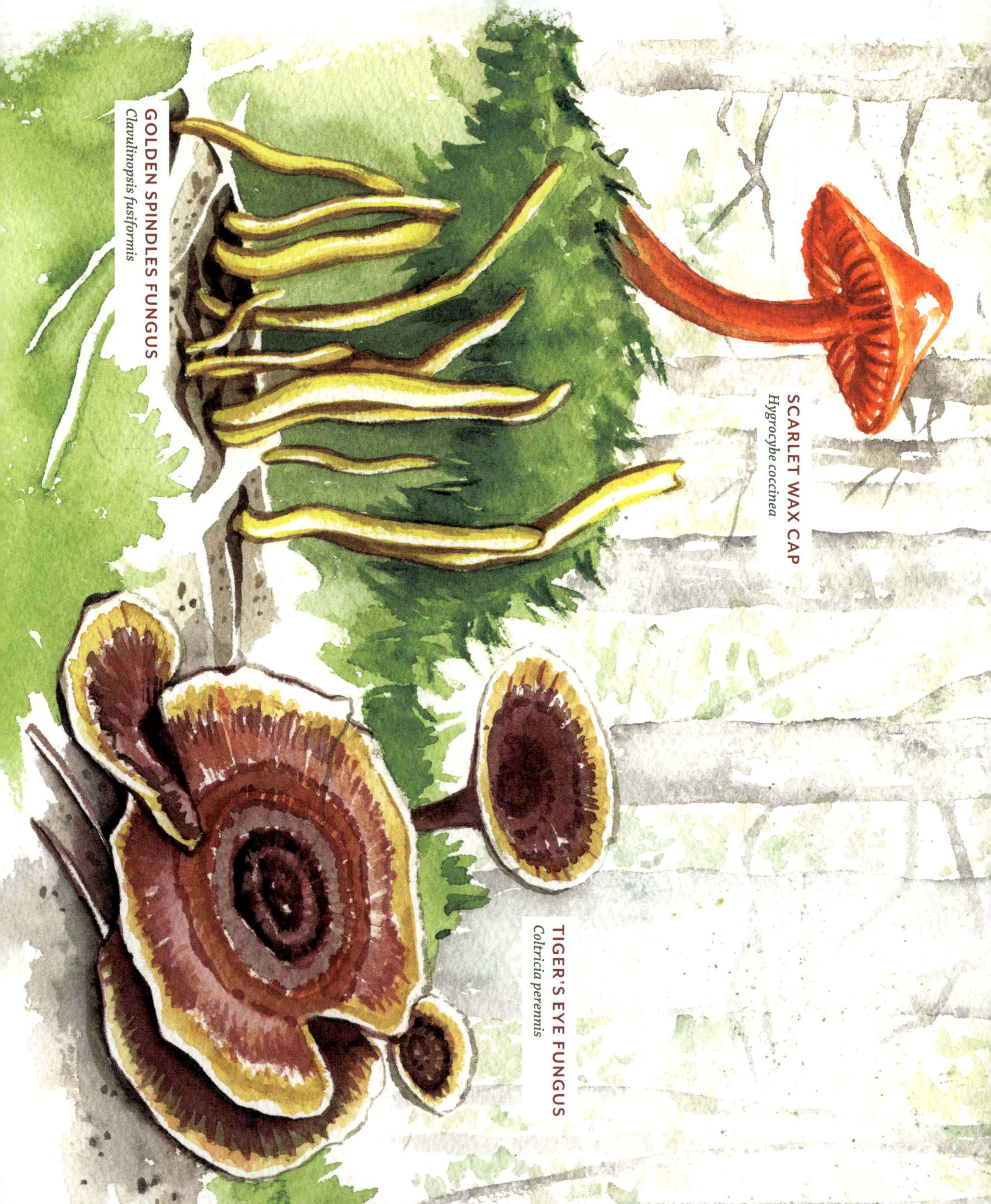

GOLDEN SPINDLES FUNGUS
Clavulinopsis fusiformis
SCARLET WAX CAP
Hygrocybe coccinea
TIGER'S EYE FUNGUS
Coltricia perennis

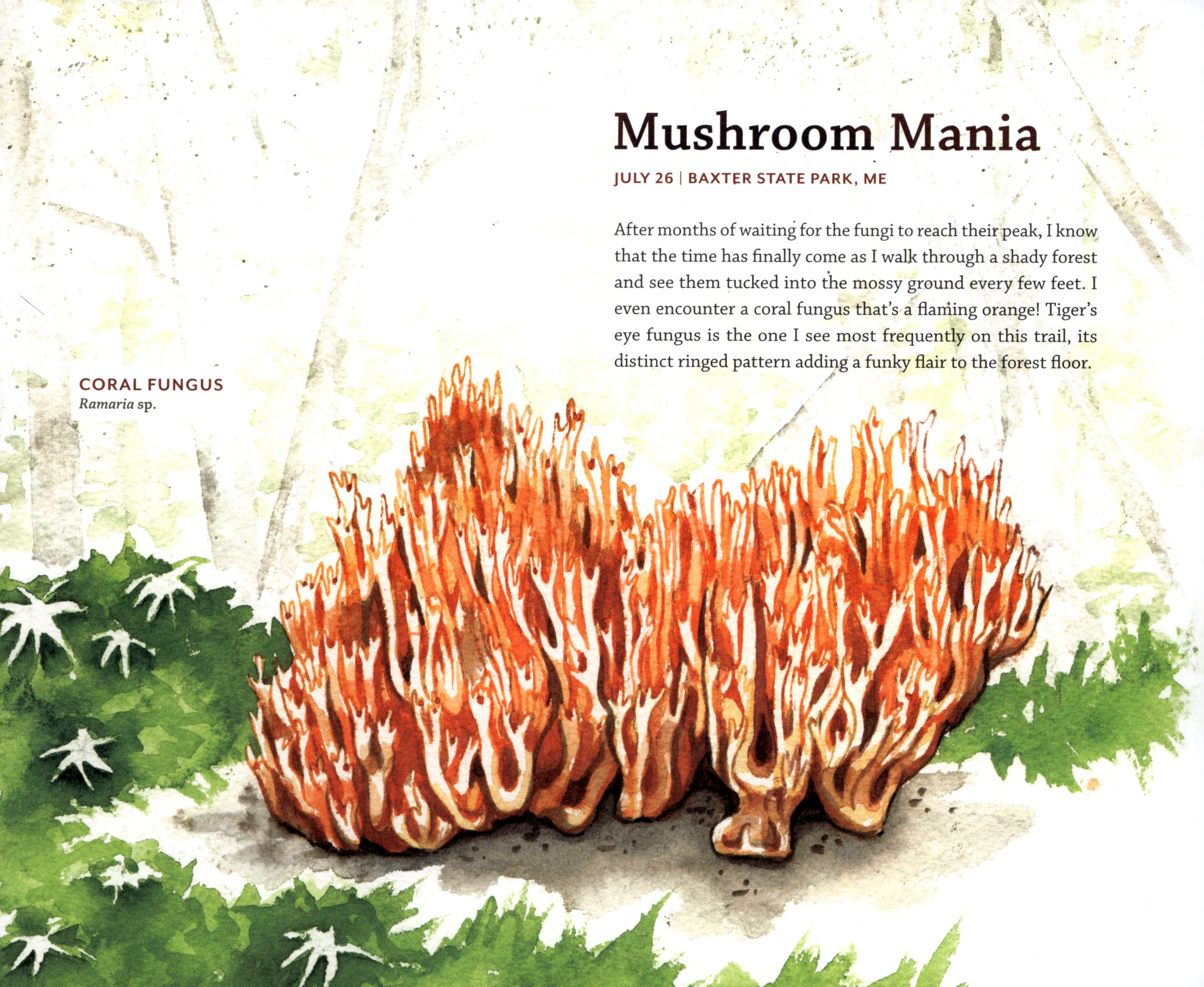

Mushroom Mania

JULY 26 | BAXTER STATE PARK, ME

After months of waiting for the fungi to reach their peak, I know that the time has finally come as I walk through a shady forest and see them tucked into the mossy ground every few feet. I even encounter a coral fungus that's a flaming orange! Tiger's eye fungus is the one I see most frequently on this trail, its distinct ringed pattern adding a funky flair to the forest floor.

Lessons from the Knife Edge

While planning our trip to Maine's Baxter State Park, I imagined myself paddling around the lakes, bird-watching, and searching for tiny creatures at my usual leisurely pace. But when a couple of friends caught wind of our plan, they suggested we meet up to hike the perilous "Knife Edge" of Mount Katahdin.

Now, there are two different routes to ascend Mount Katahdin: one follows the Appalachian Trail and the other traces the Knife Edge, a 1.1-mile path that traverses the massif's eastern ridgeline to Baxter Peak, Katahdin's highest point. The videos I'd seen online made the Knife Edge route look as horrifying as it sounds. Some stretches of the trail are only four feet wide, with drops of 2,000 feet on either side. While the adrenaline-seekers in the group eagerly lobbied for the Knife Edge, they ultimately left the decision up to me. Considering my fear of heights, and my aversion to risk-taking in general, I was the wild card.

We're often told to trust our gut, but what do you do if your gut is a hypervigilant, overactive self-preservationist? For the first half of my life, the years that should have been the most joyful and carefree, I let my fearful intuitions rule my every move. I was terrified to swim in lakes and rivers on family camping trips because I was convinced that snapping turtles would nibble my toes or leeches would bleed me dry. When I transferred to a new school in ninth grade, I ate my first few lunches in a bathroom stall because I couldn't bring myself to face a cafeteria full of strange new faces. We had a wood-burning stove in our home and every time we went away for the weekend, I'd picture us returning to a house that had been reduced to smoldering ashes. In short, I was scared of everything.

But living like that was exhausting. So when I turned eighteen, I set a new intention. With friends holding my hands and telling me stories to distract me from the pain, I got two words and a mathematical symbol tattooed on my left ankle: LOVE > FEAR. The expression was inspired by 1 John 4:18, a Bible verse that reads, "There is no fear in love, but perfect love casts out fear." (Being a good girl, I'd gotten my parents' approval ahead of time—key to that was the biblical theme.)

In the decade since getting that tattoo, I've slowly chipped away at the cage that once held me captive. I started small. Each day, I would do just one thing that scared me. At the outset this was easy, since even simple tasks—like making a phone call or having a meeting with my professors—filled me with nervous energy. In time, however, I started to see results. Eventually, I was initiating meetings with magazine art directors to show them examples of my illustration work; I was teaching multiday watercolor workshops; and I was hiking solo in mountains that were crawling with black bears. Yes, I would keep a cool water bottle with me during those meetings, covertly pressing it against my wrist to bring down my rising temperature. And yes, I would constantly look over my shoulder during those hikes to ensure I wasn't being stalked by curious bears or creepy humans. But these were huge steps forward all the same.

We pull up to the lean-to shelter at Nesowadnehunk Field Campground and my stomach is tight. Tomorrow, we'll hike either the safer path or the Knife Edge. I've deliberated for weeks, and although I'm still uncertain, I give my blessing for the latter—contingent on safe weather conditions. Strong winds, lightning strikes, and thick fog have claimed dozens of lives on this route. Secretly, I hope that the weather will take a turn for the worse, forcing us to opt for the safer trail to the summit.

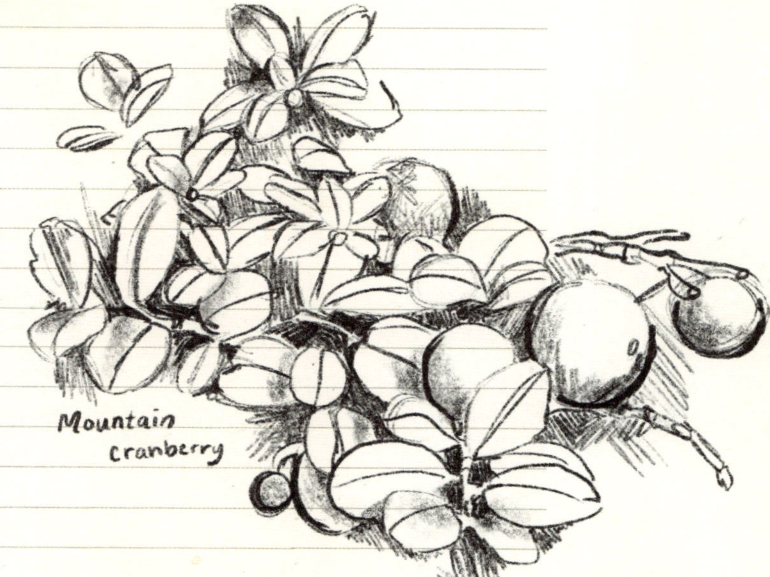

Mountain Cranberry

To my dismay, the forecast remains perfectly clear. We pile into the car at four in the morning, then bump through the darkness down a long gravel road, reaching the trailhead by daybreak. The trail starts innocently enough, with a gentle incline lined on either side with blueberry bushes bursting with fruit. My mind is preoccupied with the section of narrow trail that lies ahead, but I remind myself to look around and pay attention to my surroundings. As we cross through a band of stunted fir trees and then emerge above tree line, dark-eyed juncos cry out from their rocky perches, their short call notes a perfect imitation of glass marbles clacking together. I glimpse tiny plants tucked between the boulders—the fragile white blossoms of three-toothed cinquefoil and the shiny reds and waxy greens of mountain cranberry. Although both plants appear too delicate to exist in these conditions, they actually thrive in exposed, rocky areas with harsh weather and little topsoil.

We huff and puff our way up to the top of Pamola Peak, where we finally lay eyes on the ominous path that leads to Katahdin's summit. I peer through binoculars at a few silhouettes of ant-sized humans creeping their way along the ridge and take a deep breath to calm the knot in my stomach. The intensity of the trail immediately ratchets up a notch when we come upon a steep, thirty-foot rock crevice called the Chimney. This section requires us to descend backward, like we're climbing down a ladder. As I slide my feet blindly against the steep rock in search of footholds, I can see that the path gives way to nothingness just a few feet to each side of me. "You won't fall, you won't fall," I silently chant.

When we emerge from the Chimney and onto the thin ridge, I immediately crouch down, intending to crawl along on my hands and knees. It feels safer that way, closer to the earth. But one of our friends explains that the more upright you stand, the more stable and balanced you'll be, ready to catch yourself with the other leg if you do trip. Reluctantly, I rise to my feet. Gingerly placing one foot in front of the other, I inch my way across the ridge, my mental fortitude tested with each step.

As the rocky trail gradually widens and we approach Baxter Peak, relief and gratitude course through my veins. We sprawl out and prepare a gourmet picnic of tortillas stuffed with chicken sausage and cheddar cheese. The sunshine soothes our aching muscles, and smiles spread wide across our faces.

At the bottom of the mountain, we come upon a stunning lake called Chimney Pond, Mount Katahdin reflected in its glass-like surface. Sitting on the shore, we tug off our shoes and peel off our sweaty socks. I catch a glimpse of my little ankle tattoo, so poorly done that the letters now blur together and make it hard to read. I've often regretted it and have even considered having it removed because it seems so embarrassingly cliché, but it's funny how the sentiment has held true throughout these past ten years: I can't fully love my life if fear is consuming all my energy. I wiggle my toes in the cool mountain air and know that although I'm far from fearless, I have come further than I ever thought possible.

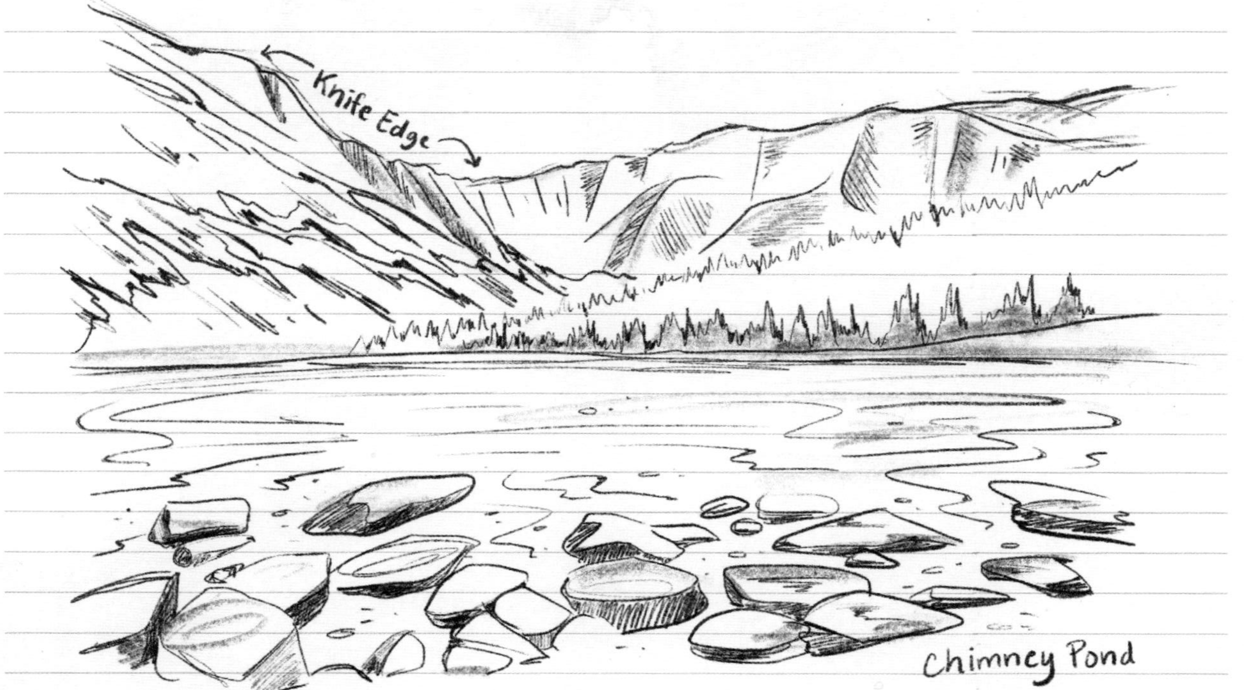

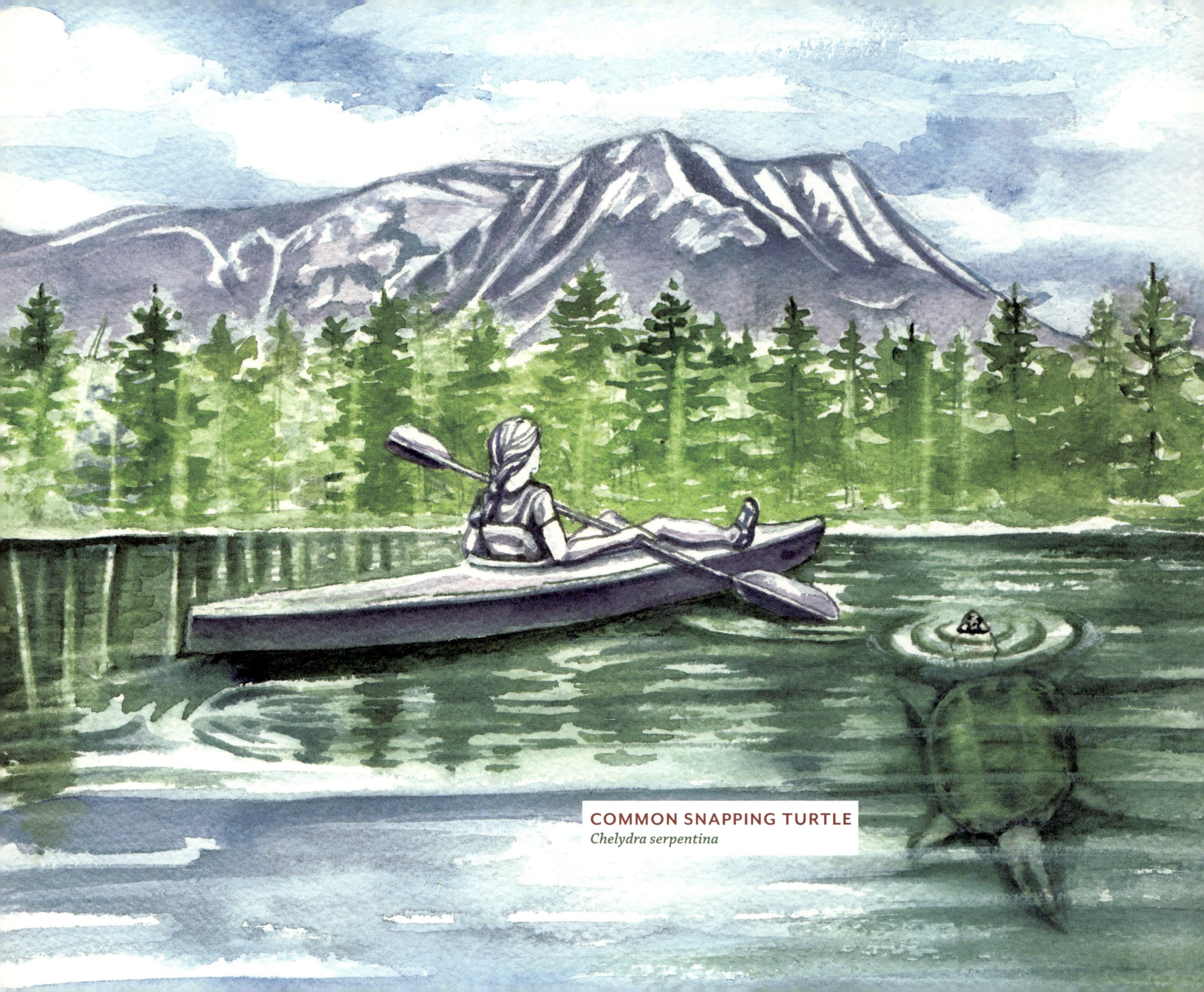

COMMON SNAPPING TURTLE
Chelydra serpentina

Lake Lurking

The tip of my kayak parts the reeds with a gentle swoosh as I glide through wetlands at the edge of the lake. Behind me, the sleek shape of a loon emerges silently from the glassy waters to scan its surroundings. In the distance, Mount Katahdin stands lonely and proud.

Suddenly, a powerful *thump, thump* on the bottom of my boat yanks me out of my reverie. As I peer into the water, searching for the culprit, a hulking snapping turtle swims out from underneath the kayak, his wide, paddle-like hands sending the underwater plants swirling into slow-motion chaos. His smooth head surfaces a few feet from me, and he holds it there for thirty seconds as he takes in long, deep breaths. He sounds just like an old man snoring! I hold my breath as he takes his, then he languidly retreats into the depths once again. Snapping turtles are nocturnal, usually spending their days at the bottoms of lakes and wetlands. I am grateful that this one took a break from his napping to come say hello.

As I skirt along the edge of the wetland bog, wonder after wonder is unveiled: three baby painted turtles sun themselves on a log, roundleaf sundew grows so thick that it creates a sparkling carpet on the moss, and the pretty pink blossoms of Fraser's marsh St. John's wort stand out among the lakeside grasses.

Floating bladderwort takes the prize for the most intriguing find of the day, due to its benign ferocity. This five-inch-tall flower uses a pontoon-like structure to stand on the water's surface, with delicate, rootlike tentacles dragging underneath. But don't let its innocent appearance fool you: this is a carnivorous plant, and those underwater tentacles hold traps—tiny spring-loaded bladders that release a sweet taste, which lures microscopic planktonic organisms. By constantly pumping water out of the bladders, the plant maintains an air pressure that's much lower than the surrounding water pressure. Once the prey gets close to a bladder, invisible hairs on the bladder trigger it to open and suck in the victim—at a velocity that's higher than that of any other known plant or animal predator on earth. On average, the traps snap shut in just half a millisecond.

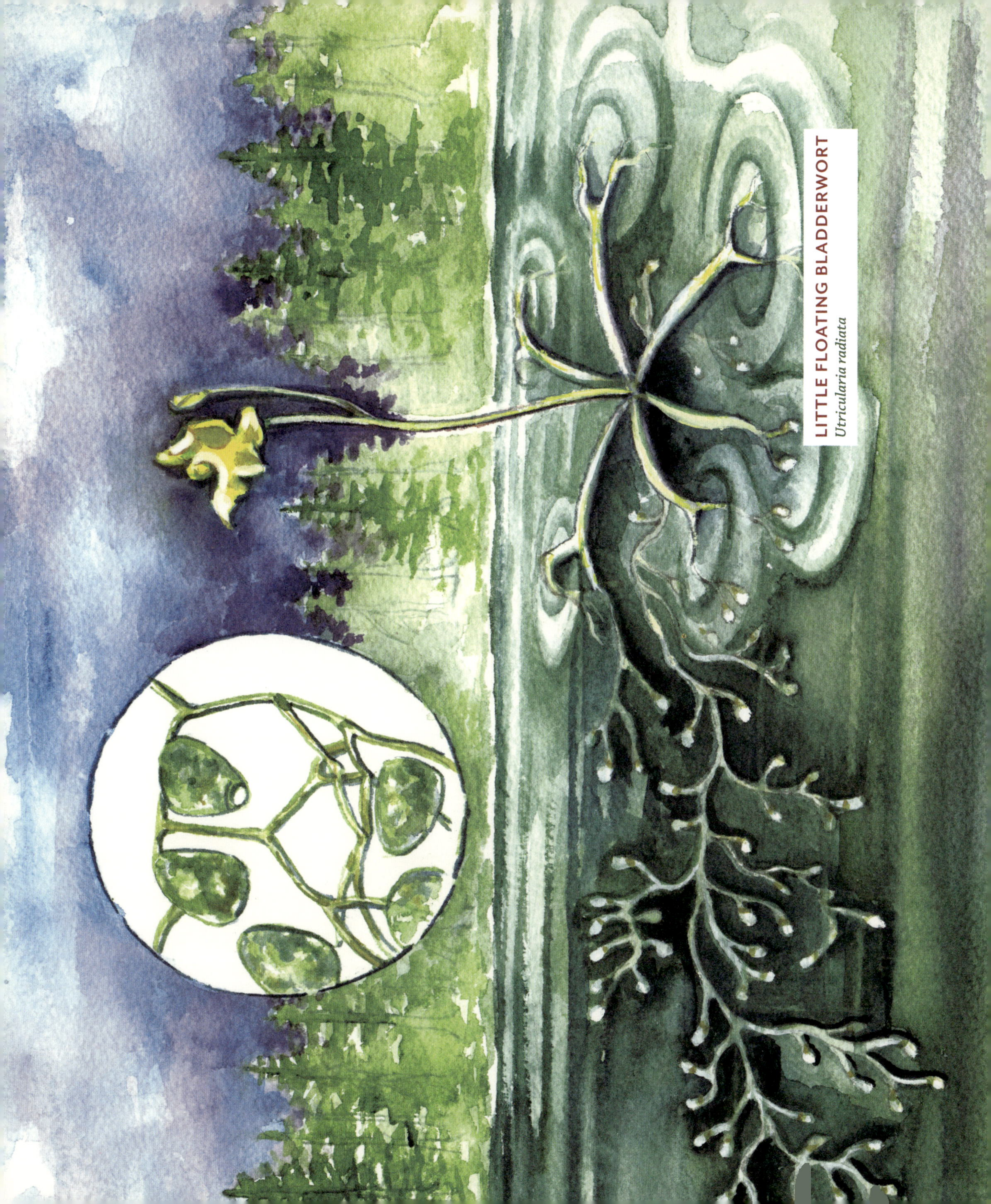

LITTLE FLOATING BLADDERWORT
Utricularia radiata

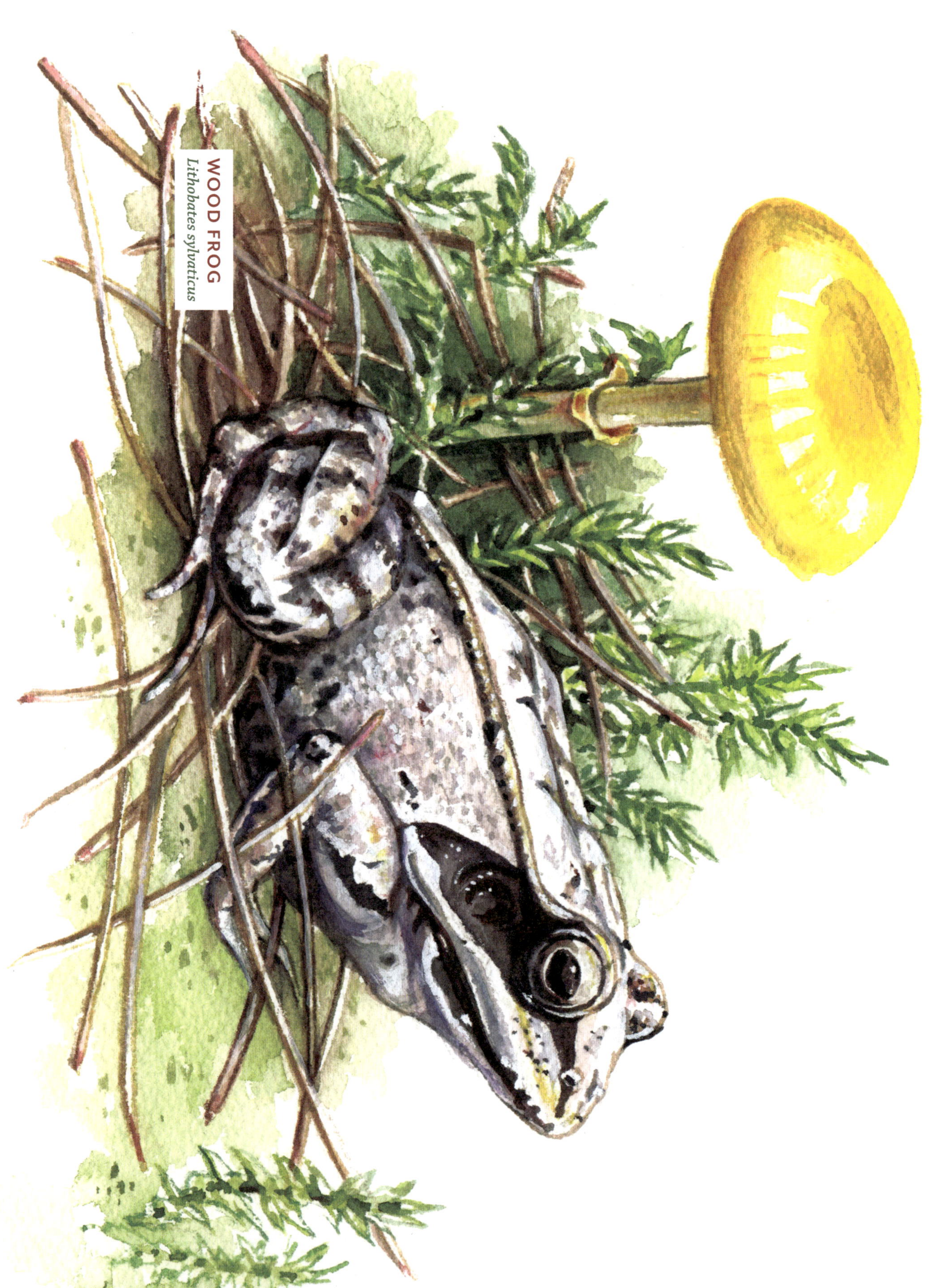

WOOD FROG
Lithobates sylvaticus

Bouncing Bandit

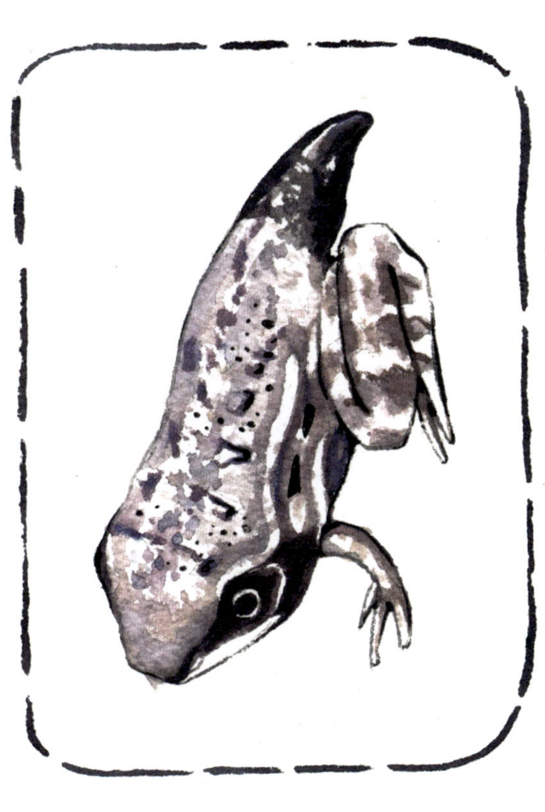

The juvenile wood frog that I found in Pennsylvania

I spy a clearing in the woods with a fallen log and a soft pine-needle floor: the perfect lunch spot. As I swallow bites of sandwich and apple, a faint stirring on the ground nearby draws my attention. A wood frog is poised just a few feet from me, the sunlight reflecting off its shiny pinkish-tan skin. It remains there for the rest of my lunch break, giving me ample opportunity to study it in detail.

Back in Pennsylvania in June, I found the teeniest wood frog, one that still had part of its dark, hornlike tail attached from its tadpole days. Even juveniles possess the distinctive bandit mask, which provides a helpful identifier. Wood frogs spend their adult lives in the forest, returning to ephemeral pools in early spring to mate and lay their eggs. These temporary pools are the safest place for eggs since there aren't any fish to eat them before they can hatch. During the winter, wood frogs actually freeze. Their hearts stop beating, they stop breathing, and ice even forms in the spaces between their cells. When springtime comes, the frogs thaw out and resume their amphibian activities.

woodpecker sightings

4/30/22	Wetlands near Greenway Trail Franklin, NC	Downy	It kept angling itself behind the tree so that I couldn't watch it while it ate - def. on purpose!
5/27/22	South River, Waynesboro, Virginia	Northern flicker	courtship dance! Male + female w/ yellow tails displayed
6/16/22	Plessinger Trail, Cowans Gap State Park, PA	Pileated	watched it aggressively preening each wing
7/19/22	Hike back from Upper Cherry Pond; Pondicherry NWR, NH	Black-backed	2 of them were squabbling loudly as they pecked

NATURE JOURNAL IDEAS

STEALTH FLOAT

One of the best ways to get a good look at ducks, beavers, muskrats, and other water-dwelling creatures is from a boat. Wildlife is less easily startled when you're floating rather than walking. Take a canoe, paddleboard, or kayak out on a lake or calm river and try to make your paddling movements as fluid as possible. You're especially likely to see activity in the riparian foliage at the edge of the water—keep your eyes peeled!

GONE SHROOMIN'

In many places, mushrooms can be found during all seasons. Go for a walk with your eyes fixed on the decaying wood along your path. Look for the funkiest specimen you can find and take a photo or sketch it.

CREATURE CONNECTIONS

Just like building a friendship, developing a deeper understanding of a wild critter requires many encounters, during which you learn a little more about them each time. Pick an animal you're likely to come across more than once—it can be something as common as a songbird or a squirrel—and begin a log of your encounters with it. First, just write down its name, where you saw it, and the date. When you run into that species again, look for one new piece of information that you can add to your log. Was it eating something? Did it make a noise? Did this individual appear to be in a different life stage than the individual you saw before? Has its plumage or fur changed colors with the season? Over the years, your observations will come together to paint a fuller picture of this creature.

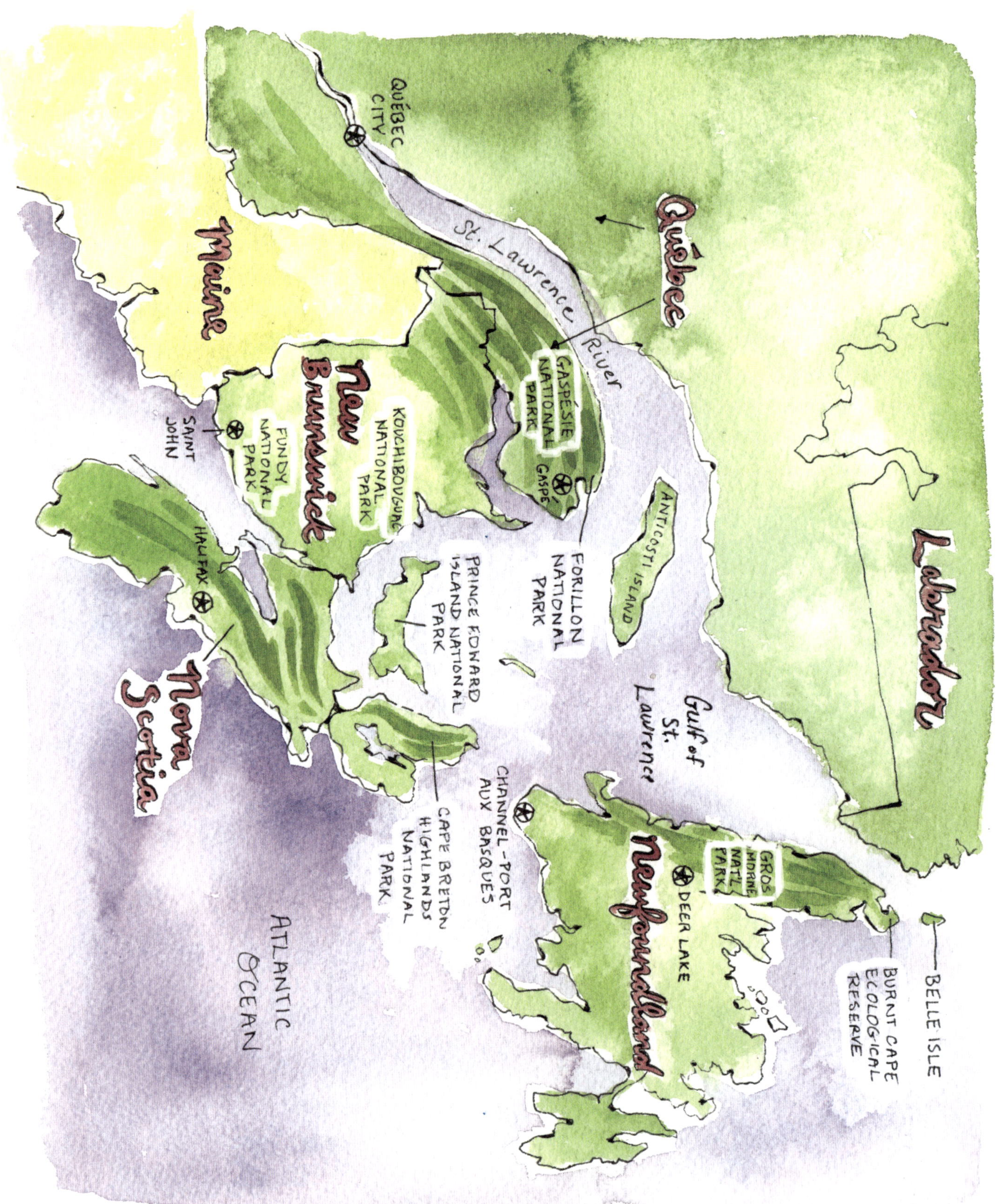

Maine
Québec City
St. Lawrence River
Québec
Labrador
New Brunswick
Gaspésie National Park
Gaspé
Kouchibouguac National Park
Fundy National Park
Saint John
Forillon National Park
Anticosti Island
Prince Edward Island National Park
Halifax
Nova Scotia
Gulf of St. Lawrence
Cape Breton Highlands National Park
Channel-Port aux Basques
Newfoundland
Deer Lake
Gros Morne Nat'l Park
Burnt Cape Ecological Reserve
Belle Isle
Atlantic Ocean

Canadian APPALACHIANS

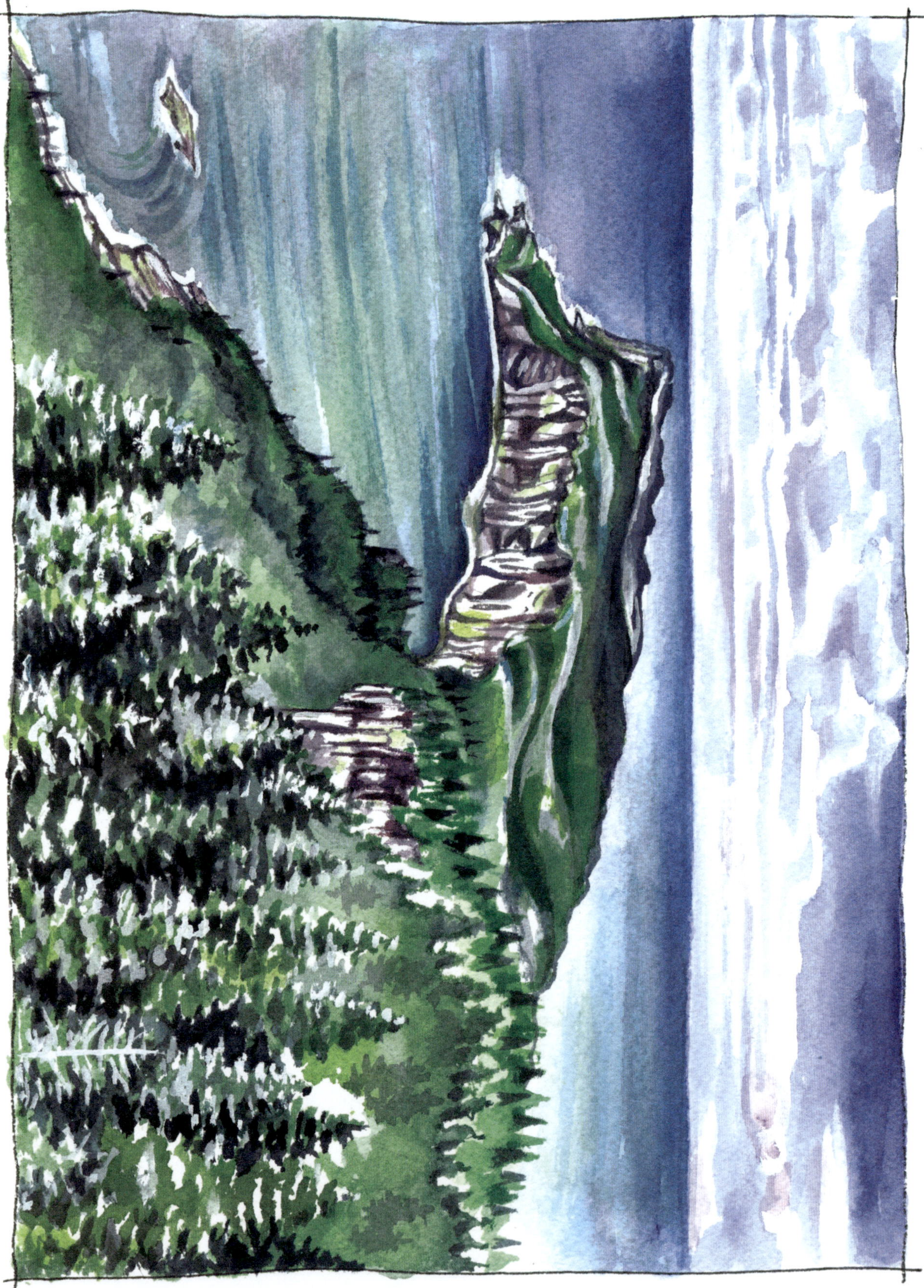

BEYOND THE BORDER

AUGUST 1

After spending four months immersing ourselves in the Appalachian Mountains of the United States, we drive across the border into New Brunswick, on our way to the Canadian Appalachians of Québec. I had always assumed that the land north of Maine was all forest, just perpetual swaths, but the view outside our window as we cross through New Brunswick is far from deep wilderness. Rolling fields produce a variety of crops, the farming landscape interrupted occasionally by small and midsize towns. It looks like Pennsylvania!

Eventually, we pass over a bridge into the French-speaking province of Québec. Whereas before the road signs had been in both English and French, here they are exclusively in French. The road takes us higher and higher into the Chic-Choc Mountains—a subrange of the Appalachians—and through the middle of Gaspésie National Park, where our surroundings transition to evergreen forests and clear, fast-moving rivers. Now *this* is the boreal forest that I'd imagined. The road spits us out on the other side of the park and skirts between the rugged Appalachian Mountains and the massive St. Lawrence River, a waterway so wide that we can't see the opposite shore.

It's strange to see a new side of these mountains. The Appalachians that I'm familiar with are landlocked and far from the sea, both culturally and geographically. But here, fishing boats bob in the distance, a windsurfer rides the waves, and seagulls fill the sky. Just as I thought I was gaining a deeper understanding of the range, a new layer unveils itself.

OPPOSITE: *Cap-Bon-Ami, Forillon National Park, QC* NEXT SPREAD: *Forillon National Park, QC*

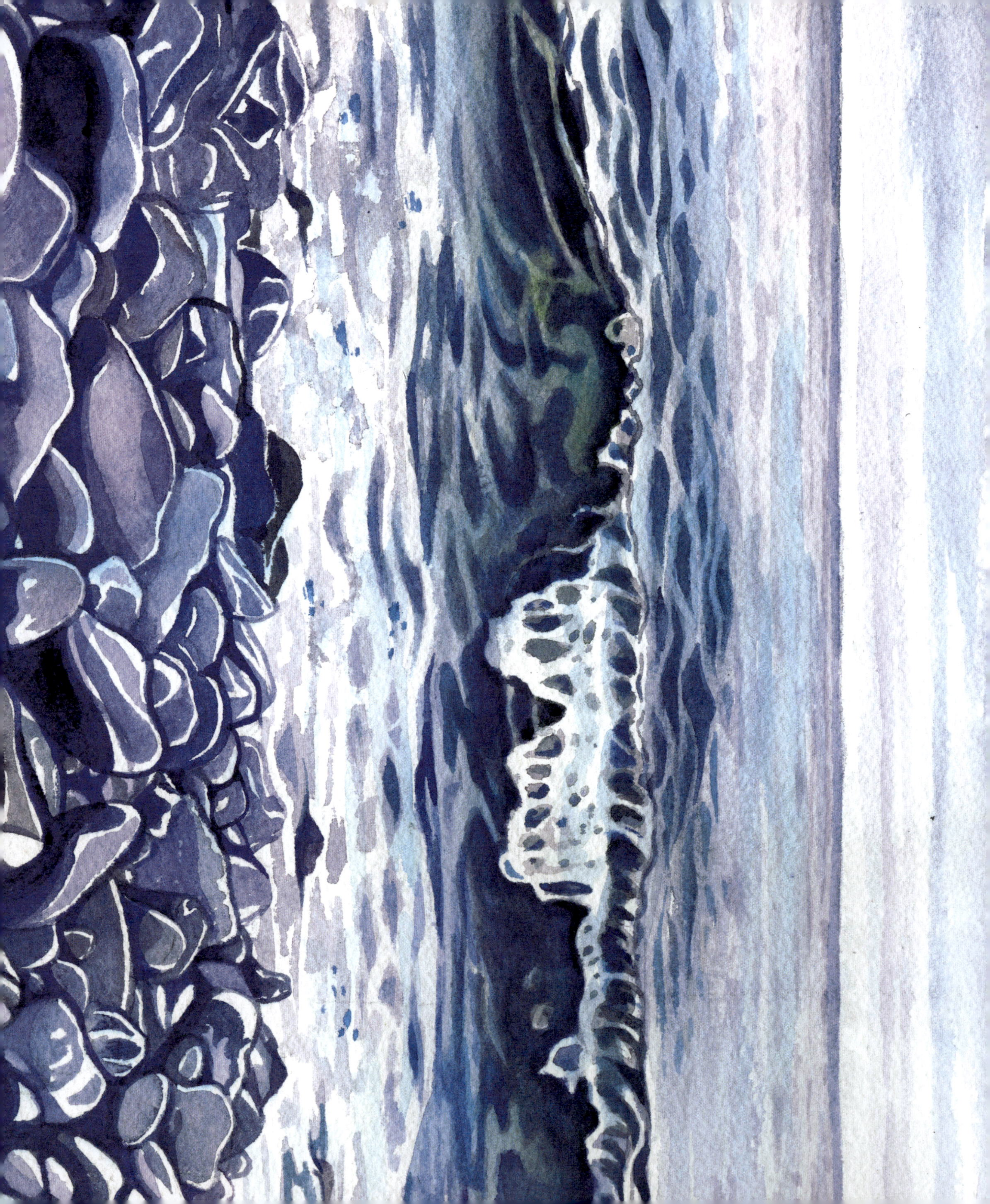

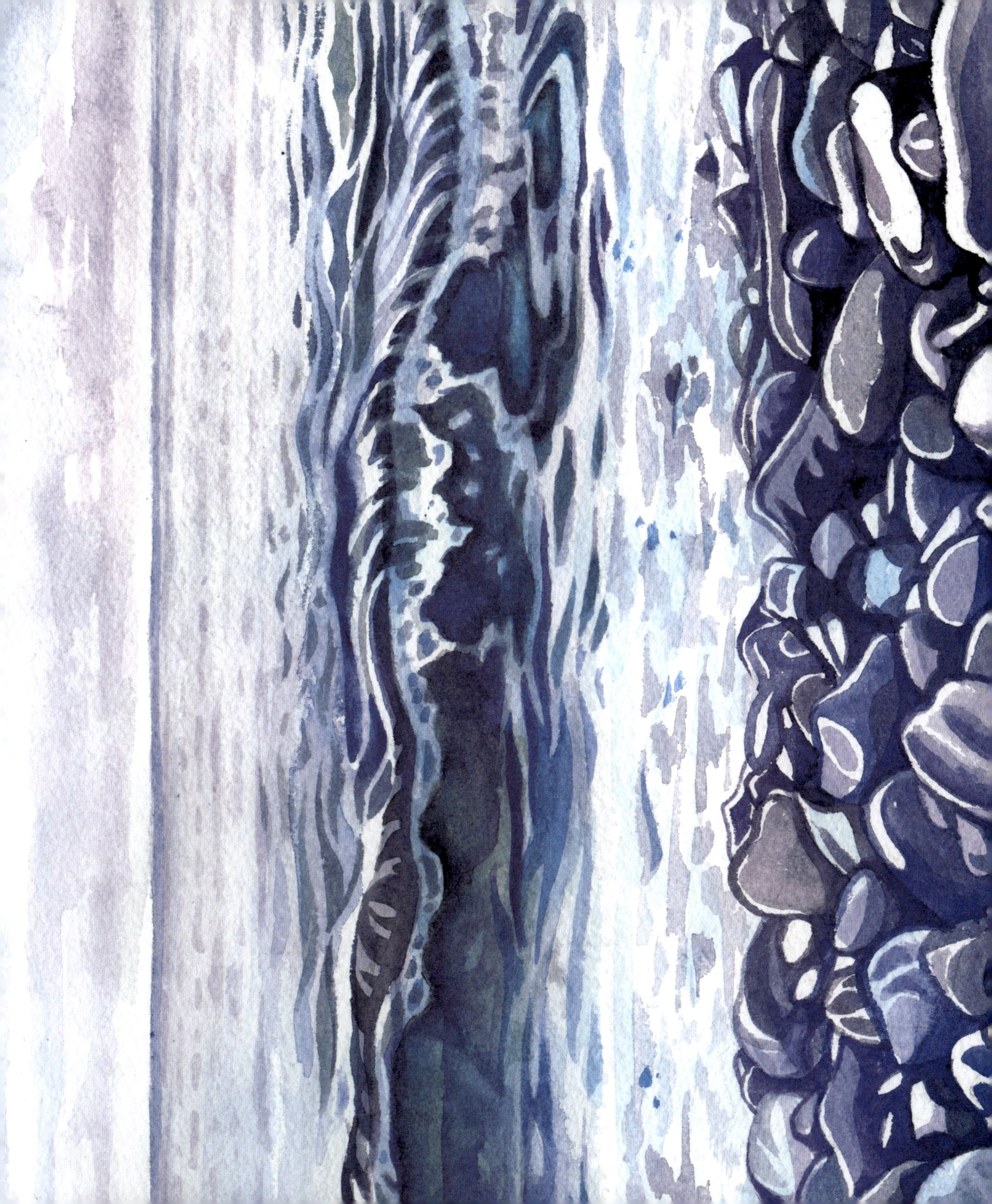

Pint-Size Pollinators

AUGUST 6 | SAINT-MAXIME-DU-MONT-LOUIS RIVER TRAIL, QC

I hear the familiar buzzing before I see it: a delicate female ruby-throated hummingbird zooms into view and sucks at a service-berry, then disappears. It's fun to admire her in a wild setting, not at a sugar-water feeder like the one my parents have always kept out on their back porch. I take a seat to wait for her return. Eventually, two new hummingbirds whir into view, this time feasting on the nectar of the native spotted jewelweed. It turns out that hummingbirds are the main pollinator of jewelweed. The bird's vibrating body causes the flower to wobble, which distributes pollen onto the bird's face.

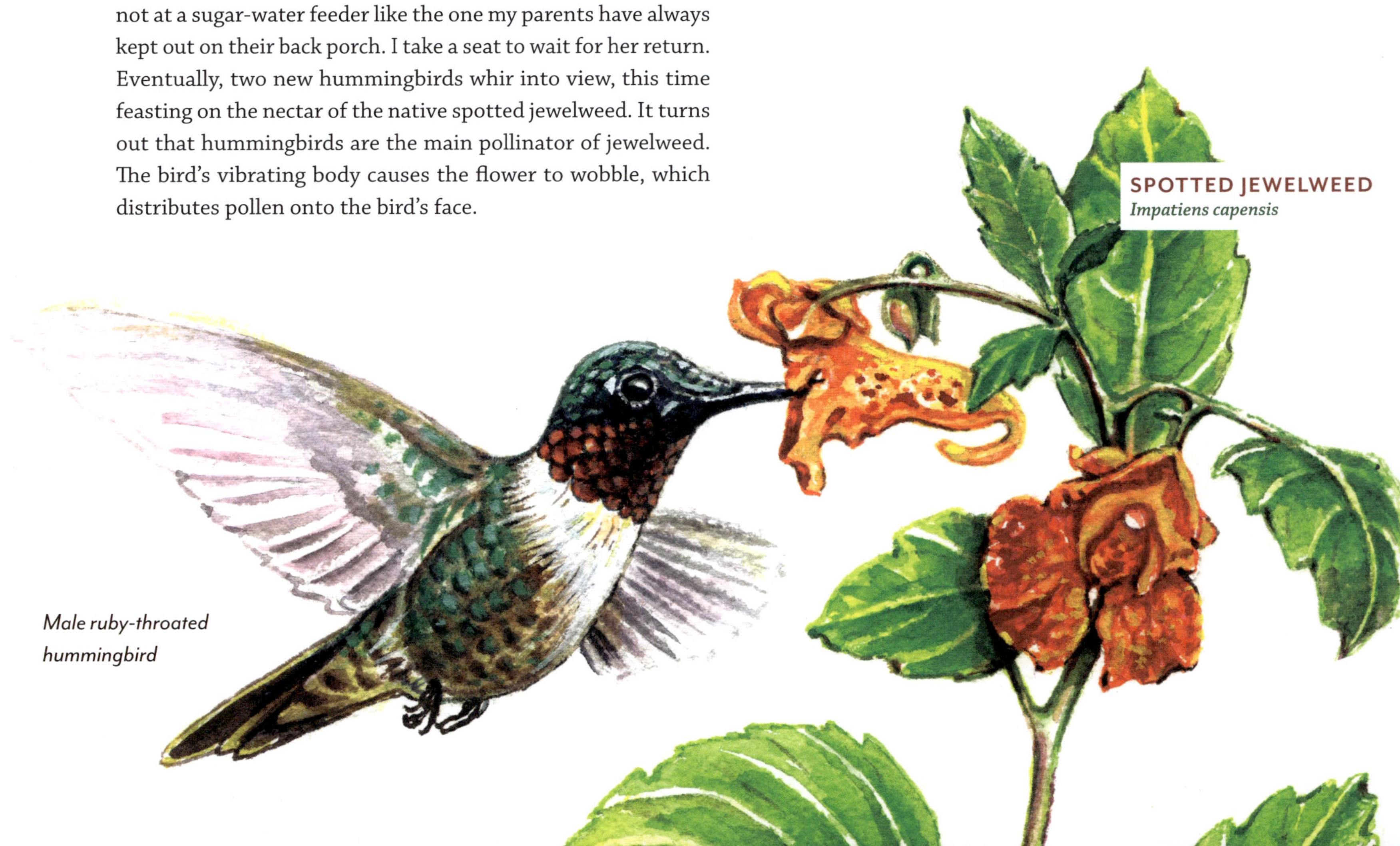

Male ruby-throated hummingbird

SPOTTED JEWELWEED
Impatiens capensis

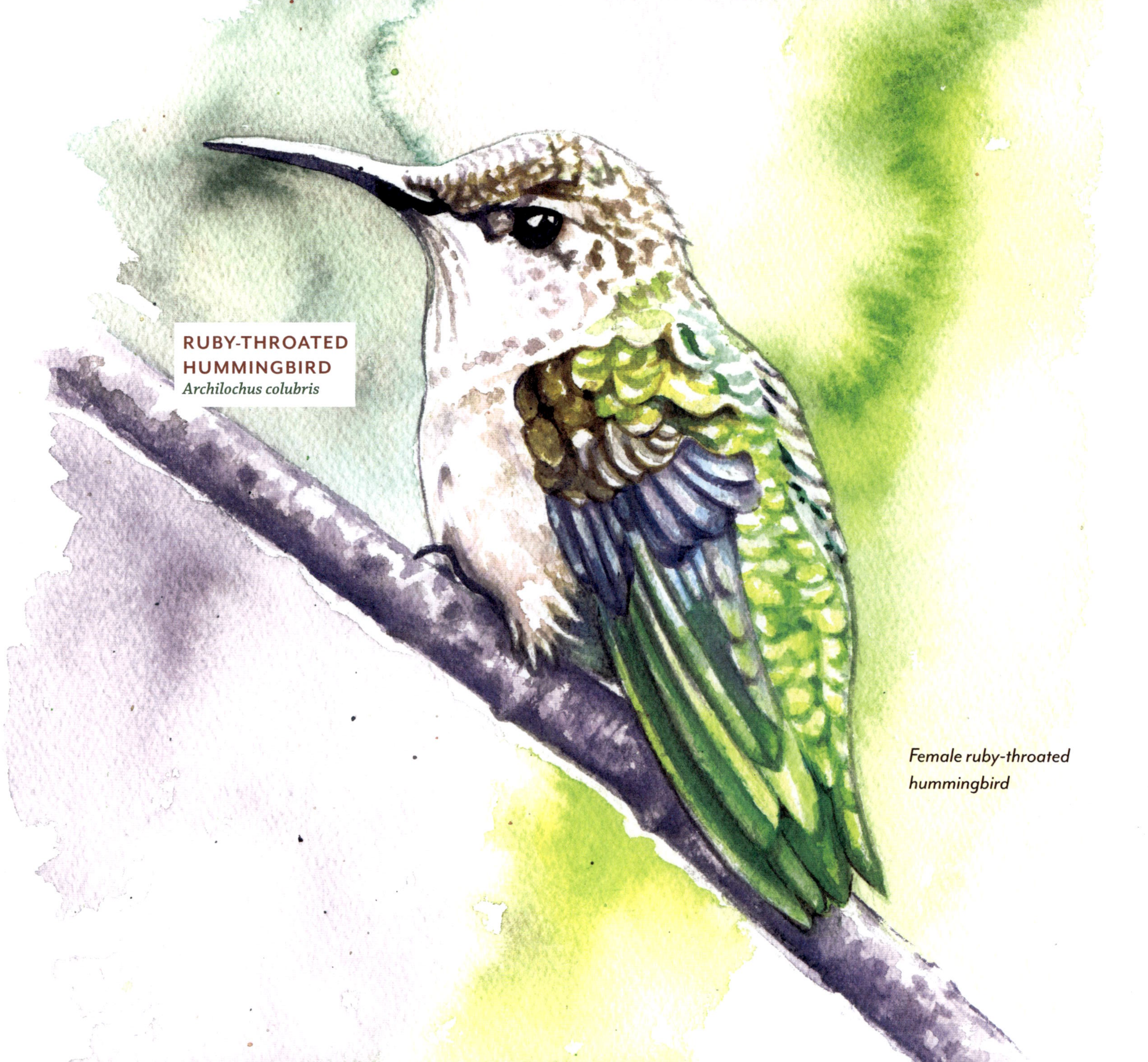

**RUBY-THROATED
HUMMINGBIRD**
Archilochus colubris

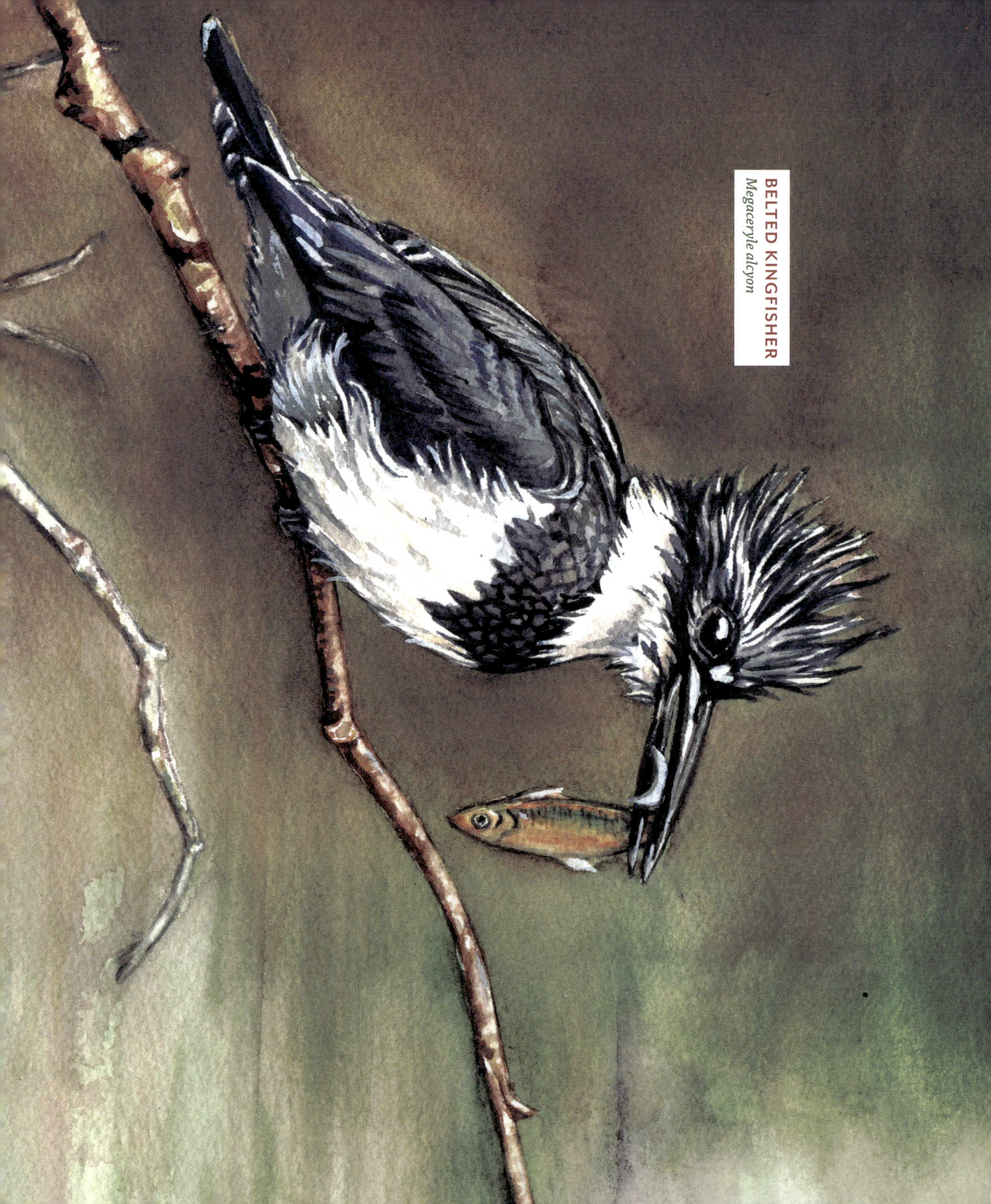
BELTED KINGFISHER
Megaceryle alcyon

King of the River

AUGUST 10 | SAINT-MAXIME-DU-MONT-LOUIS, QC

Belted kingfishers perch above the Mont-Louis River every afternoon, their keen gazes fixed on the water below. They sit perfectly still . . . until they lock eyes on their prey. Then they dive-bomb into the water and emerge straight back up to their perch with a squirming fish in their beak, their wet head feathers spiked like the hair of a '90s punk rocker.

Today as I watch the kingfishers, a family walks past and a small child asks me what I'm looking at. "I'm taking pictures of the birds," I say in French. I want to tell them about the kingfisher specifically, but I don't know the word for it. I make an attempt, saying in French, "A bird who fishes, like a king!" To which the mother responds, "Ah, *un martin-pêcheur!*" I nod my head excitedly, pretty sure that we're talking about the same bird now. We exchange smiles and they continue on. A great blue heron also fishes nearby, and his strategy is to keep his skinny legs totally still so that fish will swim right up to him. In one swift motion, he'll stretch out his long neck and nab one. I walk back to the rental where we're staying and find that Ben is cooking up some fresh cod for dinner. Fish for everyone tonight!

Cross-Country Connections

It's my first walk in the woods of Québec and I feel strangely removed from my surroundings. Over the past few months, as we've explored new areas, I've repeatedly encountered many of the same plants and animals. Now that we've jumped much farther north, I expect that the flora and fauna will be different. But my mind is so preoccupied with trying to recall French vocabulary that I simply don't have the brain space to take in new stimuli. As I finish up my loop walk, I realize I've hardly made a single observation.

Two days later, I return to the river trail for another evening stroll. This time, I try to walk with more awareness. A fleck of orange catches my eye. Could that be spotted jewelweed? I take a closer look and yes, it's the same plant that grows all over my family's farm! Next to the jewelweed I notice a tall branch of goldenrod, another wildflower that blooms in late summer on the farm. Just like that, a whole parade of familiar faces comes into focus: pearly everlasting, serviceberry, fireweed, yarrow, pheasant back mushrooms, wild sarsaparilla. How is it that changing up just a couple of factors—crossing a political border and experiencing a foreign language—suddenly casts even recognizable things in a new light?

I feel guilty when I go entire seasons without visiting the farm, almost like I'm neglecting a family member. But now I breathe in the scent of sun-warmed wildflowers and grasses and it smells just like August in West Virginia. I walk past a construction site where a few wildflower bushes have been squished, so I pluck a handful of stems and carry them back to our rental house. The blended bouquet makes it smell like home, even though we're over 1,000 miles away.

Later in the week, we pop a tire while driving the rough gravel roads of a wildlife refuge and have to go to a mechanic. He emerges from the shop and I'm surprised to find that he looks eerily similar to my oldest brother, who is also a mechanic. Even his behavior is similar. And yet this man speaks French and lives by the sea.

Now that I'm looking for similarities instead of projecting differences, I realize that the homes and lifestyle here are familiar too. The dwellings are modest and well taken care of, many of the yards adorned with children's playsets, stacks of firewood, small vegetable gardens. As I walk along the steep dirt roads behind town, four-wheelers buzz past me and I notice that one rider is wearing a camouflage hoodie and worn leather work boots, like my brothers wear back home. I see families out fishing every morning in the brackish area where the freshwater river meets the sea.

I also begin to come across clues that reveal how closely these people identify with the Appalachian Mountains. There's a maple syrup company called Appalaches Nature (Appalachian Nature), a rock band based out of Montréal named Les Appalaches (The Appalachians), and a scenic highway known as la Route de la chaîne des Appalaches (the Appalachian Range Route). No matter how out of place I feel as I try to communicate with the locals in French, a sense of connection begins to outweigh the differences. We all share the same Mountain Mama.

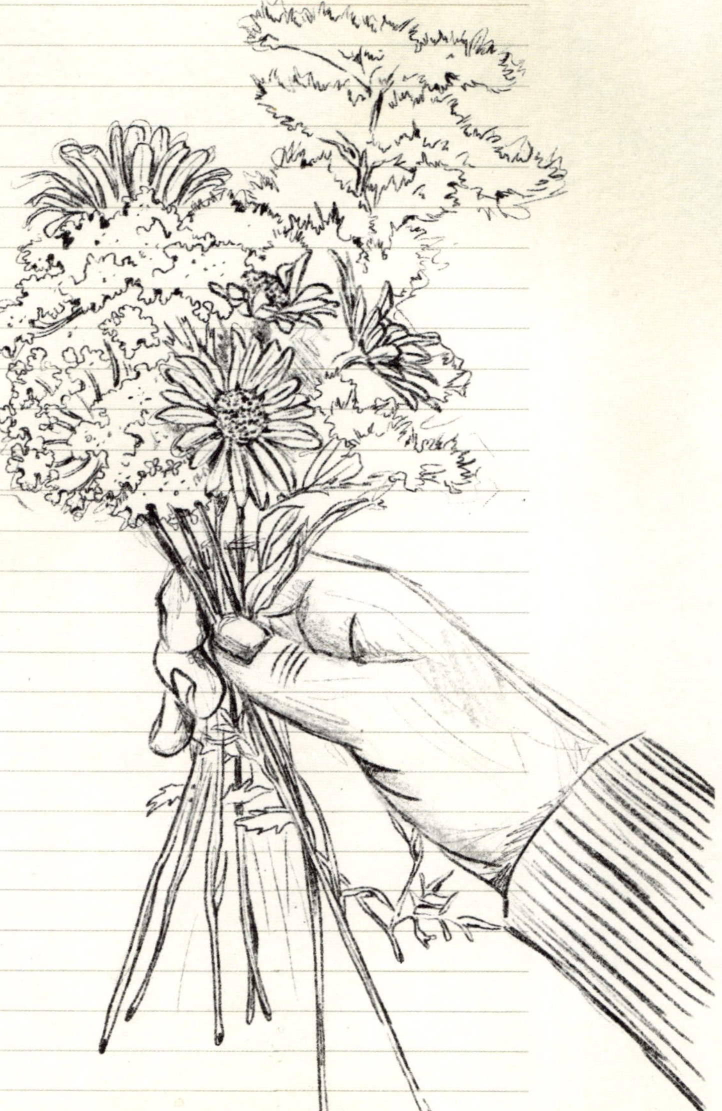

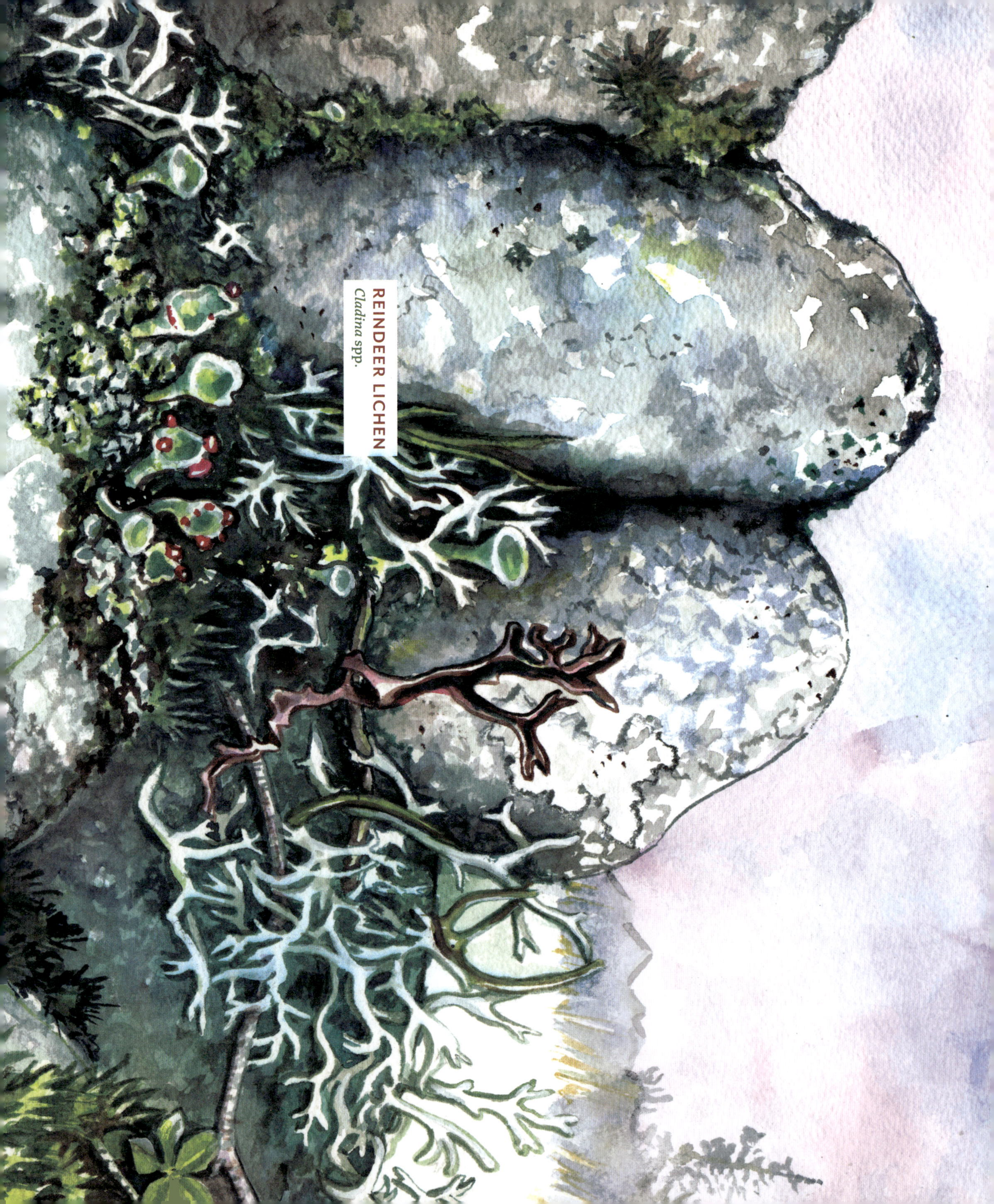
REINDEER LICHEN
Cladina spp.

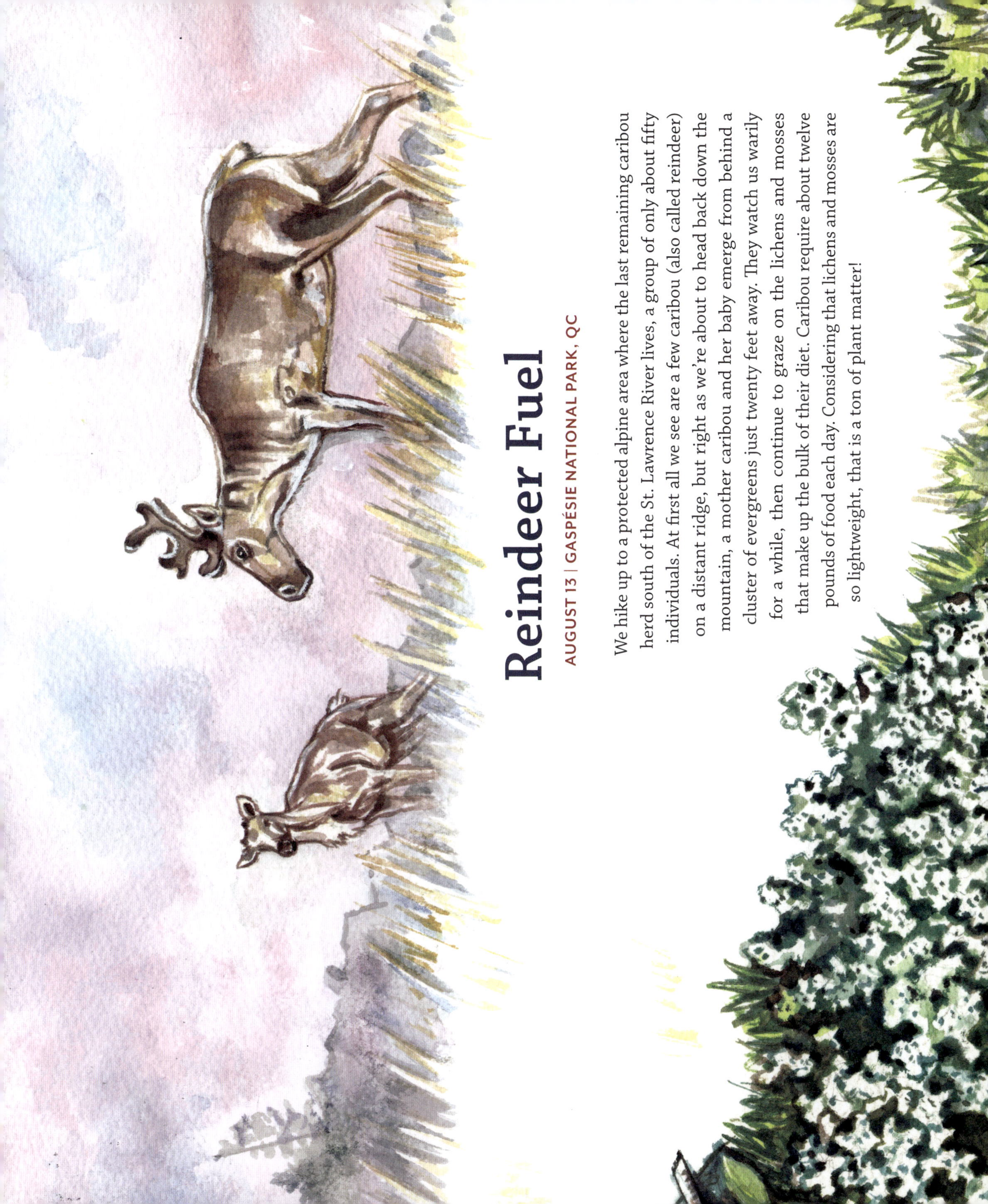

Reindeer Fuel

We hike up to a protected alpine area where the last remaining caribou herd south of the St. Lawrence River lives, a group of only about fifty individuals. At first all we see are a few caribou (also called reindeer) on a distant ridge, but right as we're about to head back down the mountain, a mother caribou and her baby emerge from behind a cluster of evergreens just twenty feet away. They watch us warily for a while, then continue to graze on the lichens and mosses that make up the bulk of their diet. Caribou require about twelve pounds of food each day. Considering that lichens and mosses are so lightweight, that is a ton of plant matter!

A Trip through Taiga

It's a quiet golden hour in the taiga. The path weaves through a stand of black spruce, the sandy soil absorbing every sound. *Taiga* is another name for northern or boreal forest, a biome normally found much farther north in Québec. But lucky for us, an outlier pocket of taiga exists here on the salt spits that border the Atlantic Ocean.

Spongy, light green reindeer lichen forms a crust over the sand. Between the lichen, I'm surprised to find cranberries, blueberries, bunchberries, common funnel mushrooms, and running-pine also growing in this dry landscape. An adorable reddish-brown vole makes a quick appearance before scampering off. I call out "Vole patrol!" to make sure Ben keeps an eye out for them too. High up in a tree, a dark-eyed junco nibbles on a caterpillar snack.

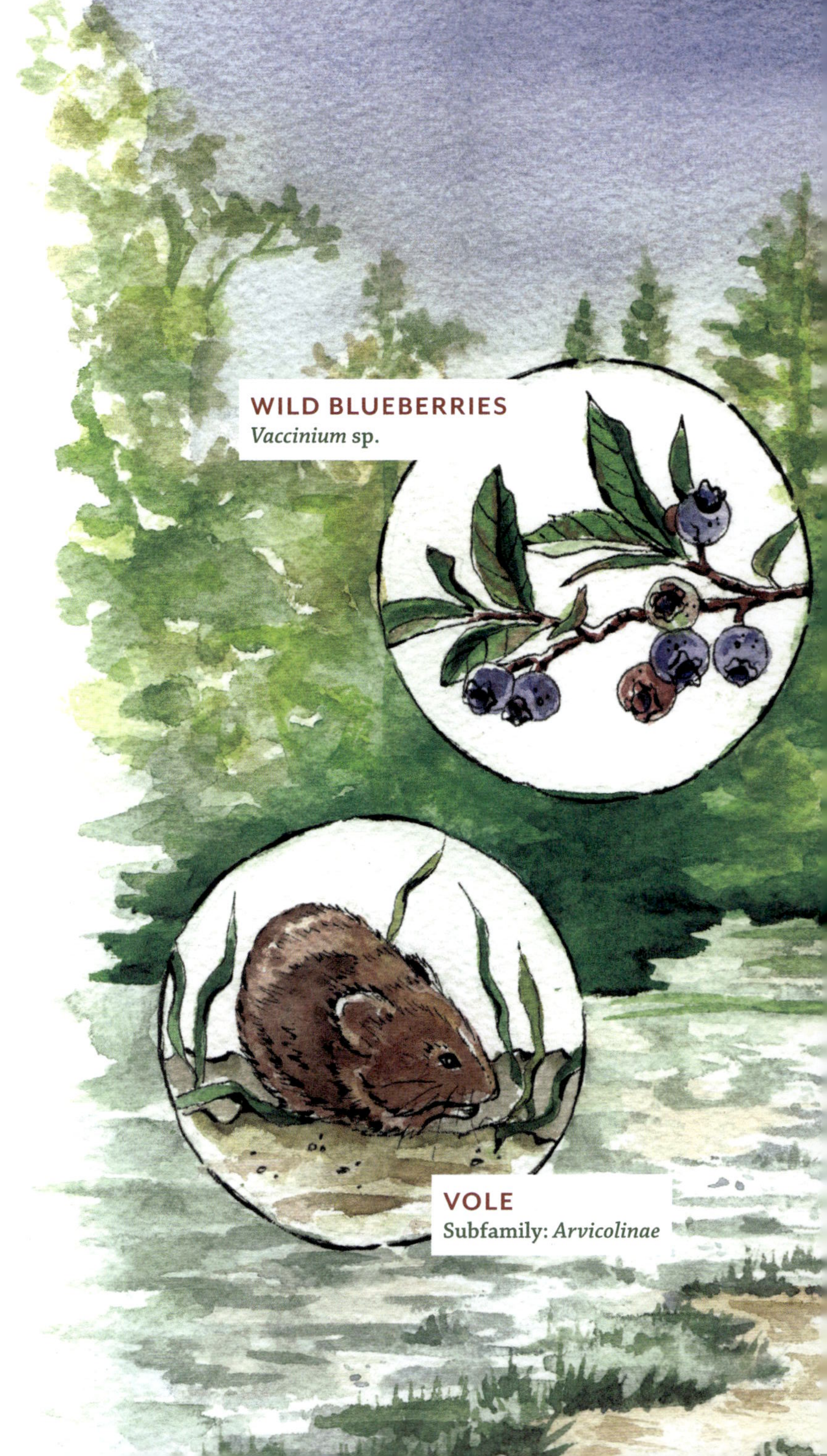

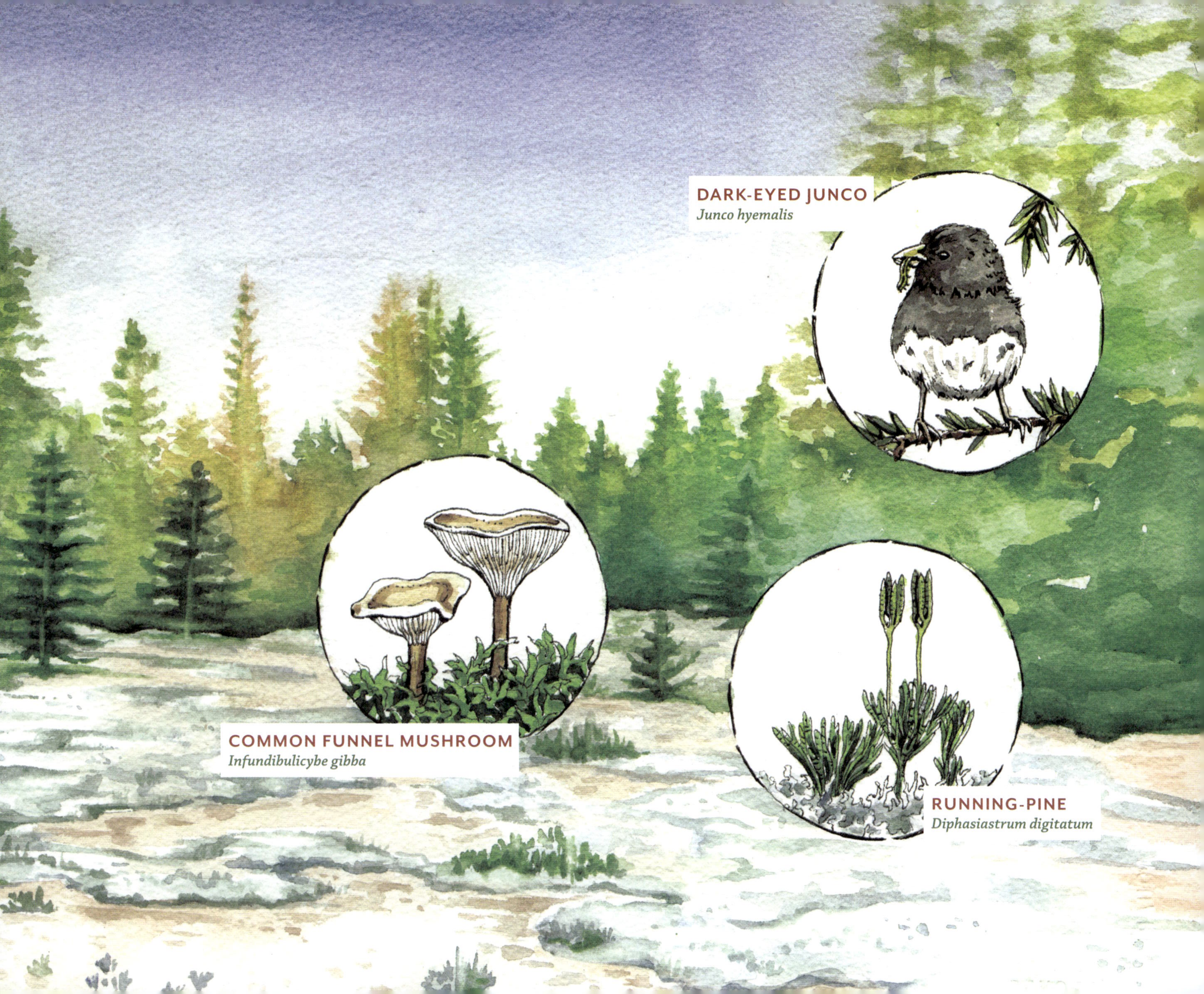

DARK-EYED JUNCO
Junco hyemalis
COMMON FUNNEL MUSHROOM
Infundibulicybe gibba
RUNNING-PINE
Diphasiastrum digitatum

Resilient Beauty

We start out from our campsite early to embark on a full-day eleven-mile hike up and around Mont-Albert, a mountain composed of a unique reddish rock called serpentine. The heavy metals found in serpentine limit the variety of plants that can grow here to only about a dozen species, one of them a hardy hot-pink flower called alpine catchfly. When we reach the North Summit, I take inventory of the other ten or so people resting at the top alongside us, our hearts pounding in sync from the exertion, the same wind whipping through our hair, all of us experiencing the shared sense of awe that accompanies a 360-degree mountain view. While the true summit is currently off-limits due to caribou protections, the only wildlife we notice up here is a spur-throated grasshopper with pretty reddish markings on its legs.

On our descent, we cross through an eight-square-mile bog that sits on a plateau just below both summits, a stark contrast to the dry, rocky landscape present elsewhere on the mountain. The sound of trickling water emanates from every direction as it drains from the mossy wetlands into a web of picturesque ponds and cascading waterfalls, eventually converging to form a rushing river at the base of the peak.

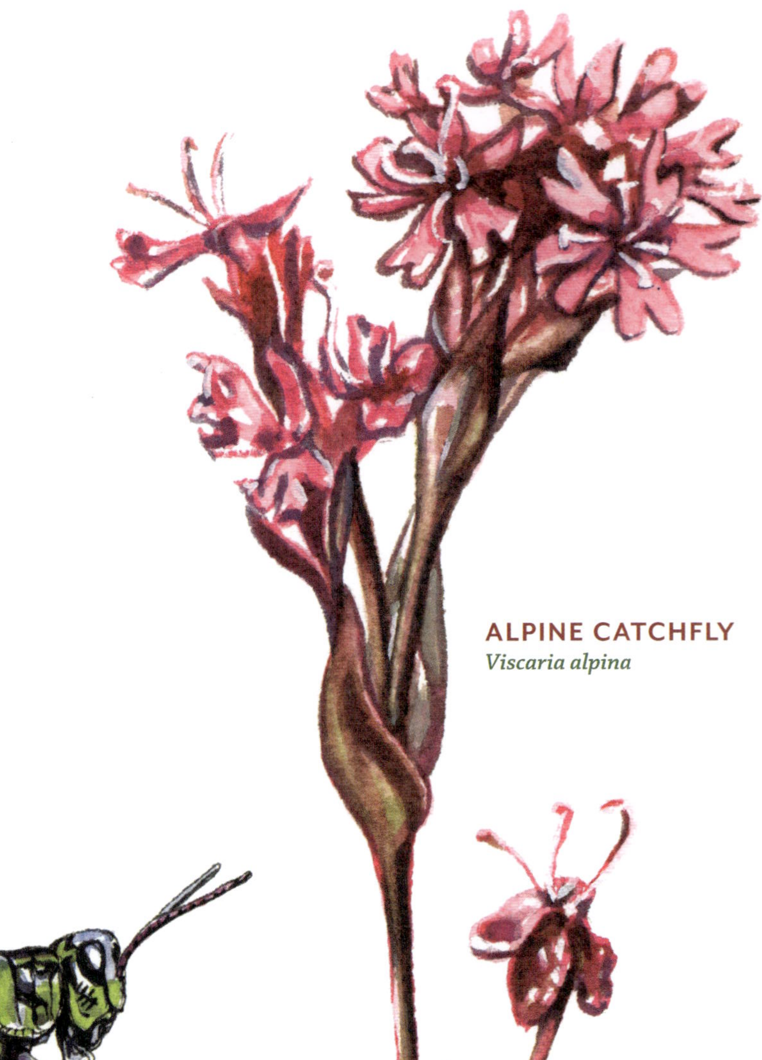

ALPINE CATCHFLY
Viscaria alpina

NORTH AMERICAN SPUR-THROATED GRASSHOPPER
Melanoplus sp.

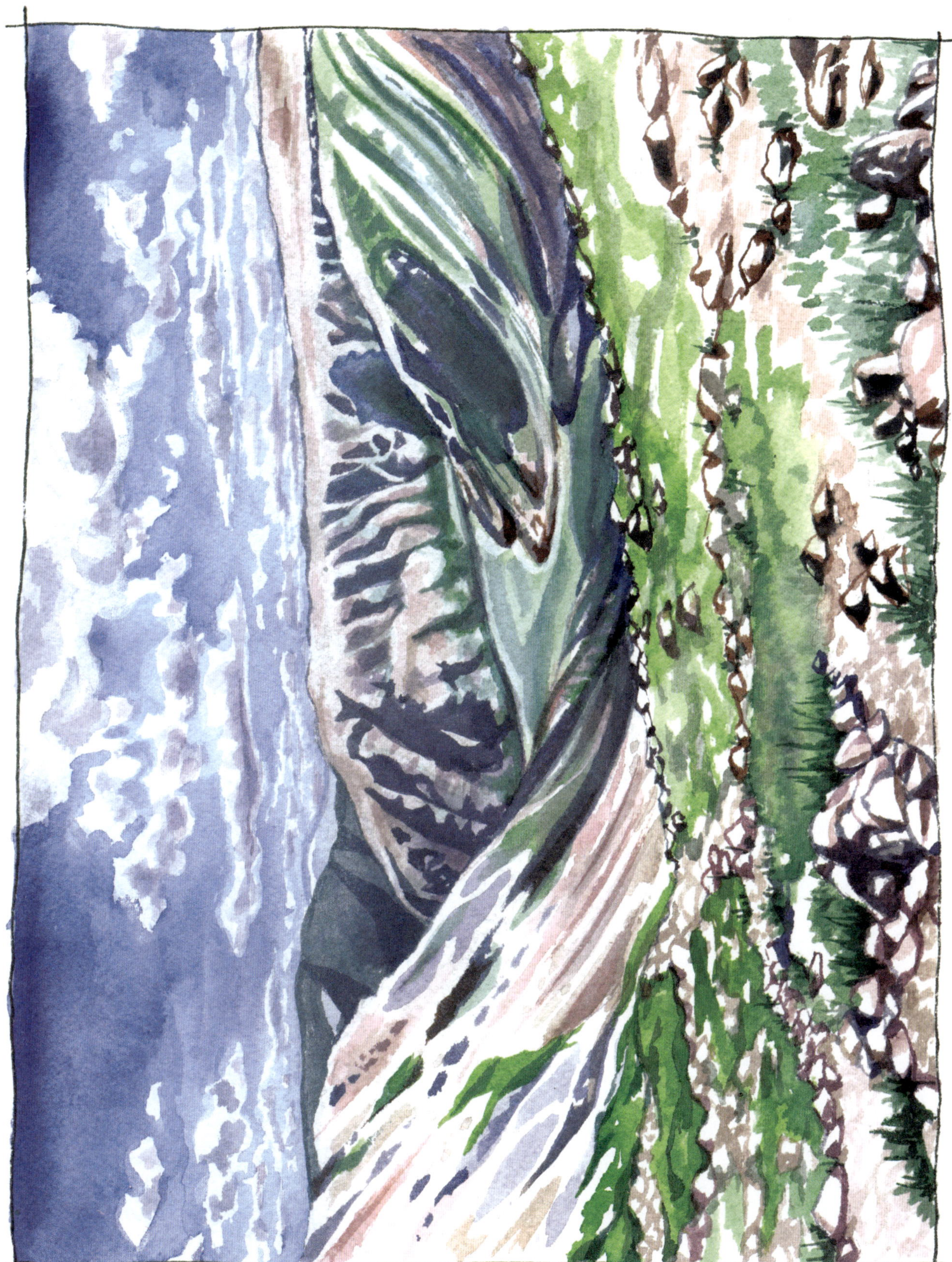

View from Mont-Albert, Gaspésie National Park, QC

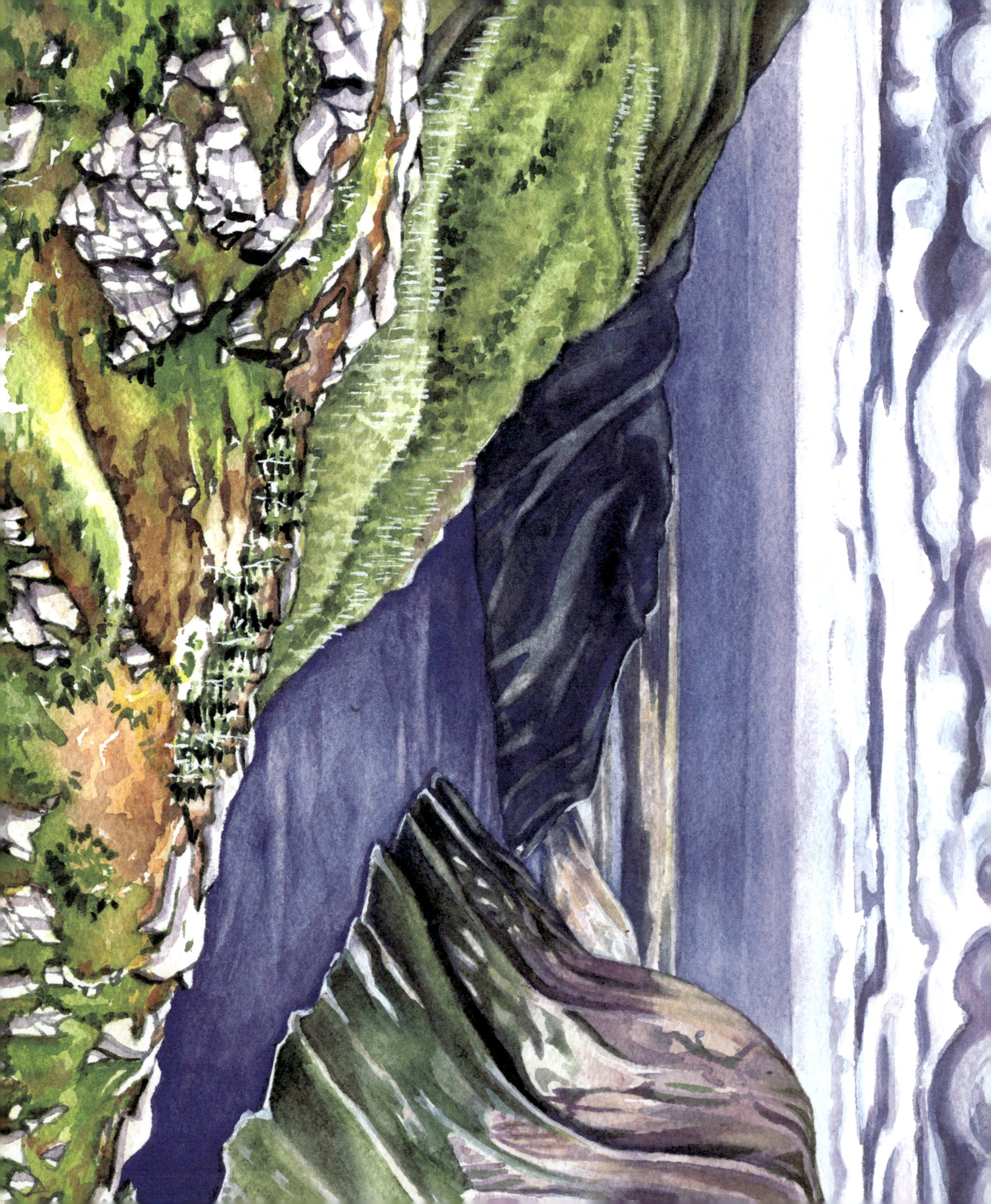

EN ROUTE TO NEWFOUNDLAND

SEPTEMBER 1

It takes two full travel days to get to Newfoundland, our final destination. First, an eleven-and-a-half-hour drive takes us back down through the Gaspé Peninsula, into New Brunswick, and then out to the eastern tip of Nova Scotia. We spend the night there, and the next morning we catch a seven-hour ferry through the Atlantic Ocean to reach Newfoundland. We decide that it would be fun to get dressed up in the one nice outfit that we each brought with us, just like people used to do for boat and airline travel back in the day. While standing with our faces to the gusting wind on the top deck, we see big white puffs of sea-foam erupting from the water—whales!

In early evening, land comes into view: a rocky peninsula dotted with a scattering of small, colorfully painted homes. The boat unloads and we begin the four-hour drive to Rocky Harbour, along the way nearly hitting a moose that weaves frantically between our car and the one in front of us. Although it's a close call, I'm grateful for this glimpse of such a massive, awkward-looking creature. Moose were introduced to Newfoundland about a hundred years ago to offer big-game hunting opportunities. Today, the island has the highest density of moose anywhere in the world.

It's almost midnight by the time we approach Rocky Harbour, a coastal town that's centrally located within Gros Morne National Park. It's been a long couple of days, but I can hardly sleep as I think about what we might find in this remote section of the Appalachian Mountains.

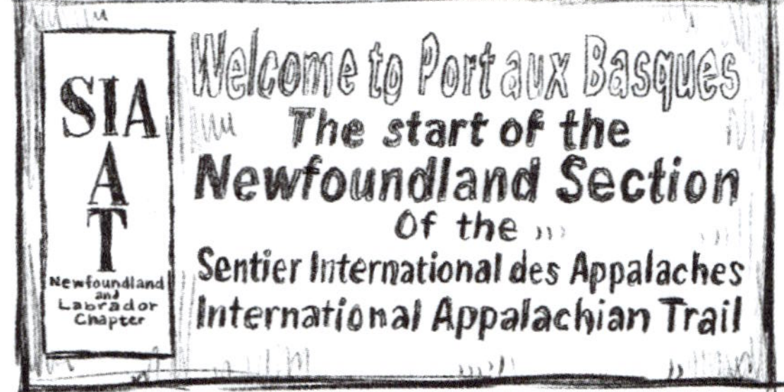

LEFT: *Ten Mile Pond, NL*

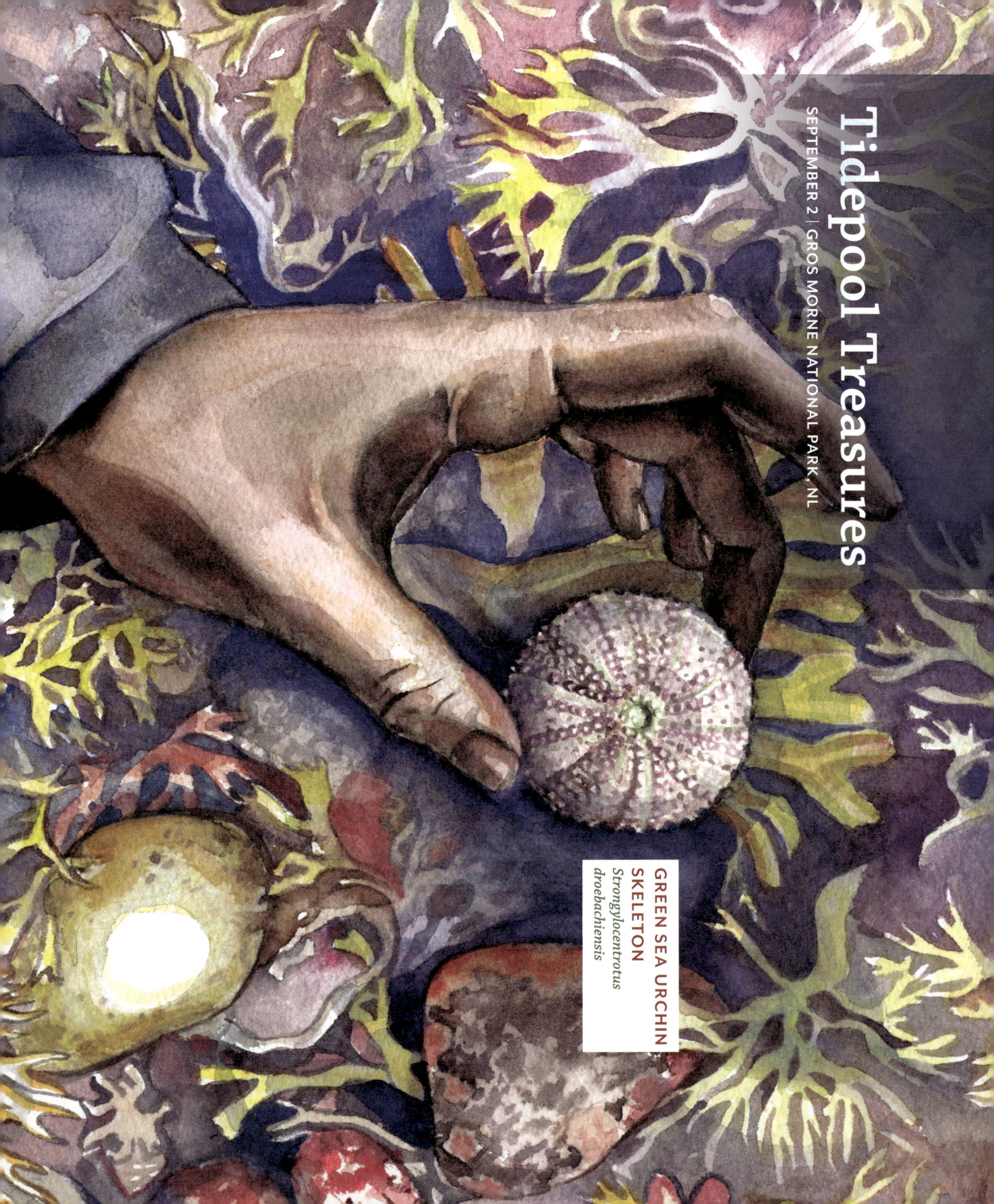
Tidepool Treasures
SEPTEMBER 2 | GROS MORNE NATIONAL PARK, NL
GREEN SEA URCHIN SKELETON
Strongylocentrotus droebachiensis

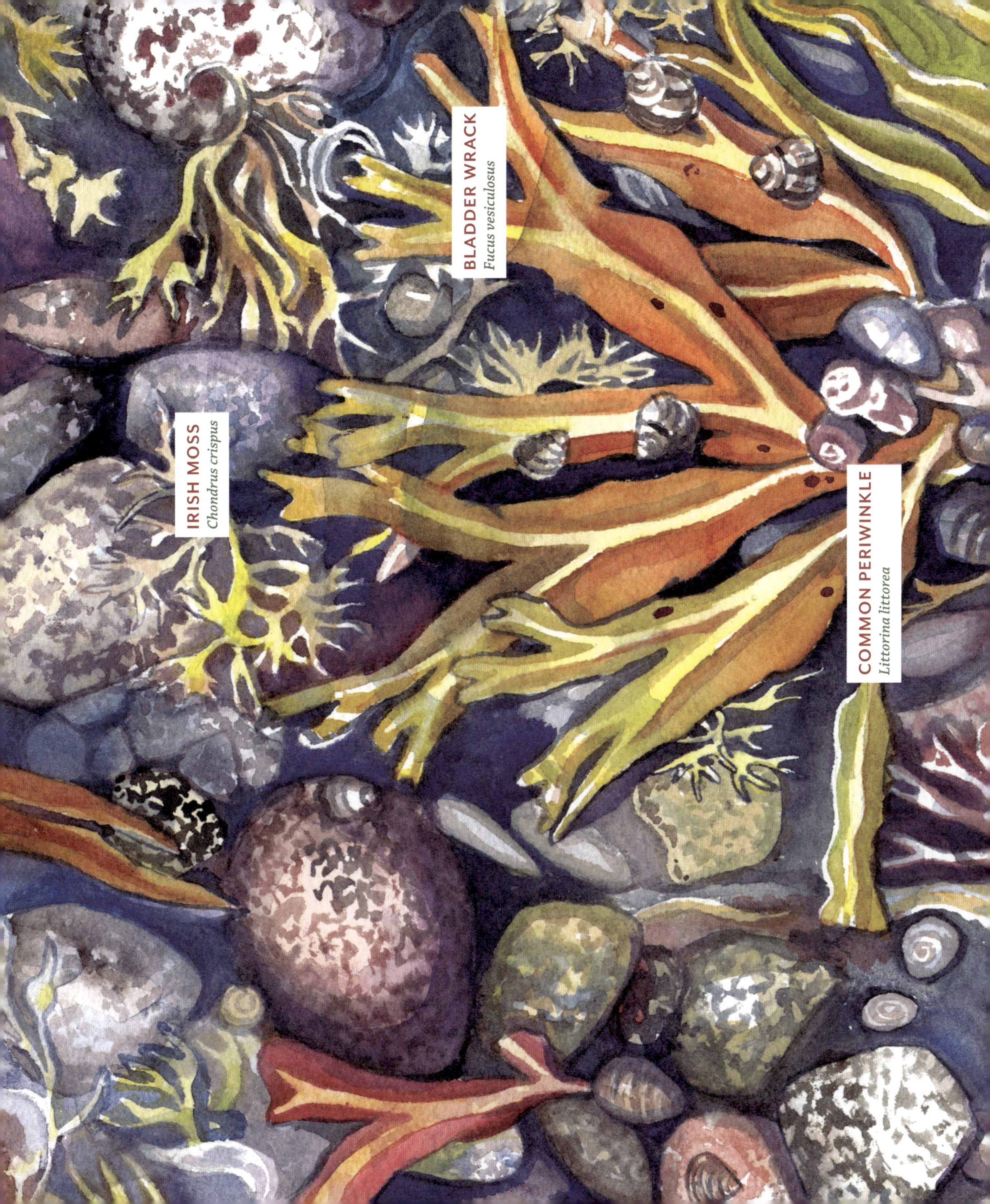
BLADDER WRACK
Fucus vesiculosus
IRISH MOSS
Chondrus crispus
COMMON PERIWINKLE
Littorina littorea

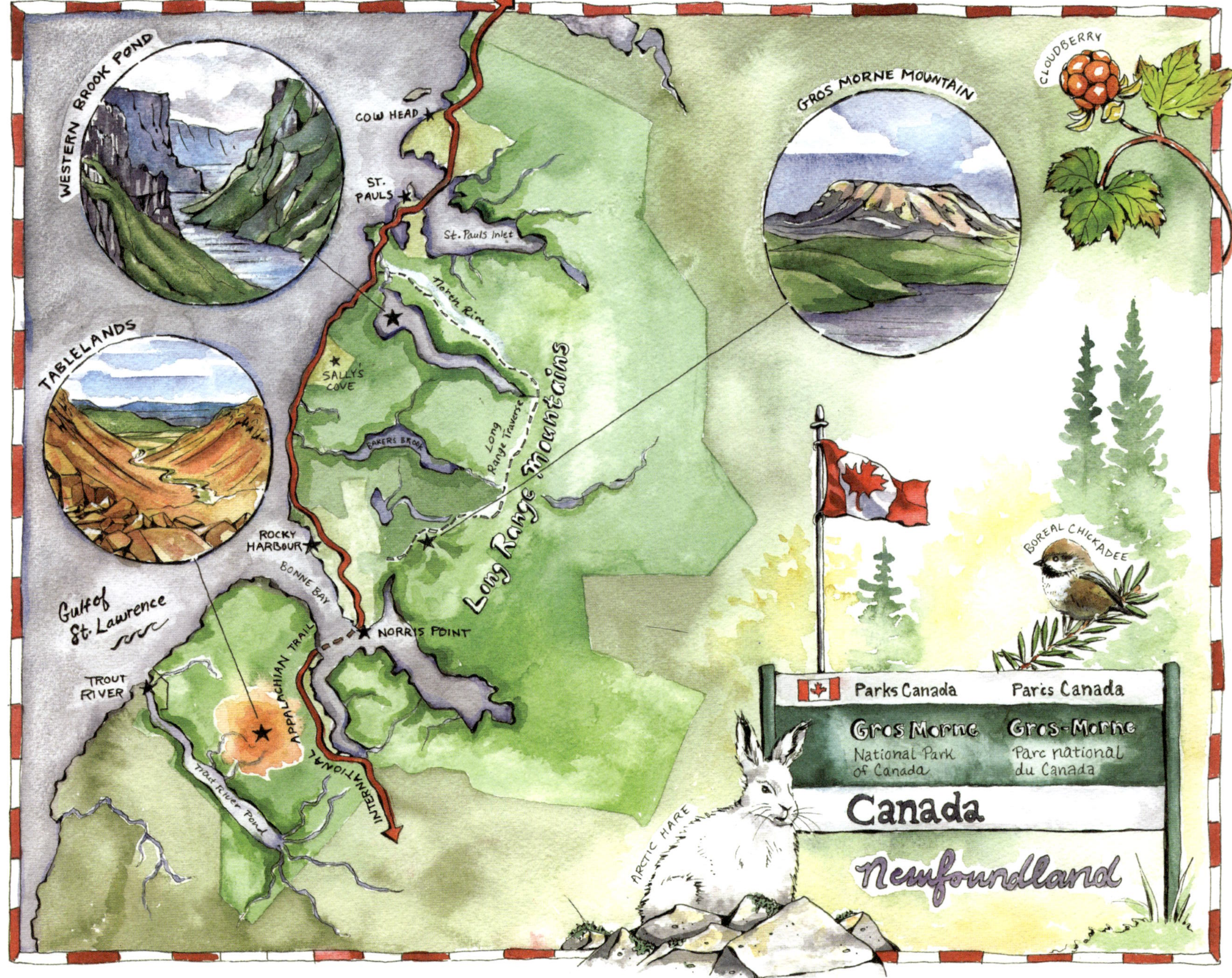

WESTERN BROOK POND
COW HEAD
ST. PAULS
St. Pauls Inlet
North Rim
SALLY'S COVE
Bakers Brook
Long Range Traverse
Long Range Mountains
TABLELANDS
Gulf of St. Lawrence
ROCKY HARBOUR
BONNE BAY
TROUT RIVER
APPALACHIAN TRAIL
INTERNATIONAL
Trout River Pond
NORRIS POINT
GROS MORNE MOUNTAIN
CLOUDBERRY
BOREAL CHICKADEE
Parks Canada
Parcs Canada
Gros Morne
National Park of Canada
Gros-Morne
Parc national du Canada
Canada
Newfoundland
ARCTIC HARE

GROS MORNE NATIONAL PARK

Gros Morne National Park parallels the west coast of Newfoundland and encompasses the Long Range Mountains, the northernmost range of the Appalachian Mountains. It takes its name—"big lonely" in French—from Gros Morne mountain, Newfoundland's second-highest peak. Glacial movements over time have formed a variety of striking landscapes that include freshwater fjords, crystal-clear lakes, gently carved valleys, and rounded alpine plateaus. In the southern section of the park, a fascinating geological hotspot called the Tablelands features an exposed section of the earth's mantle that you can walk across. If you venture up the park's namesake peak, you may spot a clucking flock of rock ptarmigan or a sprightly arctic hare hiding among the boulders.

Inside the Tangled Tuckamore

It's a windy afternoon along a coastal trail in Gros Morne National Park. I notice a hole in the tuckamore, a Newfoundlander term for the nearly impenetrable patches of entangled spruce trees that have been blown sideways by intense seaside winds. Back when sheep used to roam freely in many parts of Newfoundland, they would run into the tuckamore during storms. I can't help but see it as a mysterious and shadowy parallel world. What goes on inside this natural fortress?

Now is my chance to find out. I duck my head and nervously crawl through the arched entryway that an animal likely created. Inside, I can no longer hear the howling wind and crashing waves that were so all-consuming just a moment ago. It's very dark, the interlaced tree branches forming a gnarled shelter above my head. Despite the relative quiet, the tuckamore feels alive and menacing, with sharp branches that threaten to reach out and grab my clothing if I fall out of its favor. I startle and jump as a red squirrel lets out a shrill shriek from a branch directly above my head, and when I turn to look at him, he flicks his tail and yells again in an angry, territorial fashion.

I feel out of my element and think about turning around, but something draws me deeper inside. I follow that impulse and soon the narrow tunnel opens up into a large room, tall enough for me to stand up fully. My eyes adjust to the darkness and I see all kinds of comforting, familiar plants. Ferns decorate the floor and thick beds of moss create a cozy backdrop for ghost pipes and an array of mushrooms. All of it seems to have been perfectly planted by an elfin landscaper.

I take a seat on the soft forest floor and begin my invisibility ritual: I hold perfectly still and quiet my breath so that it's inaudible even to me. I imagine myself shrinking into my surroundings, melting into the moss like a decaying mushroom. Once I've minimized my human presence, I wait and see what will appear. This is one of the most freeing experiences for me, as someone who has been prodded her whole life to speak more loudly, participate more in the classroom and workplace, and be more assertive. Here in the wild, my quiet nature isn't a flaw—it's my superpower, a portal through which I can encounter other living creatures without scaring them away.

Many minutes of stillness swim by, and slowly birds begin to forage again and flies give off a murmuring, meditative buzz. I'm careful to move only my eyes as I scan my surroundings. Spiderwebs glisten in the golden light. Wispy old man's beard lichen drapes from the uppermost limbs. Now that I've been here awhile, I'm

no longer uneasy; the tuckamore feels like a sanctuary.

An hour passes. Suddenly, I hear a few footsteps outside and a man pokes his head through the entrance of my newfound home. He squints as he looks around, his eyes getting used to the low light. "Hello there!" I say in a friendly tone, trying not to startle him. Alas, his eyes widen as he takes in this stranger hunched gnomelike among the mushrooms and moss, and he leaps backward to make a quick exit. The sanctuary settles again, and I feel humbled to be part of the stillness.

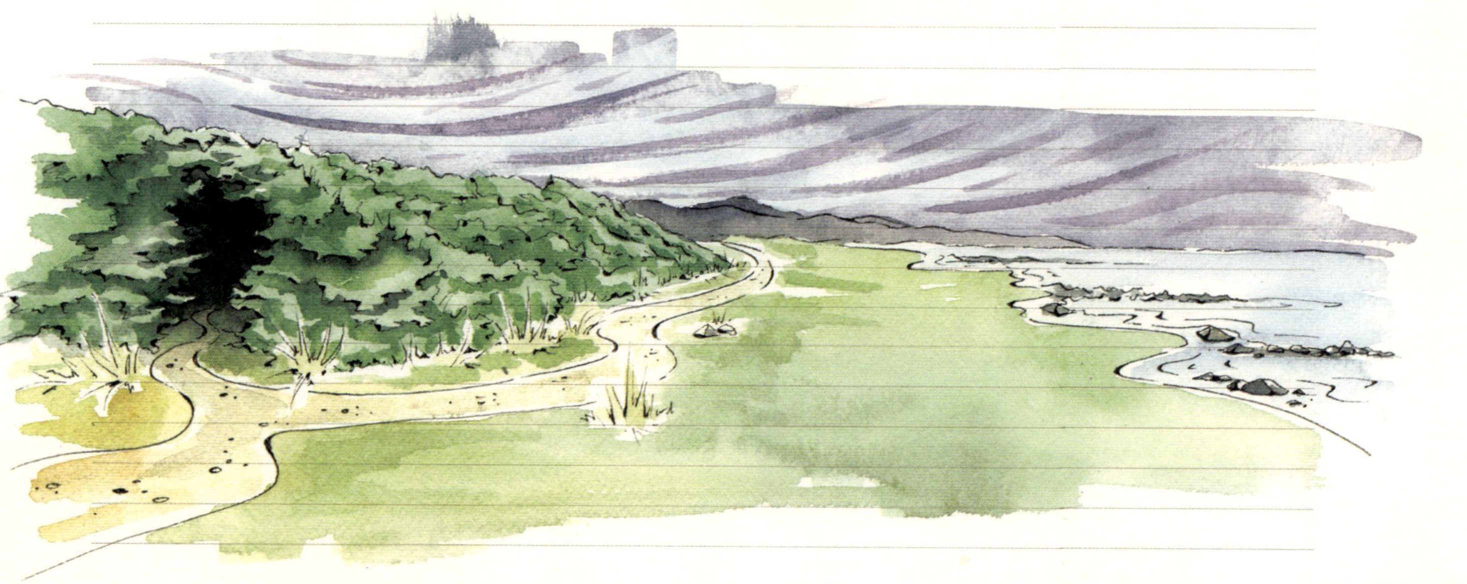

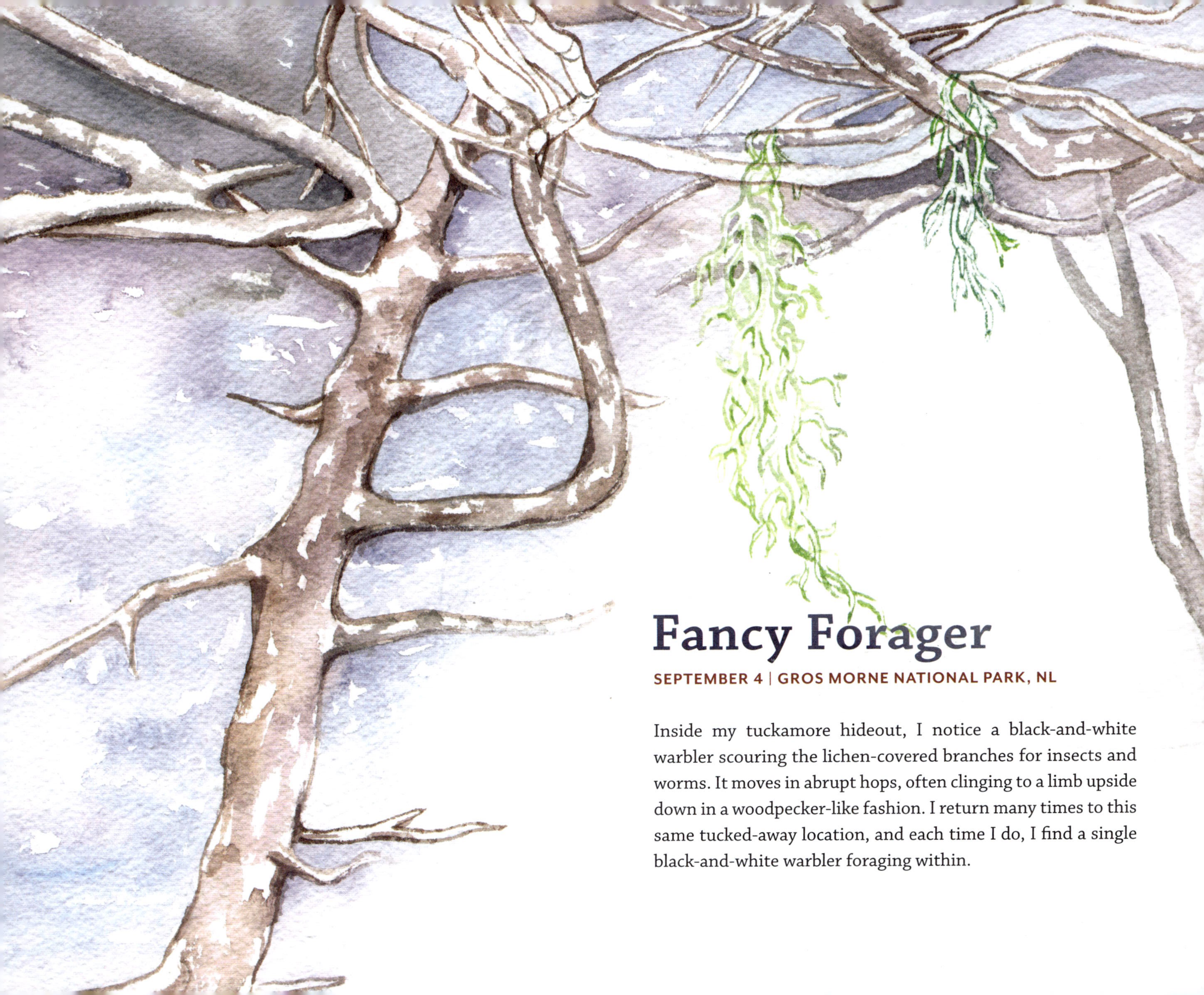

Fancy Forager

SEPTEMBER 4 | GROS MORNE NATIONAL PARK, NL

Inside my tuckamore hideout, I notice a black-and-white warbler scouring the lichen-covered branches for insects and worms. It moves in abrupt hops, often clinging to a limb upside down in a woodpecker-like fashion. I return many times to this same tucked-away location, and each time I do, I find a single black-and-white warbler foraging within.

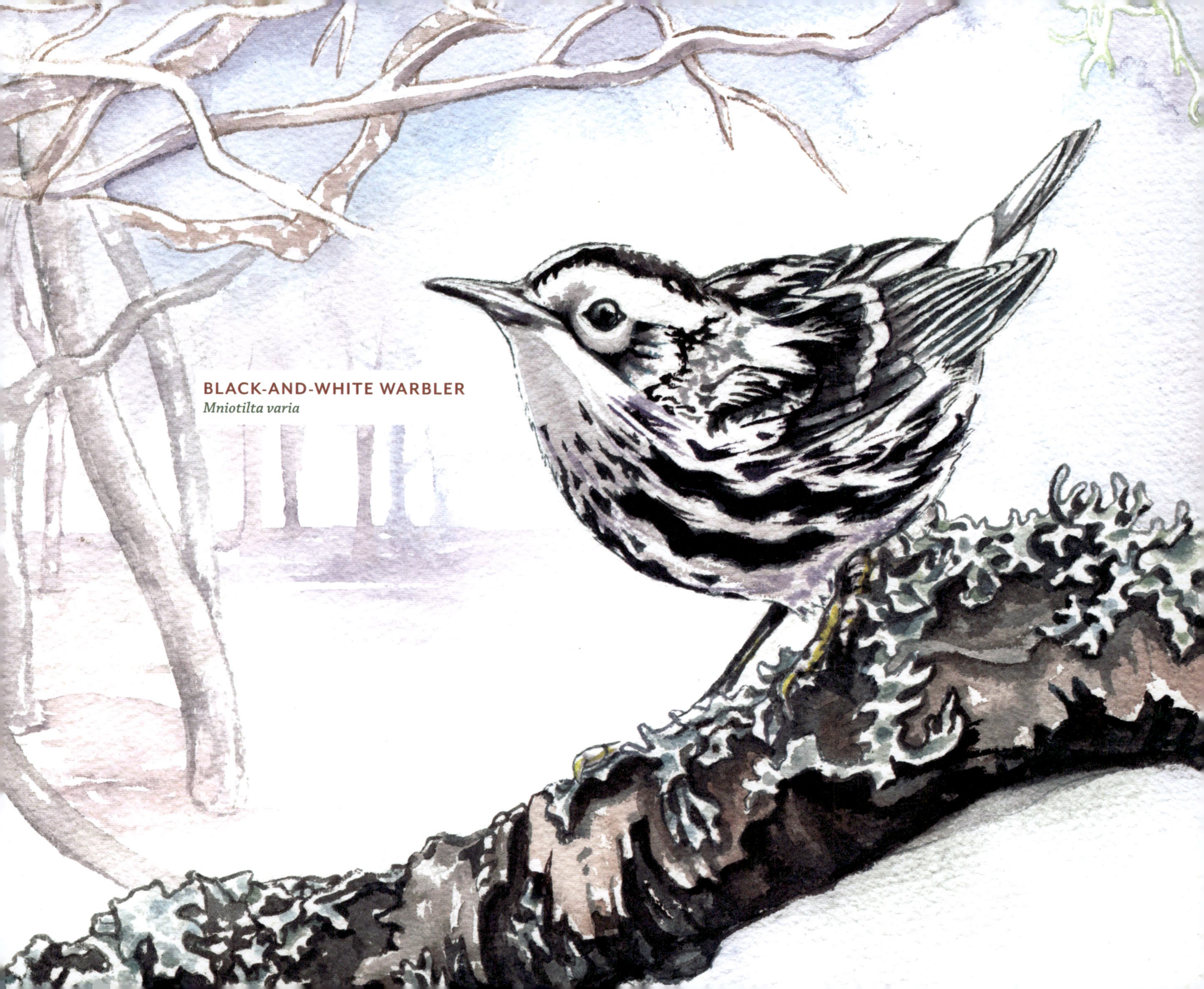
BLACK-AND-WHITE WARBLER
Mniotilta varia

Flesh-Eating Florals

It feels like we're hiking across Mars as we traverse the orange-tinted landscape, one of just two places in the world where the upper part of the earth's mantle is exposed and easily accessible. Located at the southern end of Gros Morne National Park, the Tablelands is characterized by an iron-heavy rock called peridotite that was pushed to the surface by tectonic activity long ago. Because this area contributed greatly to proving the theory that the earth is covered in tectonic plates, it's been preserved as a UNESCO World Heritage Site.

At first, the only plant life I see emerging from this inhospitable soil is a scattering of hardy nonnative weeds that line the trail. But wait—clumps of purple pitcher plants appear to be thriving here! This meat-eating plant traps insects in the sticky hairs that line its tube-shaped leaves, the "pitchers," and eventually the insects slip down inside to where the plant can dissolve and ingest them—the key to the pitcher plant's survival in nutrient-poor areas like the Tablelands. Researchers have also observed pitcher plants consuming much larger critters—including unfortunate eastern red-spotted newts—that wriggle their way, unsuspecting, into the pitchers.

What insects have these plants eaten recently? A park ranger nearby offers to let us use an empty eyedropper to draw a small sample and find out. I squeeze the bulb, gently insert the end into one of the pitchers, and watch as the clear plastic tube fills

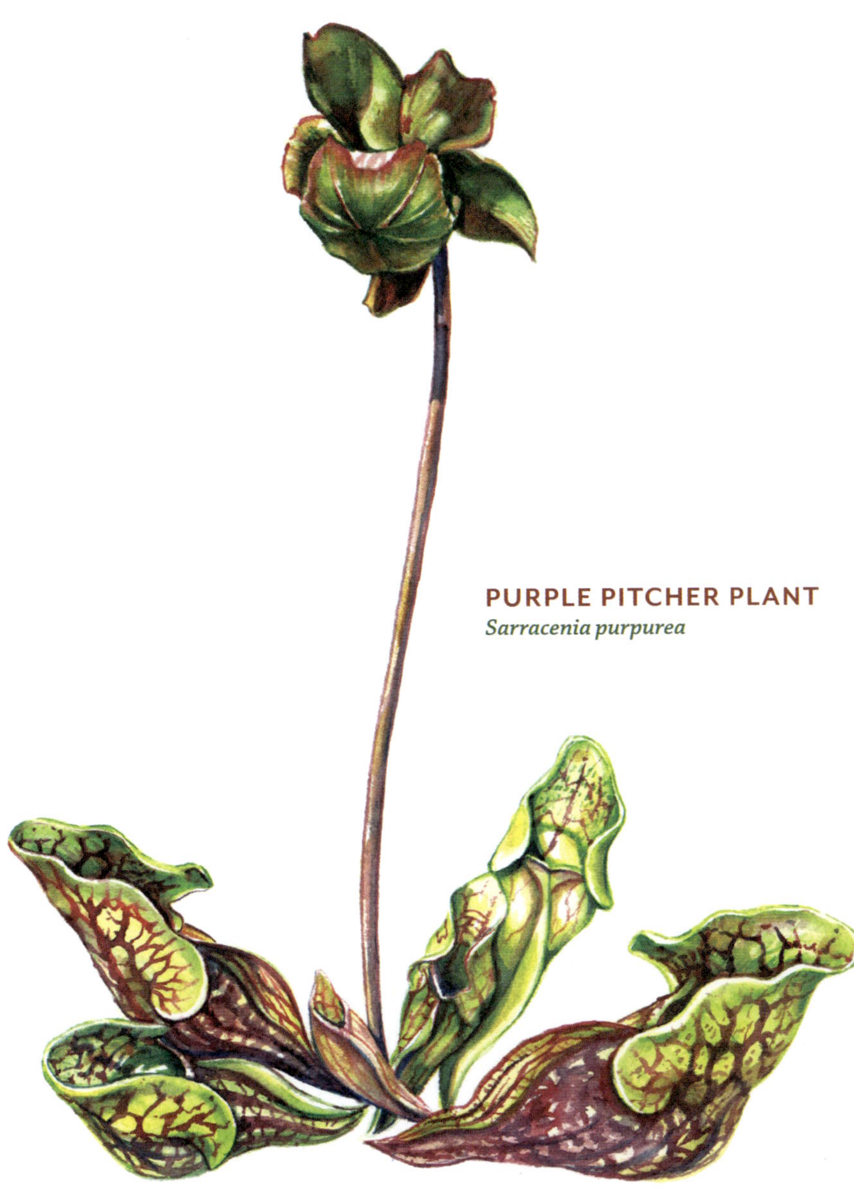

PURPLE PITCHER PLANT
Sarracenia purpurea

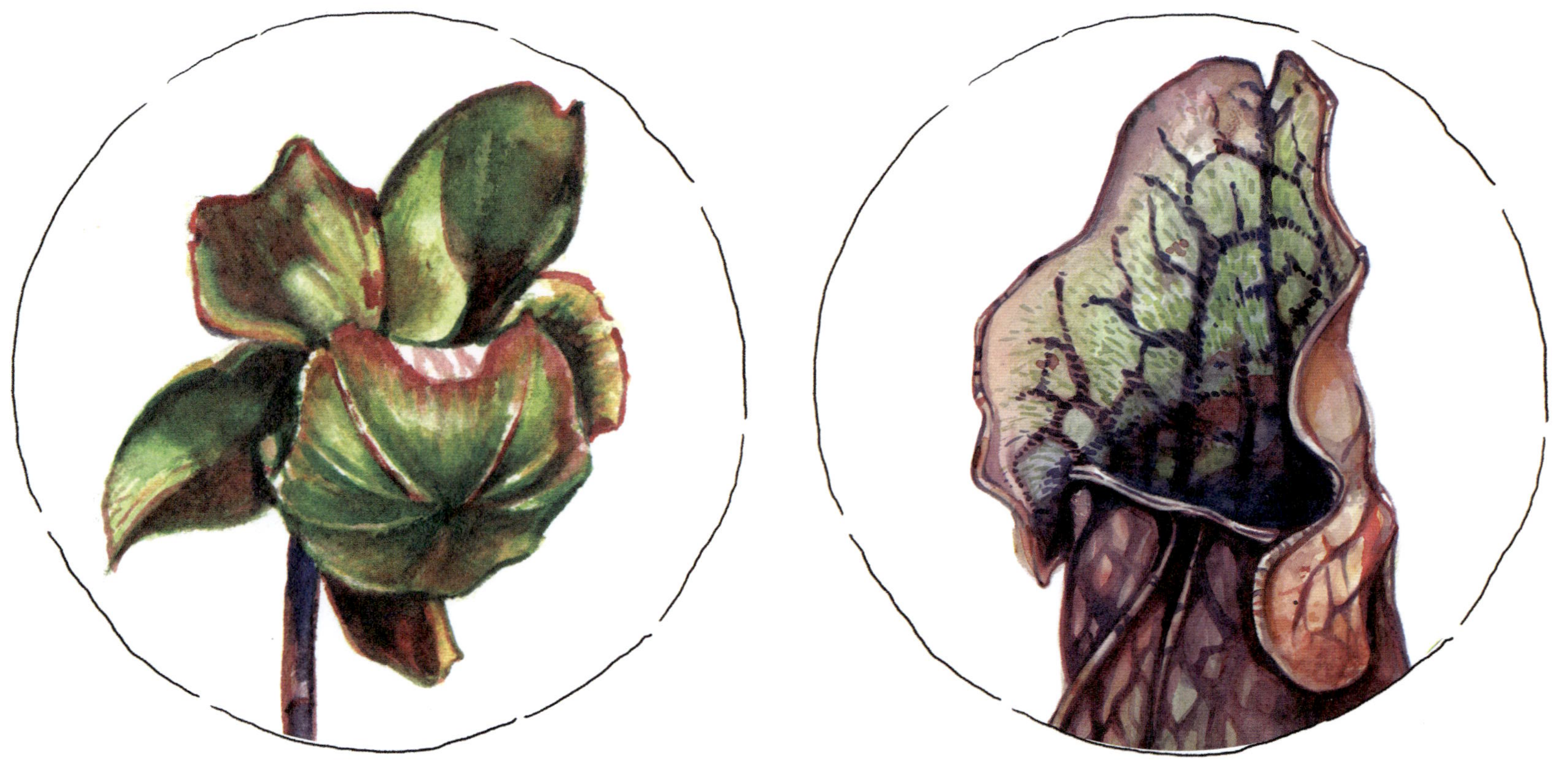

with a thick, greenish liquid. The green color indicates dissolved caterpillars. Yum!

Another pitcher contains tiny white worms: midge larvae. Certain species of midge flies and mosquitoes enjoy a symbiotic relationship with pitcher plants. The insects will lay their eggs in the rainwater that collects inside the pitchers. Once the eggs hatch, the babies will feed on other insects that have been trapped inside. This is helpful for the pitcher plant because it breaks down the nutrients into smaller, more easily absorbed particles. Eventually, the flies and mosquitoes will grow wings and make their great escape.

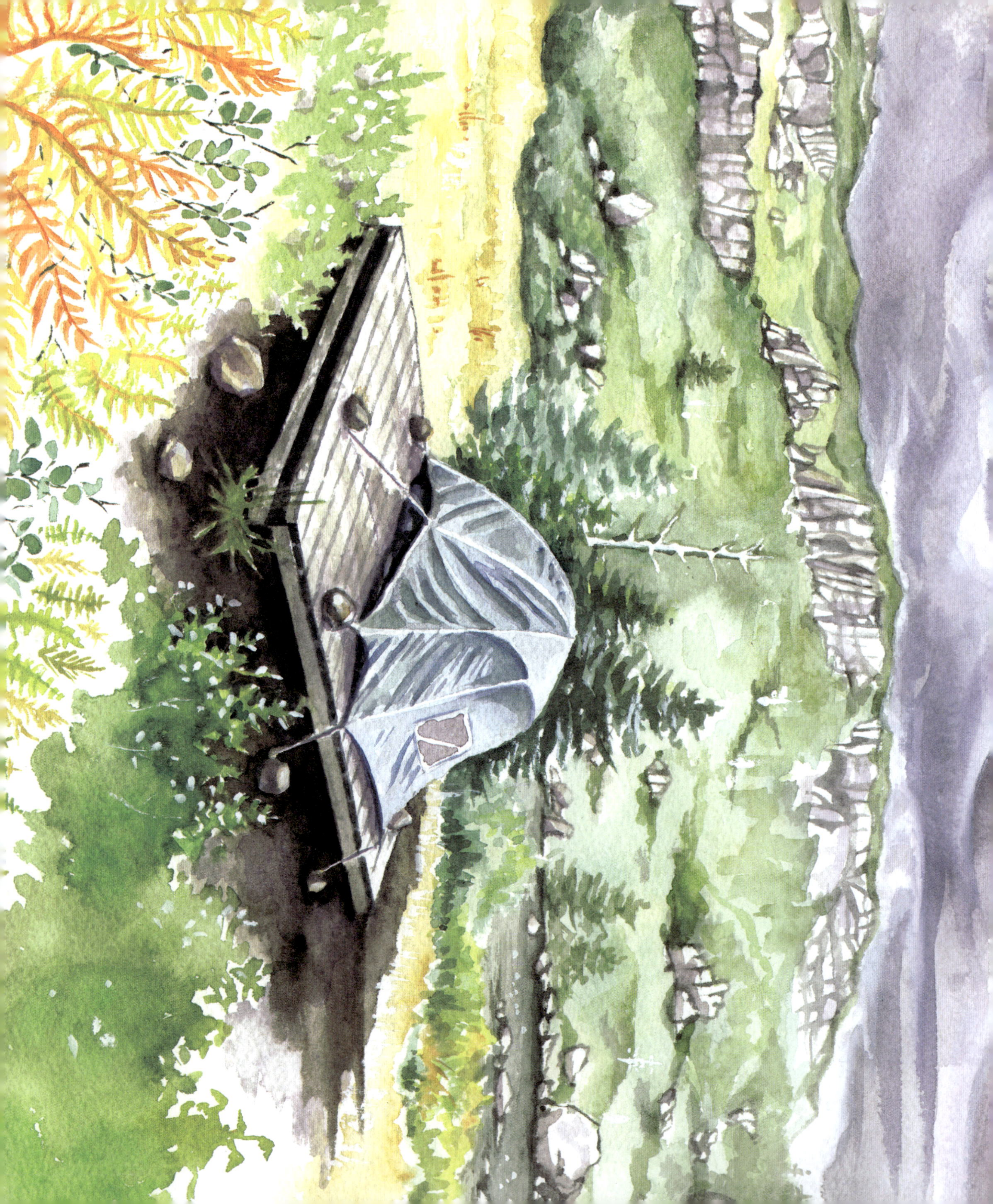

TRAVERSING THE LONG RANGE

We board the tour boat with thirty other rain-jacket-clad folks, 2,000-foot cliffs on either side beckoning us into the misty mouth of Western Brook Pond. During our hour-long ride, the sun extends a furtive greeting between gaps in the gray clouds and at one point a full rainbow arcs across the glacier-carved fjord. Now filled with pristine freshwater and separated from the ocean, Western Brook Pond reveals its saltwater history in the form of seashells, whalebones, and other fossils that archeologists have unearthed.

As the boat reaches a dead end and pulls up to a lonesome dock, Ben and I shoulder our heavy backpacks and move to disembark. The rest of the boat guests appear shocked as we explain that we're getting off here to complete a three-to-five-day trek through the backcountry. "I hope there are clear trails out there!" one lady remarks. We pause and then explain that there aren't any marked trails, but that we will use our map and compass to orient ourselves. "I hope you have bear spray!" another person chimes in. "Well, actually, rangers told us that the bears here aren't habituated to humans, so they'll run if they even smell us," I respond. We hop onto the dock and give all the worried onlookers big smiles, waving cheerfully to help them feel more confident in our ability to survive out here. And then . . . it's just us.

Because this is a highly protected wilderness area, only twelve hikers per day are granted permits for the route we plan to do. But since we were the only ones dropped off by the boat today, it's unlikely that our paths will cross with anyone else's over the next few days. The isolation is at once eerie and exhilarating.

We make our way along a rocky streambed and between boulders that are swathed in decadent layers of mosses, lichens, and liverworts. The route steepens, and eventually we are pulling ourselves up using roots and trees. When we finally emerge from the canyon and take a seat on the edge of the vast plateau, the fog parts just long enough for us to catch a glimpse of *the* iconic view of Western Brook Pond, one we've seen on everything from postcards to bags of coffee since arriving in Newfoundland. There's the ravine we've just ascended, flanked by cliffs on either side, and then the fjord, snaking away through even taller cliffs.

Now that we're on top of the plateau, the ground feels like one big sponge and our bootsteps make slurping sounds as we trudge through the squishy sphagnum bogs. We pass countless lakes of all sizes, each filled with crystal-clear water. A loon calls out from one of the ponds and we pause to watch her dip and dive with mesmerizing fluidity. Then the navigation

gets trickier. The landscape up here looks very similar in every direction, and it's easy to inadvertently veer onto animal trails created by moose or caribou. We check in with our map and compass every twenty minutes or so to make sure we're traveling in the right direction.

In the early evening, we arrive at a lake where a small tent platform has been constructed to keep the sensitive wetland plants from getting damaged. We find a beach nearby, where our bare footprints mingle with huge moose tracks in the white sand. I make a mermaid sand sculpture, then we sit and munch on peanuts and dried apricots. I imagine how hilarious the moose must have looked as it sunbathed at this idyllic spot.

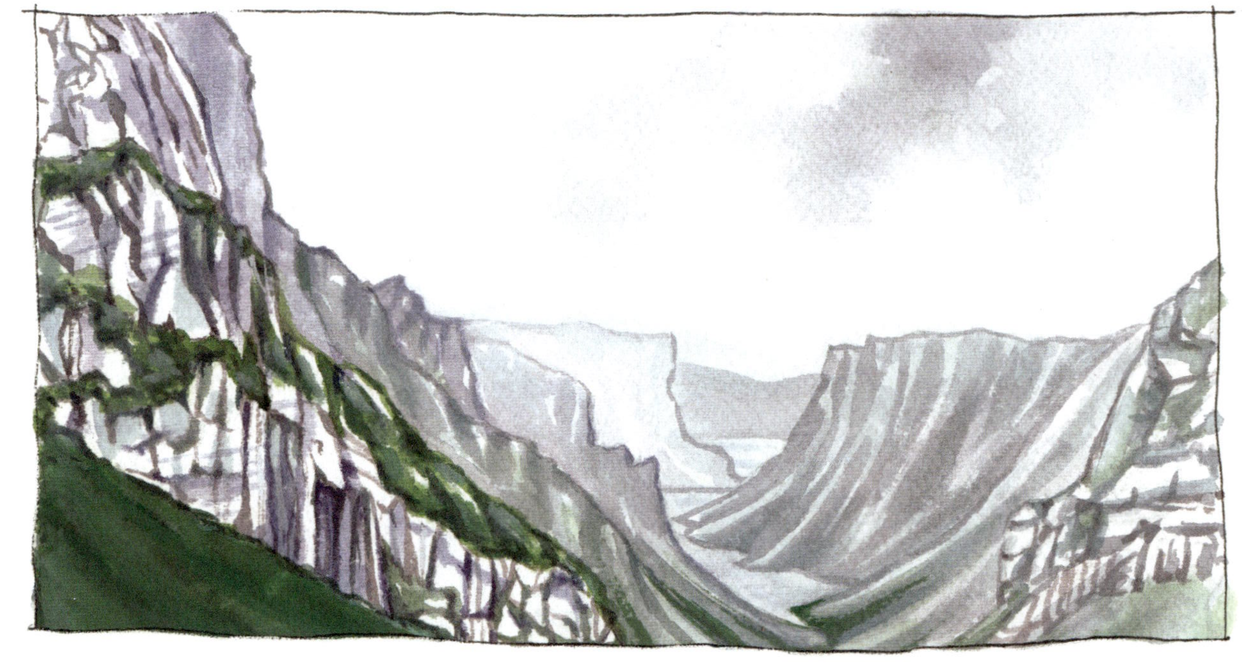

Western Brook Pond, NL

The next day is clear again, and we hike only four or five miles so that we can spend more time lounging at the next campsite while the weather is nice. Instead of rushing through this experience, why not relish the feeling of being the only humans on this wild, expansive plateau? The area has never been exploited for its natural resources, a kind of purity that's nearly impossible to find these days. Unlike many of the other designated wildernesses I've visited, which were at one time manipulated by some sort of industry, this area has only ever been utilized for small-scale hunting and fishing.

During our second night, the weather changes dramatically and we awake to freezing temperatures, insane wind gusts, and incessant rain. I have a circulation issue in my fingers and toes, so I'd packed hand warmers and gloves. Unfortunately, they're no match for the wet, frigid conditions; somehow, they make things worse. The best I can do is to tuck my hands into yesterday's muddy wool socks. My fingers are excruciatingly painful all day, and they eventually grow so numb that I can't even zip my jacket or hold a map. My fearful imagination gets the best of me; I visualize my fingertips blackened from frostbite. What if I'm unable to paint again? When our route requires us to climb down a cliff and my frozen fingers can't grip the rock well enough to hold my body weight, Ben proves himself a steady partner as he maneuvers me and my pack safely to the base.

Even though it's only one o'clock, we decide to set up camp and get warm, grateful that we brought a few extra days' worth of food so that we can take another short day. We find a nice spot in a river valley with a few trees to shelter us from the hammering winds. With my hands still useless, Ben assembles the tent, gets my bedding out, and then helps me take off my wet raingear and clothes, removing one disgustingly clammy sock at a time. Then I burrow into my down mummy sleeping bag and remain there in stillness, listening to the wind howling as it threatens to flatten our tent. Ben crawls into his sleeping bag and there we lie, a pair of snug burritos side by side. Three hours pass with neither of us falling asleep or speaking a word— the feeling of warming our bodies has become a strange sort of meditation as we notice each finger and toe slowly revive itself.

My mind wanders to all the critters I've met over these past few months that have had buddies with them: the flock of eight harlequin ducks huddled together on a cliff on the Forillon peninsula, the gang of five beautiful brook trout swimming together in a lake beneath Mont Jacques-Cartier. Even mushrooms often grow in pairs or groups. I used to think I was totally independent. But then my life merged with Ben's. He's a quiet, gentle man who loves cooking and trail running and whose favorite pastime is grocery shopping. And I now know that while, yes, I could technically sustain myself on my own, my life is infinitely warmer with a best friend by my side.

The cold wind and rain continue through the rest of our trek, but the knowledge that we only have a few days to experience this pristine wilderness helps me muster more gratitude than complaints. Eventually, we make it back to the trailhead. Yet even as I delight in the hot shower and soft blankets that follow, I find myself already nostalgic for the pure freedom of existence among the lakes and bogs and birdsong.

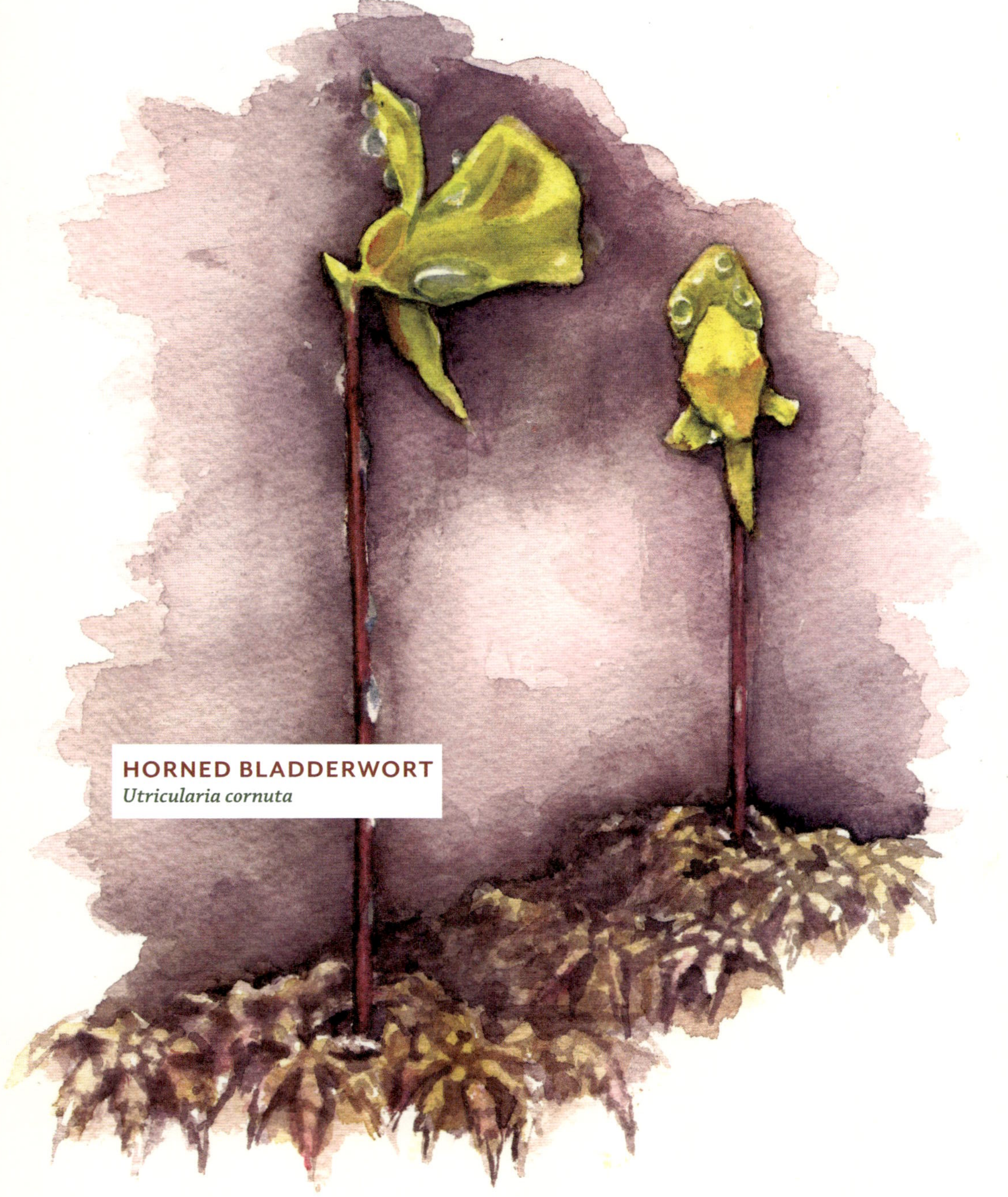

Tiny Botanical Beauties

SEPTEMBER 15–20 | LONG RANGE MOUNTAINS, NL

HORNED BLADDERWORT

This is an enchanted land. Some of my favorite carnivorous plants, which I usually think of as rare and difficult to find, are literally blanketing the landscape. Purple pitcher plants cluster in such abundance that they look like any other field of wildflowers. In a flat, muddy area that is otherwise barren, I notice a scattering of about thirty tiny yellow flowers: horned bladderwort! Unlike the floating bladderwort that I found in Maine, this carnivorous plant grows out of the mucky moss. Nearby, sundew plants grow as dense as the sphagnum moss.

SWAMP LAUREL

Even though the blooms are small, this flower's hot-pink hue catches my attention from afar. Swamp laurel is a relative of mountain laurel, a familiar friend that can be found in many parts of the Appalachian Mountains. The flowers are toxic to animals and humans, so while swamp laurel is pollinated by bees, the honey the bees produce will be poisonous to humans.

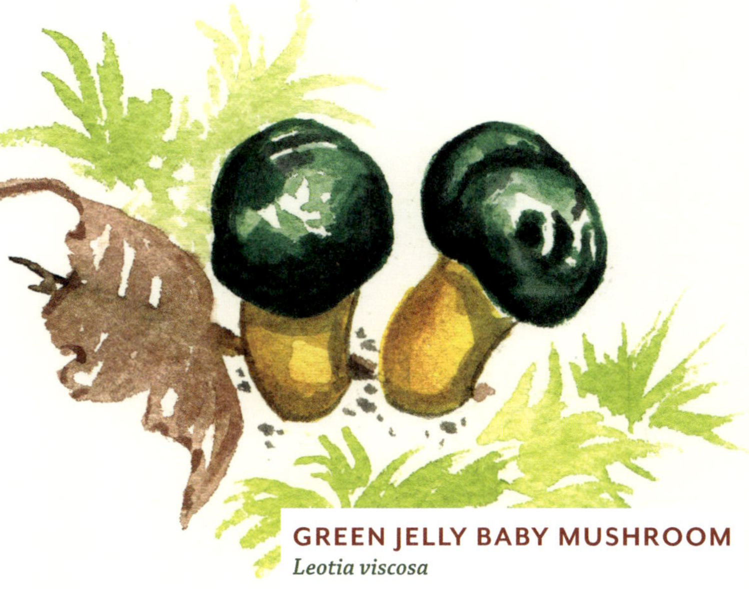

GREEN JELLY BABY MUSHROOM
Leotia viscosa

SWAMP LAUREL
Kalmia polifolia

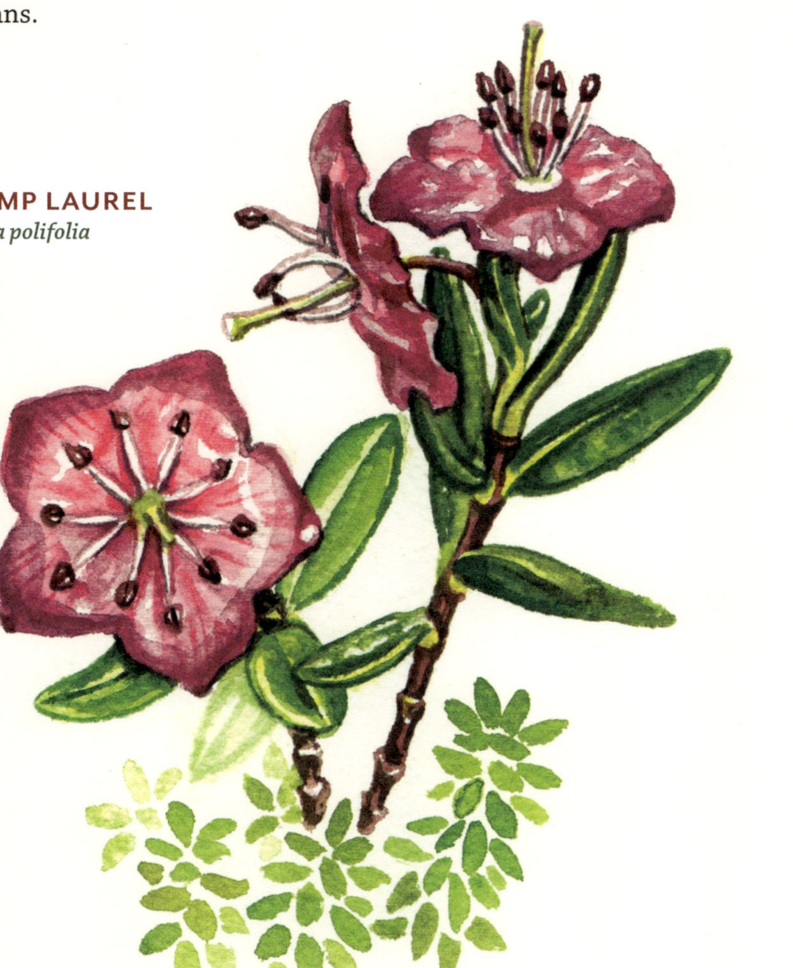

GREEN JELLY BABY MUSHROOM

I kneel to inspect a piece of wood that's tinted emerald with what I assume is green stain fungus. But as I get closer, I realize I've found something even more exciting: green jelly baby mushrooms! Each mushroom is just one centimeter tall and looks exactly like a green jelly bean perched on a bright yellow stem. This mushroom is unusual because its spore-producing structures are located directly on the slimy green upper surface of the cap, rather than on the underside in the form of pores or gills.

A Look at the Limestone Barrens

SEPTEMBER 28 | BURNT CAPE ECOLOGICAL RESERVE, NL

We walk out onto a desolate peninsula that juts into the frigid waters of the Strait of Belle Isle. This natural area resembles a construction site, the ground covered with limestone gravel as far as we can see. But the combination of the harsh arctic conditions and the calcium-rich limestone shale has created a habitat for more than 300 resilient plant species, many of which moved in after the last glaciers retreated; thirty of these species are considered rare. So what looks like a wasteland on the surface is actually one of the most important botanical sites in Newfoundland. I keep my eyes peeled, and eventually minuscule pops of color in the distance draw me in. It's too late in the season for many of the wildflowers I've read about, like yellow lady's slippers, but nevertheless I find a few hidden wonders: common juniper, sunburst lichen, and seaside bluebells.

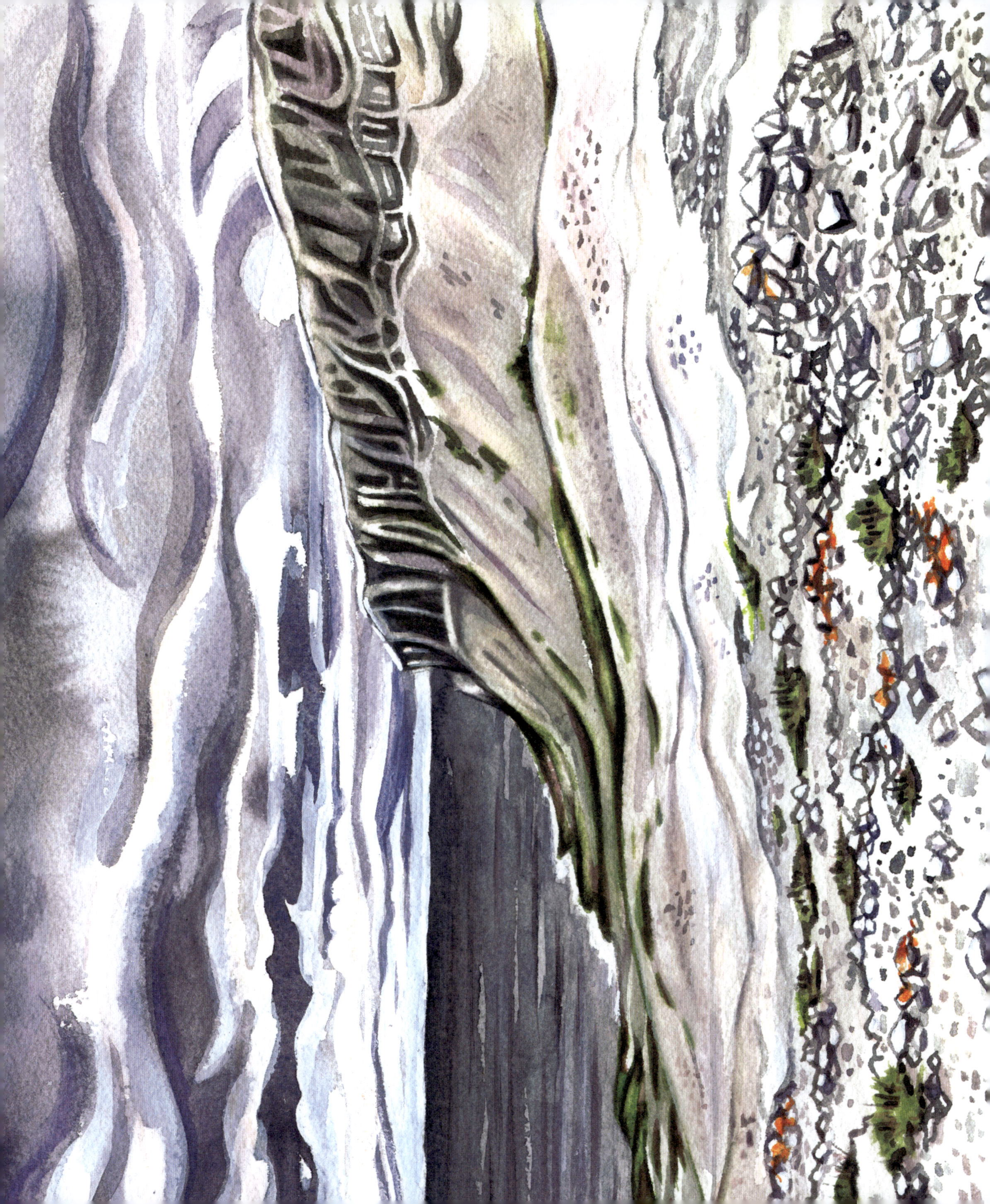

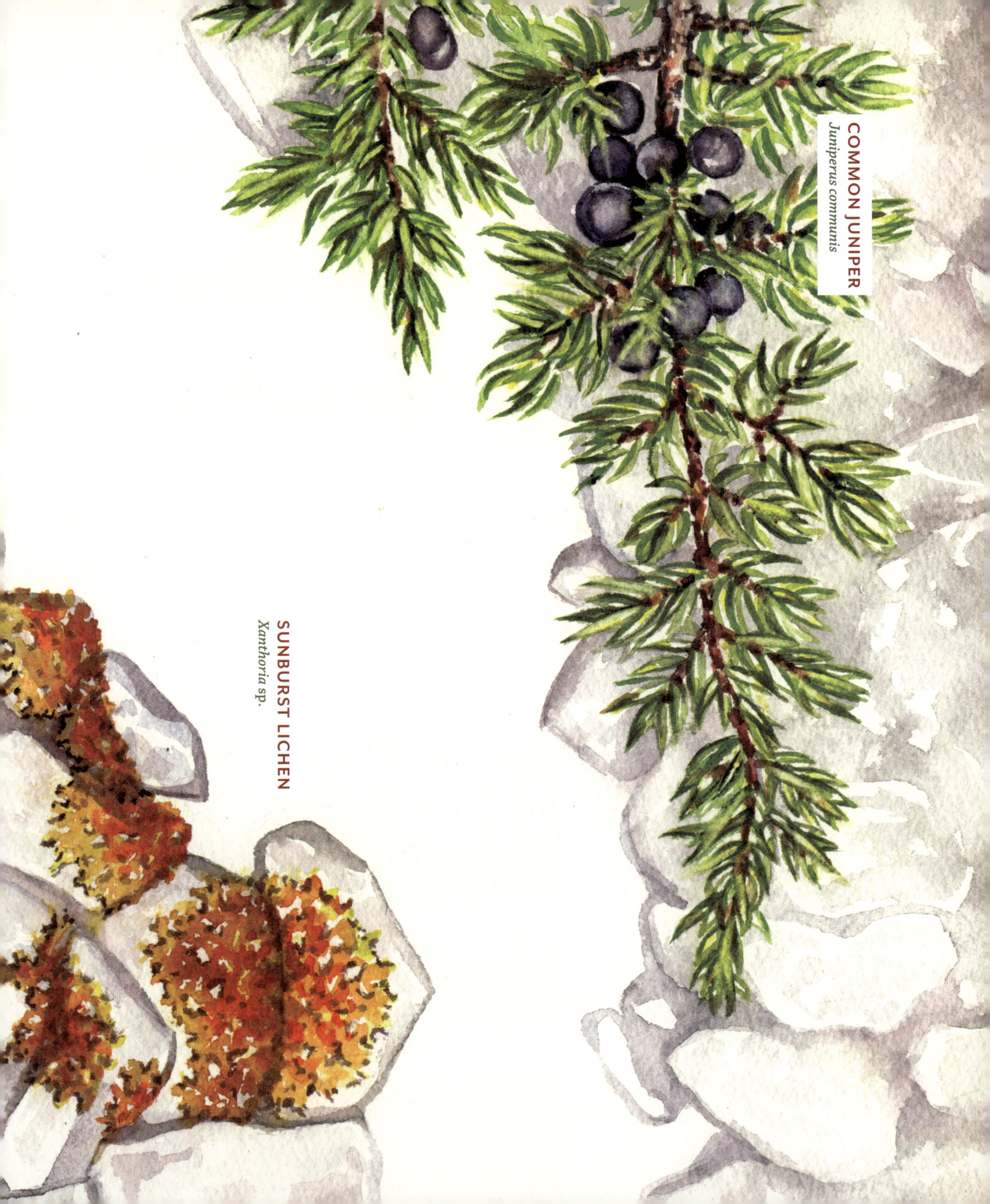
COMMON JUNIPER
Juniperus communis
SUNBURST LICHEN
Xanthoria sp.

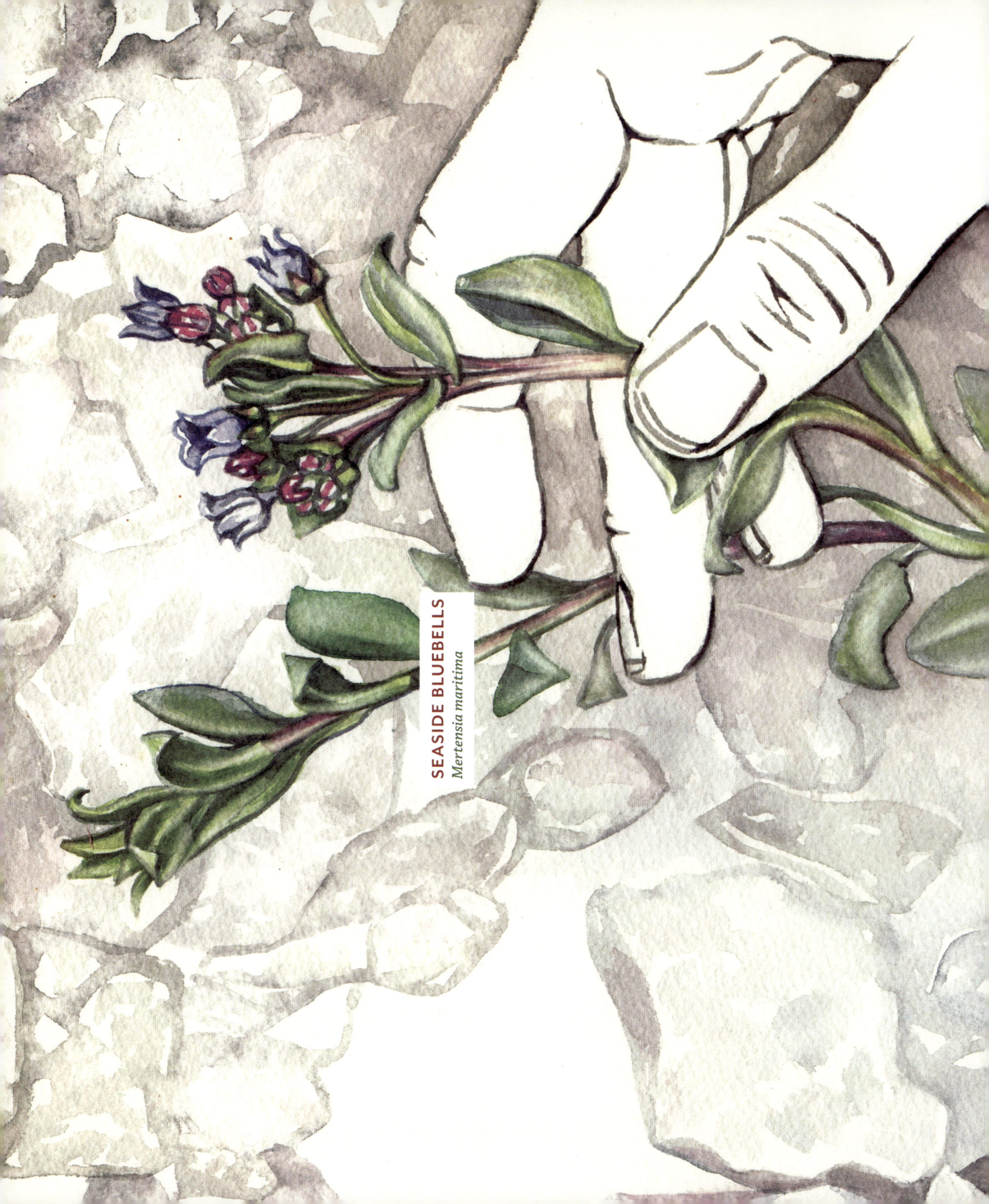
SEASIDE BLUEBELLS
Mertensia maritima

The Fox and the Fiddle

My relationship with music sprouted before I was even born, when my mom would play the banjo with her pregnant belly (me!) filling out the hollow backside of the instrument.

The Elk Creek Family Band was already well established by the time I came along, the middle child of seven. My older brothers were proficient on the mandolin, Dobro, penny whistle, and washtub bass, and my older sister played the fiddle. Our songs covered all sorts of topics, but most of the crowd favorites were animal themed: there was one about wishing we were moles in the ground, another about a herd of hogs who turned on their farmer and ate him, and yet another about the unfortunate day our dog Noodle got into a fight with a porcupine and came home with needles in her nose. We performed at small venues all over Central Appalachia—in churches and schools, at amusement parks and family reunions.

We kids got to keep most of the money from the shows, and our music financed new bicycles, a trampoline, and even a couple of family vacations. Unfortunately, I never got good at playing an instrument. This was by design. From a young age, I felt that visual art was my true passion, and my childhood logic told me that I shouldn't get sidetracked by learning to play an instrument while there were paintings to paint. Instead, I sang.

The fun and games came to a screeching halt when I entered high school. A teacher asked me to stand during an all-school meeting and then told the whole student body that our family band was going to play a show at a local coffeehouse that weekend. I sat back down, horrified. The kid next to me whispered, "Rosalie, your face is the color of that red backpack." The rest of the day, various classmates asked me about the band, barely disguising little smirks. A popular girl mocked me during gym class: "Rosalie, sing for us!" she cried out.

Then they found our family band website. It just kept getting worse.

That day after school, I burst into my dad's home office and begged him to take down the website. He laughed it off at first but soon realized that I was serious. So, with the click of a button, this website that he had lovingly crafted—featuring music videos, profiles of each band member, and a list of all the venues we'd played—vanished from the internet. I'm pretty sure that the coffeehouse gig was our last show, too. The buzz about Rosalie being in a hillbilly family band faded just as quickly as it had appeared, but for years after that, I lived with the guilt that I had been the one to disband us. My shame at being different from my peers had brought it all to an end after eighteen fun-filled years.

But it wasn't really over. Even as we all grew up and went off to college, the jam sessions continued whenever we were together. A year ago, I decided that I wanted to be able to contribute more than just my voice, so I started learning to play the fiddle—something that I'd always secretly wanted to do but was too afraid to try. Now, I regularly meet with an online fiddle teacher, and Ben has picked up some guitar to help me keep rhythm while I play.

We're staying in a basement rental in Newfoundland this month and there's a family upstairs, so we've taken to practicing our music outside so that I don't disturb them with my mediocre playing. We go to a dreamy cliffside picnic area on the outskirts of town where we can practice without anyone hearing us. It's incredibly freeing to play as loudly as we want without fear of judgment, our music accompanied by the sounds of waves crashing below and gulls cackling overhead. One day while practicing on the cliffs, I look up to see a gorgeous red fox seated about twenty feet from us, watching and listening with perked ears. We keep playing and are elated to find that he remains attentive, his eyes flicking back and forth from our faces to our instruments. When we stand to leave, he darts away and disappears behind the craggy hillside.

The next day, we return to our practice spot and again the fox appears, silently. On three separate occasions, we play our hearts out while our one-fox audience watches and listens. It's very possible that he associates humans with food and pays his visits simply to see if we will feed him (which we never do, of course), but I like to think that he's just in the mood for a little live music.

Every year, my mom asks for the same thing for her birthday: a family jam session. And all these years later, I still catch myself humming the old tune about moles in the ground on an alarmingly frequent basis. The joy of making music together was too much a part of the fabric of our family for one insecure middle child to unravel it. We'll be playing our hillbilly songs until we're old and gray, even if our only audience is a fox.

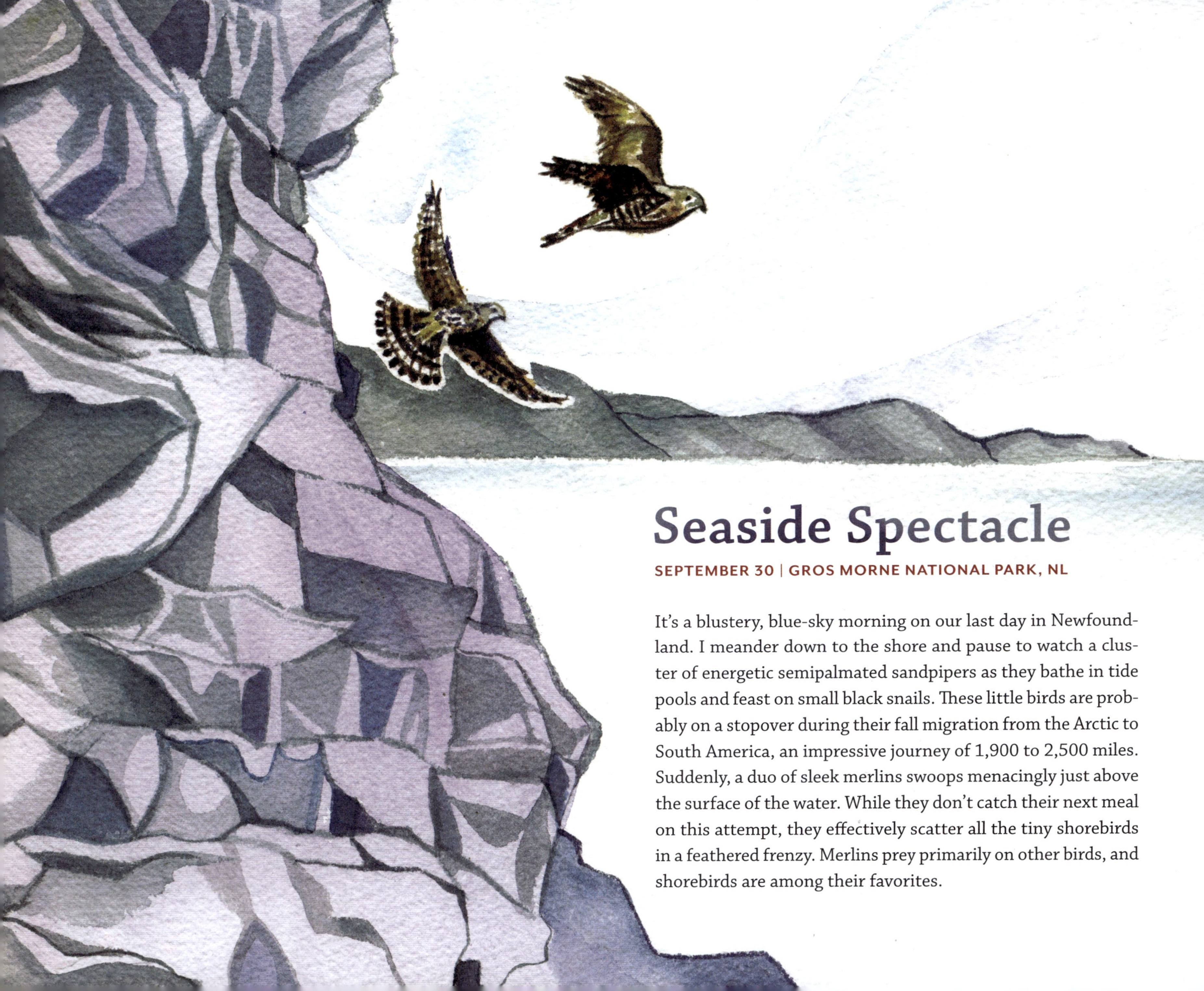

Seaside Spectacle

SEPTEMBER 30 | GROS MORNE NATIONAL PARK, NL

It's a blustery, blue-sky morning on our last day in Newfoundland. I meander down to the shore and pause to watch a cluster of energetic semipalmated sandpipers as they bathe in tide pools and feast on small black snails. These little birds are probably on a stopover during their fall migration from the Arctic to South America, an impressive journey of 1,900 to 2,500 miles. Suddenly, a duo of sleek merlins swoops menacingly just above the surface of the water. While they don't catch their next meal on this attempt, they effectively scatter all the tiny shorebirds in a feathered frenzy. Merlins prey primarily on other birds, and shorebirds are among their favorites.

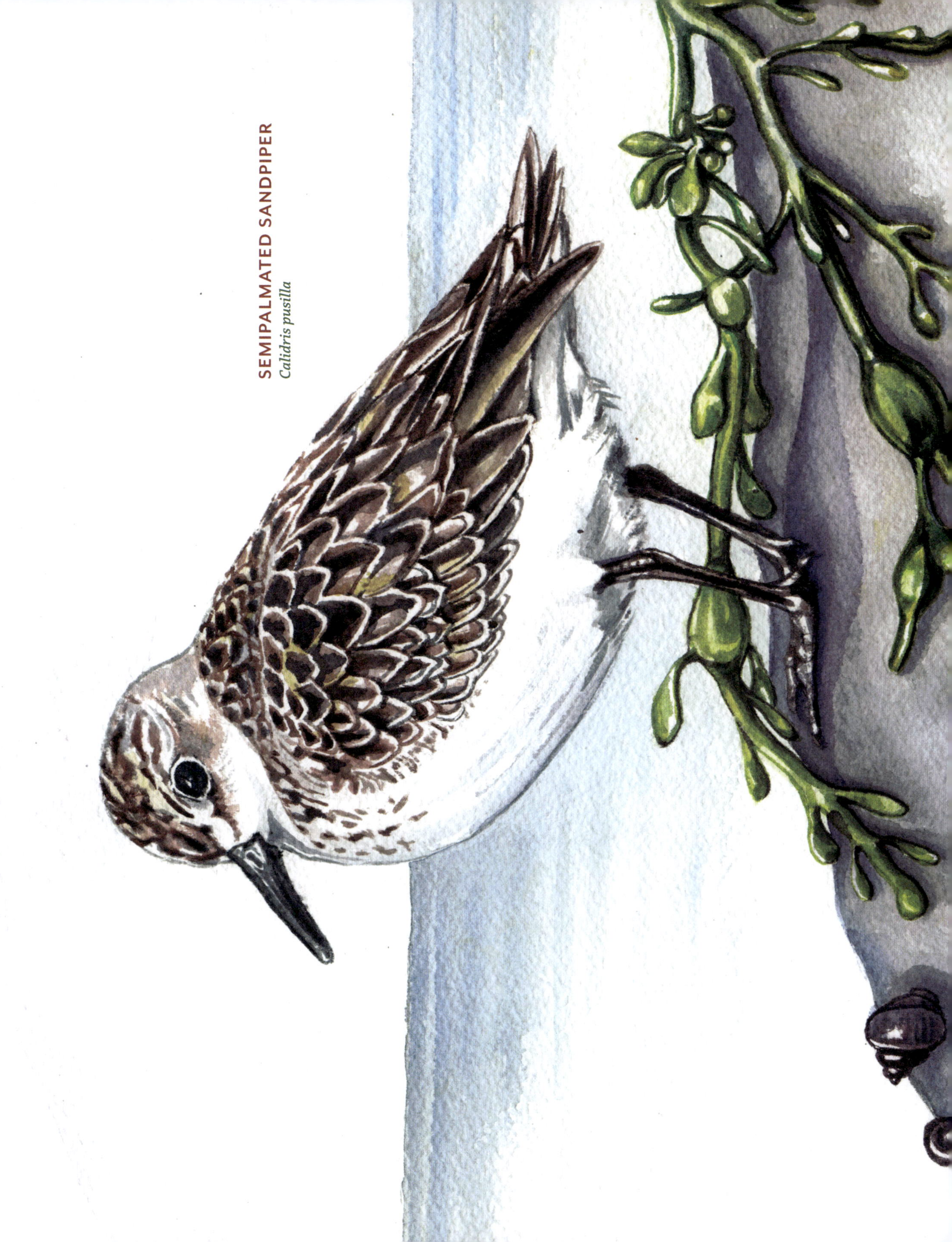

SEMIPALMATED SANDPIPER
Calidris pusilla

I Notice...

- Lime-green seaweed stands out starkly against the water-darkened sand
- A great blue heron fishes where the river water meets seawater. It doesn't waste energy taking lots of lunges, but instead holds totally still and lunges its neck into the water every 10-15 minutes.
- The gulls are flocked together in a huge group and are squawking loudly.

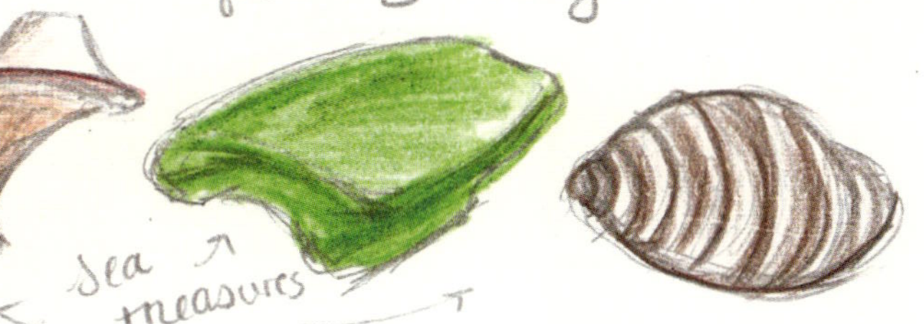

Sea treasures

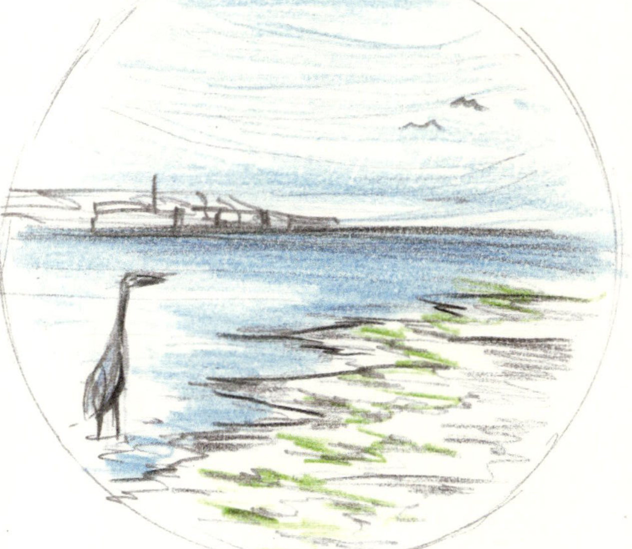

- All the birds out this morning have such different personalities!

- A family of mallards - 2 fuzzy ducklings + a mother duck - are hanging out along the river shore, rapidly nibbling at the green seaweed + digging their faces deep into it, looking for something. Babies copy every movement mom makes.

Baby mallard

★ August 10, 2022
★ The beach where the river feeds into the St. Lawrence.
★ 6:30 — 7:40 Am
★ St-Maxime-du-Mont-Louis

NATURE JOURNAL IDEAS

INVISIBILITY IMMERSION

When we're constantly on the move during our outdoor time, our footsteps and voices scare off wildlife from afar. My most exciting wildlife observations arise when I've made myself as invisible as possible, usually by sitting quietly among foliage near a body of water. Your challenge is to think of a safe place outdoors where you can practice becoming invisible. If you can, plant yourself discreetly next to a river, wetland, or lake at dawn or dusk when wildlife is most active. Try not to make any movements or sounds, waiting as long as your patience level allows. Who knows what will show up!

WASTELAND WALK

I never would have thought that the limestone barrens would be home to so many interesting plants. Think about an outdoor location near you that seems like a wasteland. Maybe it's an overgrown bank on the edge of town, an abandoned sports field, or a drainage wetland. Get creative! Walk around the area and look and listen for signs of life, keeping your mind open to unexpected discoveries.

BLOODTHIRSTY BOTANICALS

Carnivorous plants are some of my favorite finds. Do some reading to determine whether there are carnivorous plants in your region and, if so, find out where they grow. Hint: they're often in bogs! Go on a little pilgrimage and see what insect-eating plants you can find. Each plant has a specific way of capturing its prey. Make note of any insects you see trapped in the plant's leaves.

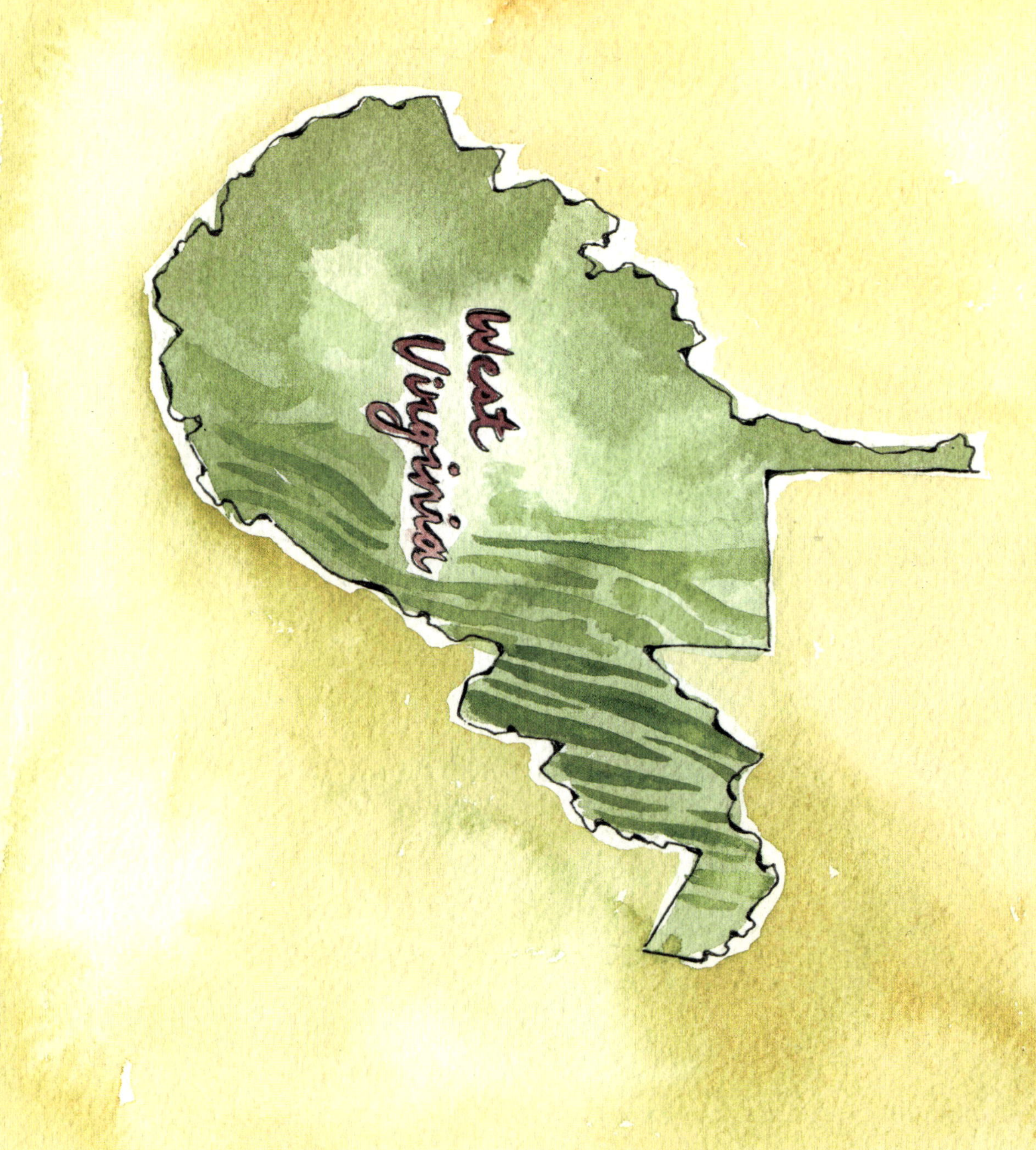

West Virginia

home AGAIN

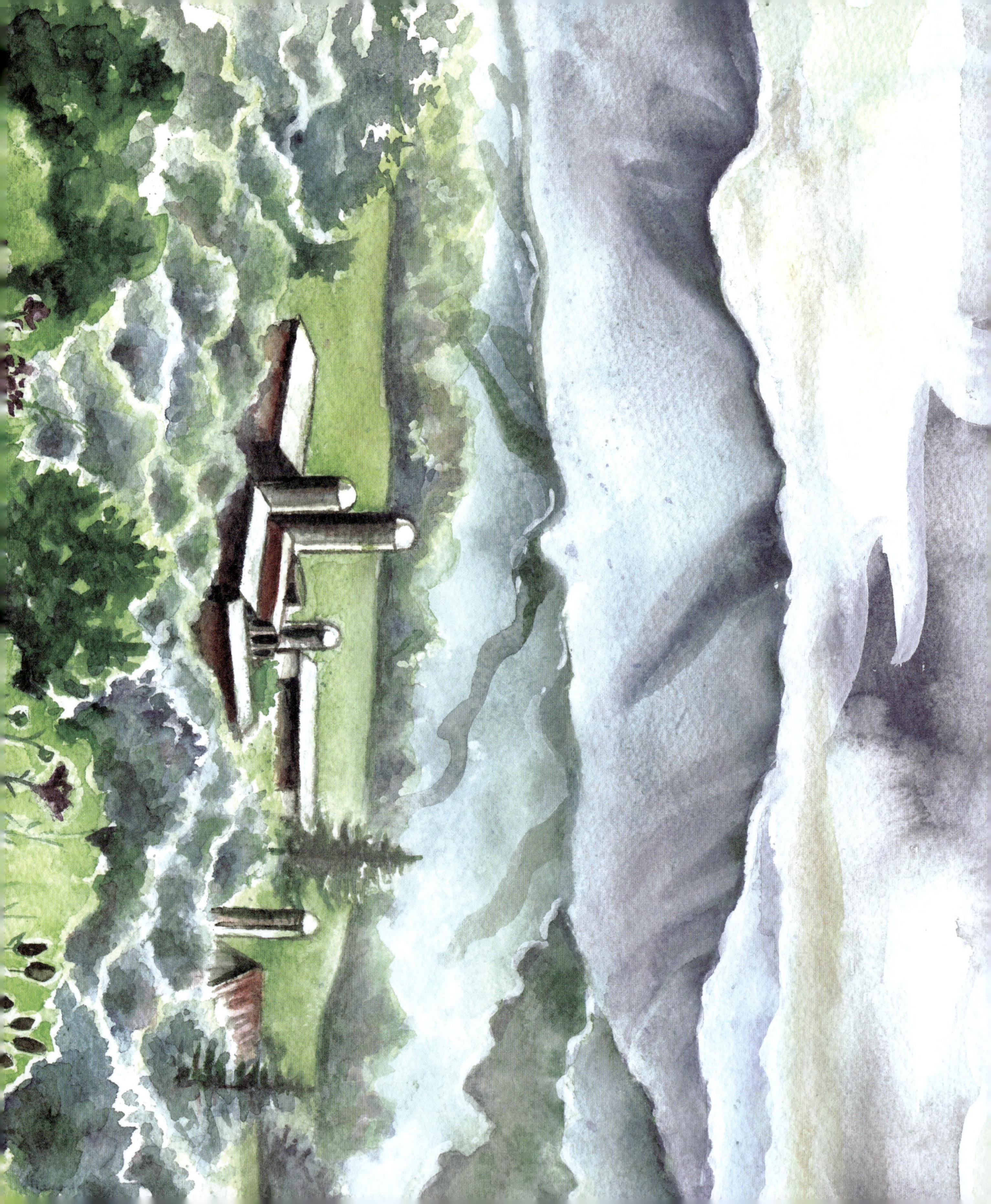

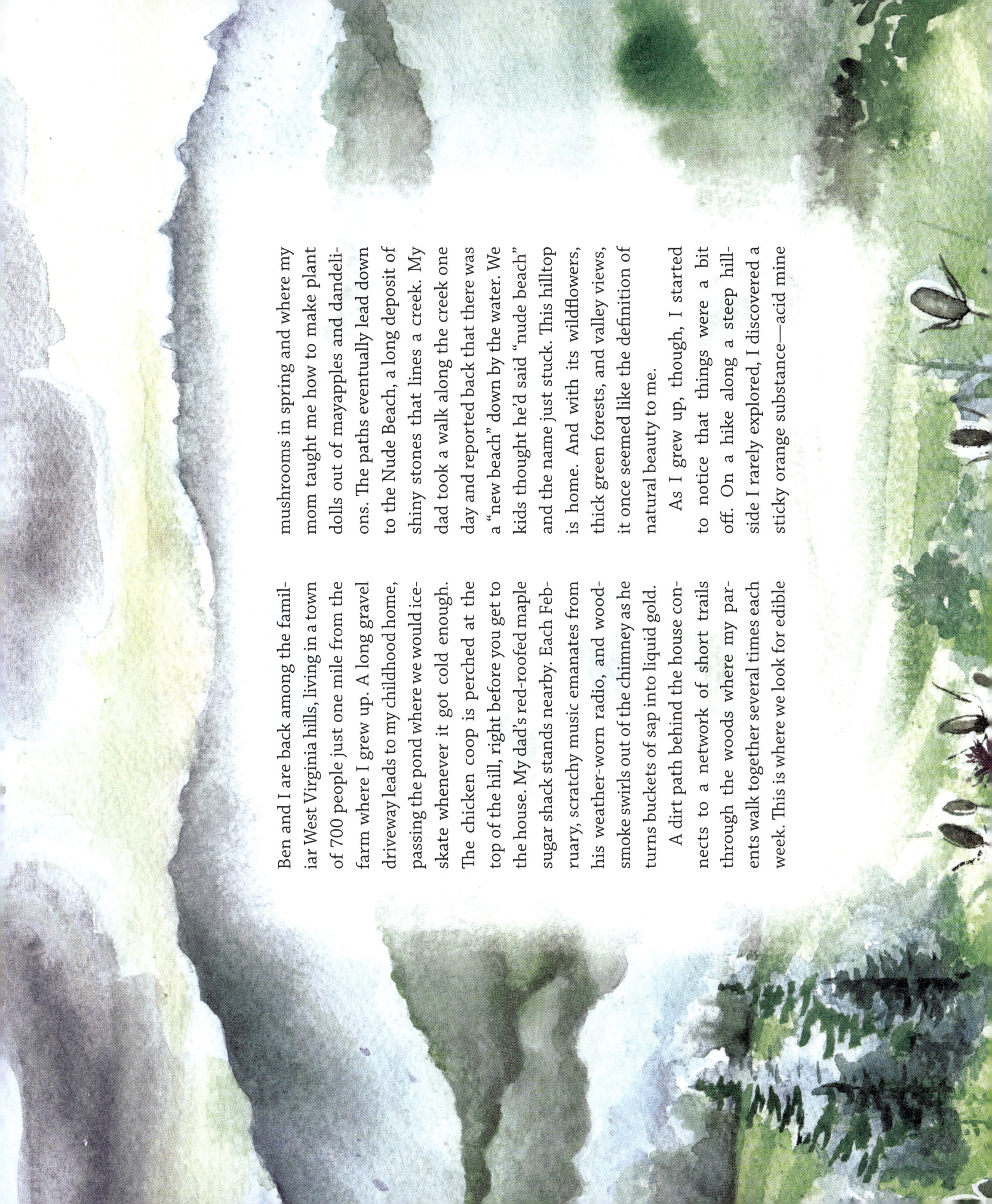

Ben and I are back among the familiar West Virginia hills, living in a town of 700 people just one mile from the farm where I grew up. A long gravel driveway leads to my childhood home, passing the pond where we would iceskate whenever it got cold enough. The chicken coop is perched at the top of the hill, right before you get to the house. My dad's red-roofed maple sugar shack stands nearby. Each February, scratchy music emanates from his weather-worn radio, and woodsmoke swirls out of the chimney as he turns buckets of sap into liquid gold.

A dirt path behind the house connects to a network of short trails through the woods where my parents walk together several times each week. This is where we look for edible mushrooms in spring and where my mom taught me how to make plant dolls out of mayapples and dandelions. The paths eventually lead down to the Nude Beach, a long deposit of shiny stones that lines a creek. My dad took a walk along the creek one day and reported back that there was a "new beach" down by the water. We kids thought he'd said "nude beach" and the name just stuck. This hilltop is home. And with its wildflowers, thick green forests, and valley views, it once seemed like the definition of natural beauty to me.

As I grew up, though, I started to notice that things were a bit off. On a hike along a steep hillside I rarely explored, I discovered a sticky orange substance—acid mine

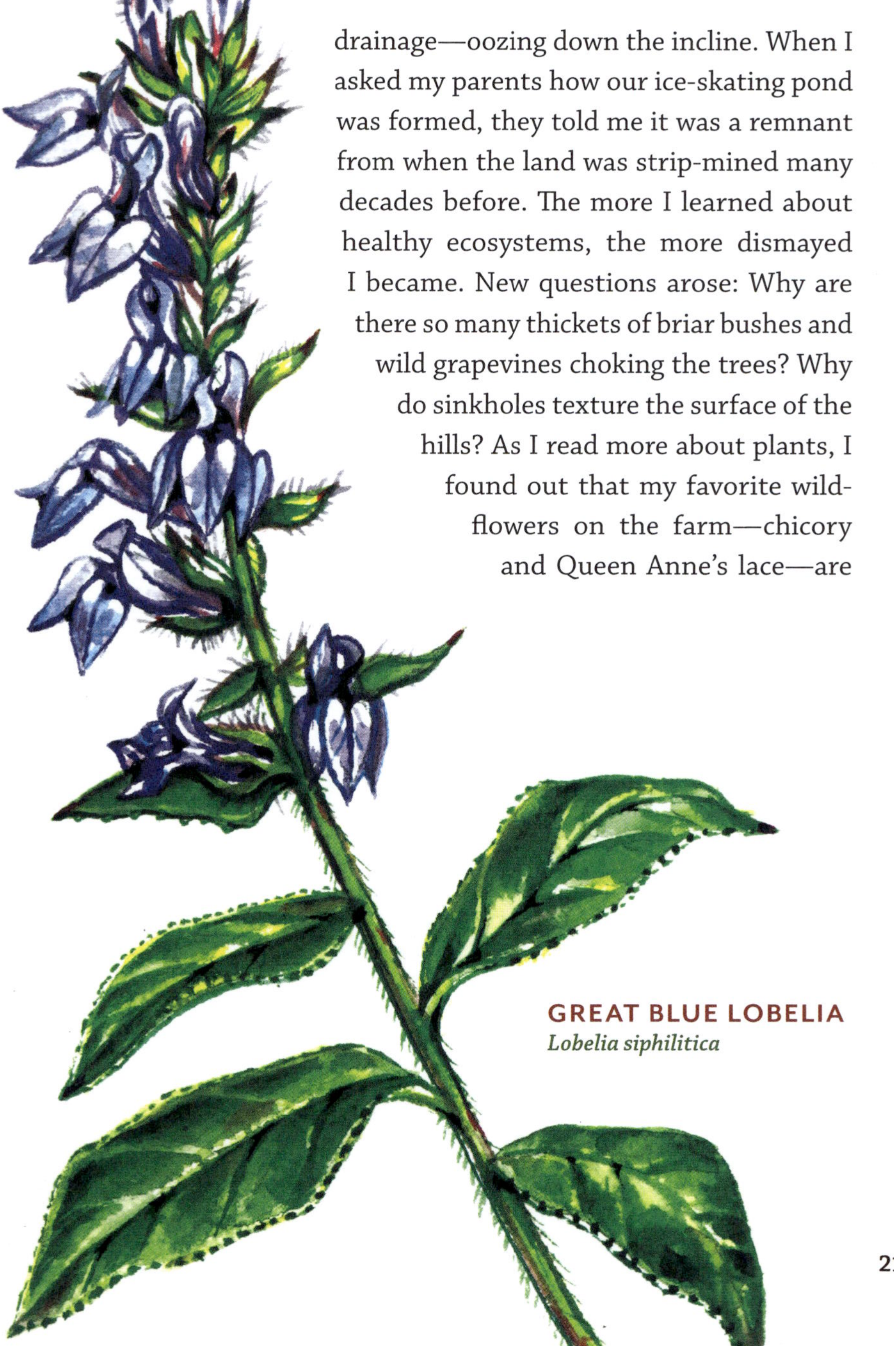

GREAT BLUE LOBELIA
Lobelia siphilitica

drainage—oozing down the incline. When I asked my parents how our ice-skating pond was formed, they told me it was a remnant from when the land was strip-mined many decades before. The more I learned about healthy ecosystems, the more dismayed I became. New questions arose: Why are there so many thickets of briar bushes and wild grapevines choking the trees? Why do sinkholes texture the surface of the hills? As I read more about plants, I found out that my favorite wild-flowers on the farm—chicory and Queen Anne's lace—are

actually nonnative weeds that compete with native plants for resources.

Zooming out, I considered the area beyond the farm. The Ohio River runs nearby, just a ten-minute drive away, and the entire waterfront is crowded with industry, smokestacks puffing dark clouds into the sky. Each day on the way to and from school, we would pass a massive carbon coke plant. I remember holding my breath to keep out the nauseating fumes.

The flora and fauna of West Virginia have inspired me to create countless paintings over the years, but I've often traveled to other parts of the state that are less scarred by industry to collect my observations. I've sought out the rugged beauty and biodiversity of protected lands like the nearly one-million acre Monongahela National Forest and New River Gorge National Park and Preserve. These places have also endured deforestation and coal mining, but they've been given the time and resources to heal. And although my parents are working hard to remediate acid mine runoff, eliminate invasive species, and cultivate native plants and trees, it's a constant uphill battle.

Now that Ben and I are back, right on the heels of experiencing truly pristine wilderness in Newfoundland, I wonder how I'll view these hills and valleys. It would be so easy to dismiss this area as having been lost to industry, but unlike before, I find that I'm curious about the resilient plants and animals that continue to call these hills home. Because as long as the birds still sing and the trees still grow, this place is worth protecting too. In search of natural surroundings that enjoy more protections, what might I have overlooked?

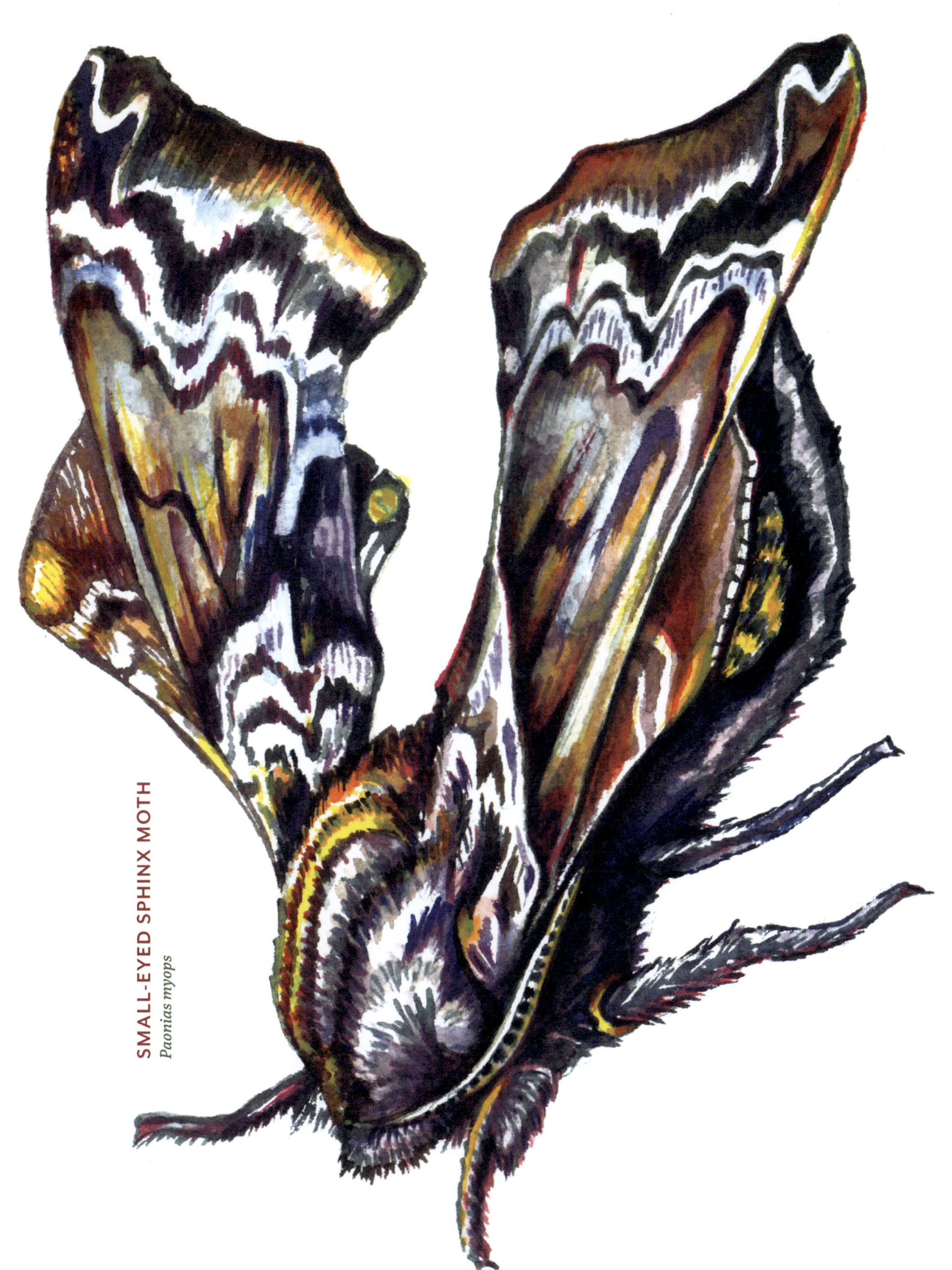

SMALL-EYED SPHINX MOTH
Paonias myops

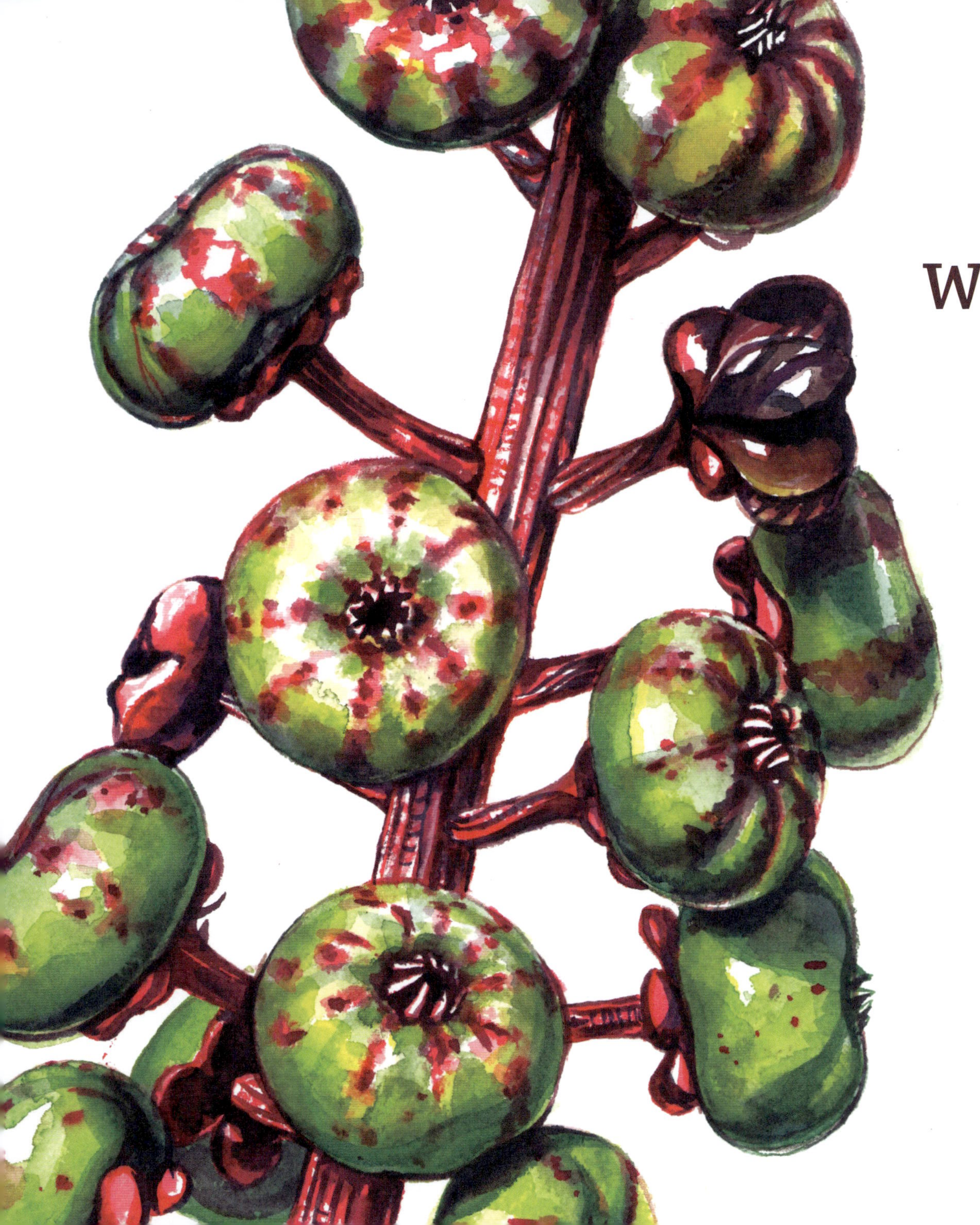

Wondrous Weed

Along a bike trail that I frequently traveled as a kid, I spot American pokeweed. I typically pass right by it since it's so common, but this time I stop and turn the branch over in my hand, admiring the pokeweed's intense colors—its fuchsia stem, the tie-dye of the unripe berries, and the royal purple of the ripe ones. While the berries are poisonous to humans, they are a significant source of food for a range of mammals, such as opossums, deer, and raccoons. They also nourish many of our region's airborne friends, among them crows, mockingbirds, cardinals, and mourning doves. Pokeweed is also an important host plant for the stunning giant leopard moth.

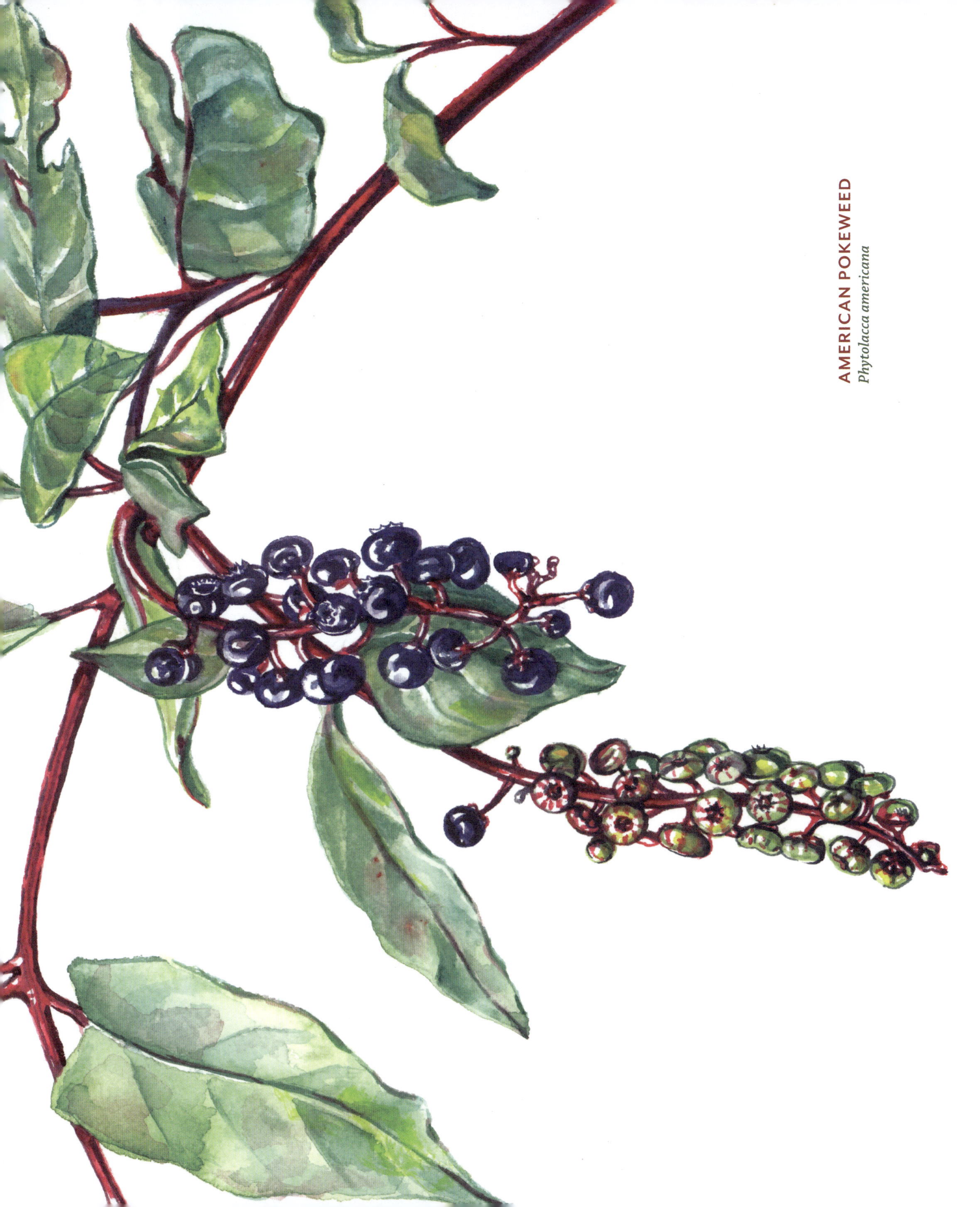
AMERICAN POKEWEED
Phytolacca americana

Dance of Decay

My birthday is in October, and autumn happens to be my favorite season. This time of year seems to echo the many contradictions that I hold in my heart: My desire to cling tightly to youth and beauty confronts my understanding that there is loveliness even in entropy. My yearning for change and newness butts up against my wish that everything I know and love stay the same.

My first taste of autumn this year came when I opened the door to a late-September morning in Newfoundland, the air smelling of leaves and damp earth as it filled and chilled my nostrils. The aspen had already turned to gold, and the wind had yanked a few of the less resilient leaves off the trees, demoting them to the forest floor with a swoosh and a swirl: the dance of decay.

Before I knew it, it was October and we were on our way home, a seven-hour ferry ride through the Atlantic followed by a twenty-three-hour drive back to West Virginia. As we passed through the countryside where I grew up, I noticed a familiar house, one in which a kid who rode my bus used to live. The roof had caved in and there was now a full-grown tree sprouting through the front porch. Had I really been alive long enough for nature to reclaim an entire house?

Mid-October arrived and a familiar melancholy washed over me during my birthday week. When I was an early teen, my birthday began to feel like an annual burden, something that I needed to get through as quickly as possible. Parties focused entirely too much attention on my shy self, far more than I could handle, and even to this day I struggle to understand why we celebrate getting older. I like being alive, and the passing of another year means one year closer to death. But perhaps that's precisely the reason for a birthday party: to counteract the grief that comes with losing another year on this vibrant planet.

This time around, on my twenty-eighth birthday, I am back home and surrounded by my family. We've gathered to celebrate four of us who have birthdays right around the same time. It is a fifth family member's birthday today too—my grandma's. She died last November from a short but intense illness. It was on this day one year ago that I shared my last conversation with her, my birthday buddy. The following month, she was gone.

On this year's shared birthday weekend, we gather in the living room to partake in the first annual family talent show. My dad is the emcee, and he uses a karaoke microphone to announce each act. A gaggle of six nieces and nephews dance to the song "Baby Shark," which is followed by a performance that involves four straight minutes of cartwheels by my nieces. One sister and her boyfriend tell ghost stories and scare all the toddlers, and another sister swirls across the floor with her fiancé in an impressive salsa routine.

Then one brother and his family get up to play a song about adding another limb to their family tree, and suddenly we all look at each other and know that they're announcing a new baby. My mom starts crying and jumps up to hug them and soon we're all crying and laughing. Clearly they've clinched first place at our homegrown talent show.

After spending so many months studying plants and animals that I encountered throughout the Appalachians, I somehow feel more in tune with the behaviors, expressions, and interactions of humans too. I notice the nervous waver in my mom's voice as she sings her song for the talent show. I watch a tear trickle down my hard-shelled oldest brother's cheek during the pregnancy announcement. I sense that my dad is walking on clouds the whole weekend, his heart filled to the brim with all his babies and their babies under one roof. People say that love is simply paying sustained attention to something. And on my journey to becoming a better observer of the plants and animals I meet each day, I think I've become more awake to life itself.

As we stand in a big circle and sing "Happy Birthday" after the talent show ends, my dad's voice rises above everyone else's in a faux-operatic style that he adopts exclusively for the birthday song. Another year has come and gone. But this time I'm not as sorrowful. I think about the average life span of many of the tiny critters that I've so enjoyed getting to know over the past months—hummingbirds live between three and five years, voles for about a year, and dragonflies for fewer than eight weeks. What a short time to enjoy this earth! Suddenly, the human life-span feels like an abundant gift.

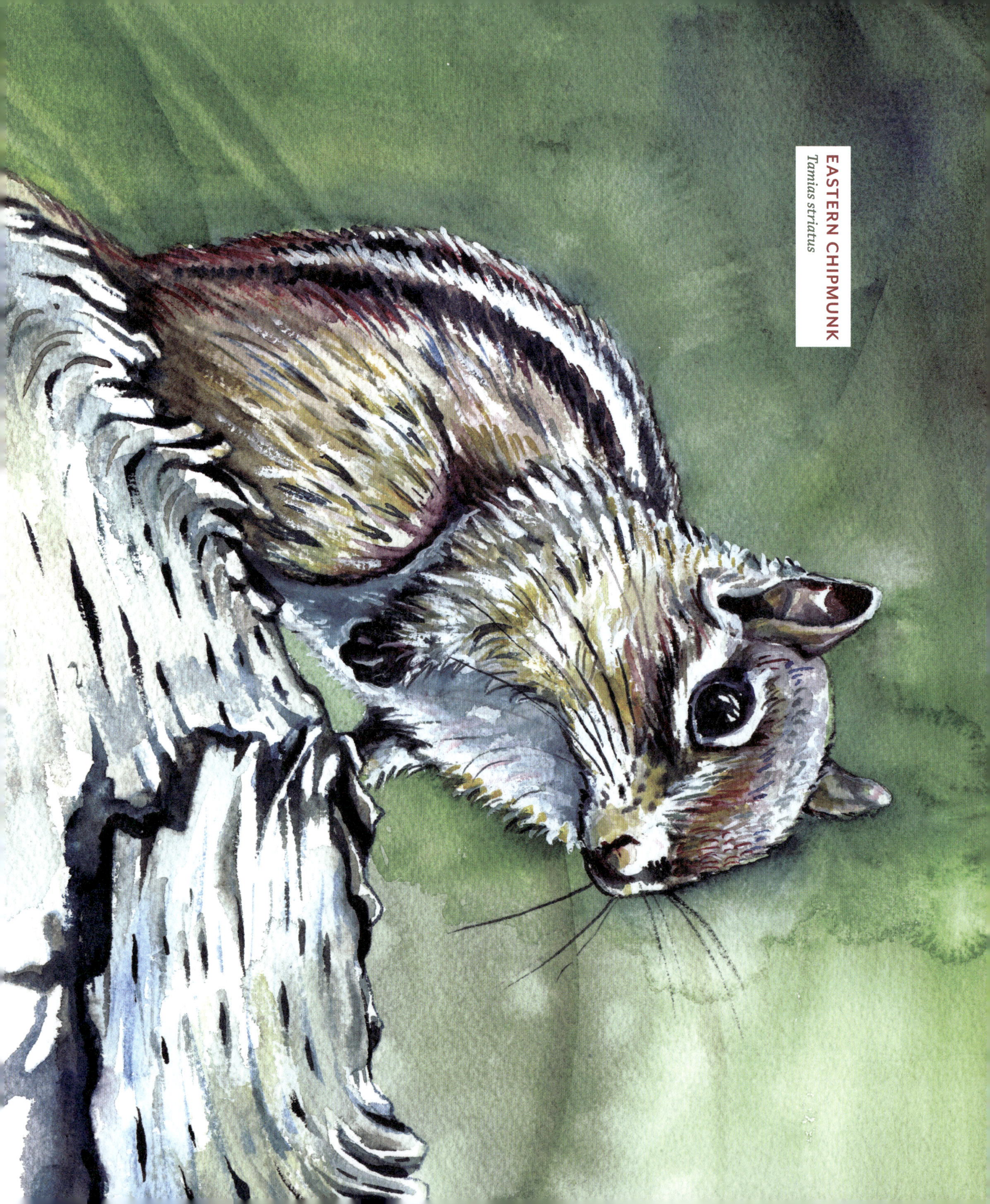
EASTERN CHIPMUNK
Tamias striatus

Chipper Chum

I bring my camera with me to the park, hoping to capture some birds in action to paint when I return home. The *peck-peck-pecking* of red-bellied and downy woodpeckers taunts me all morning, but I only spot one before it skirts around a tree trunk and out of view.

As I sit there, deflated, a perky chipmunk scurries over and gazes at me, his dainty fingers trembling ever so slightly. "You know," I say to the chipmunk, "I've never painted one of you before. Maybe you'll model for me?" Sure enough, my chipmunk friend holds still long enough for me to get my camera ready and snap a few shots, then he hurries off to continue his chipmunk duties, perhaps gathering acorns for the imminent winter. Chipmunks store their food—berries, nuts, bird eggs, mushrooms, seeds, and insects—in large caches that they cover with soil. And unlike nest-building squirrels, chipmunks dig extensive underground burrows with multiple entrances, creating a lot of easily accessible escape routes. My favorite fact about these critters is that a group of chipmunks called a "scurry."

As I continue exploring the park, I notice chipmunks everywhere. Their clucking vocalizations merge with those of the woodpeckers, a syncopated soundtrack to my walk. How did I not see them before? Did they seem so mundane that I tuned them out completely?

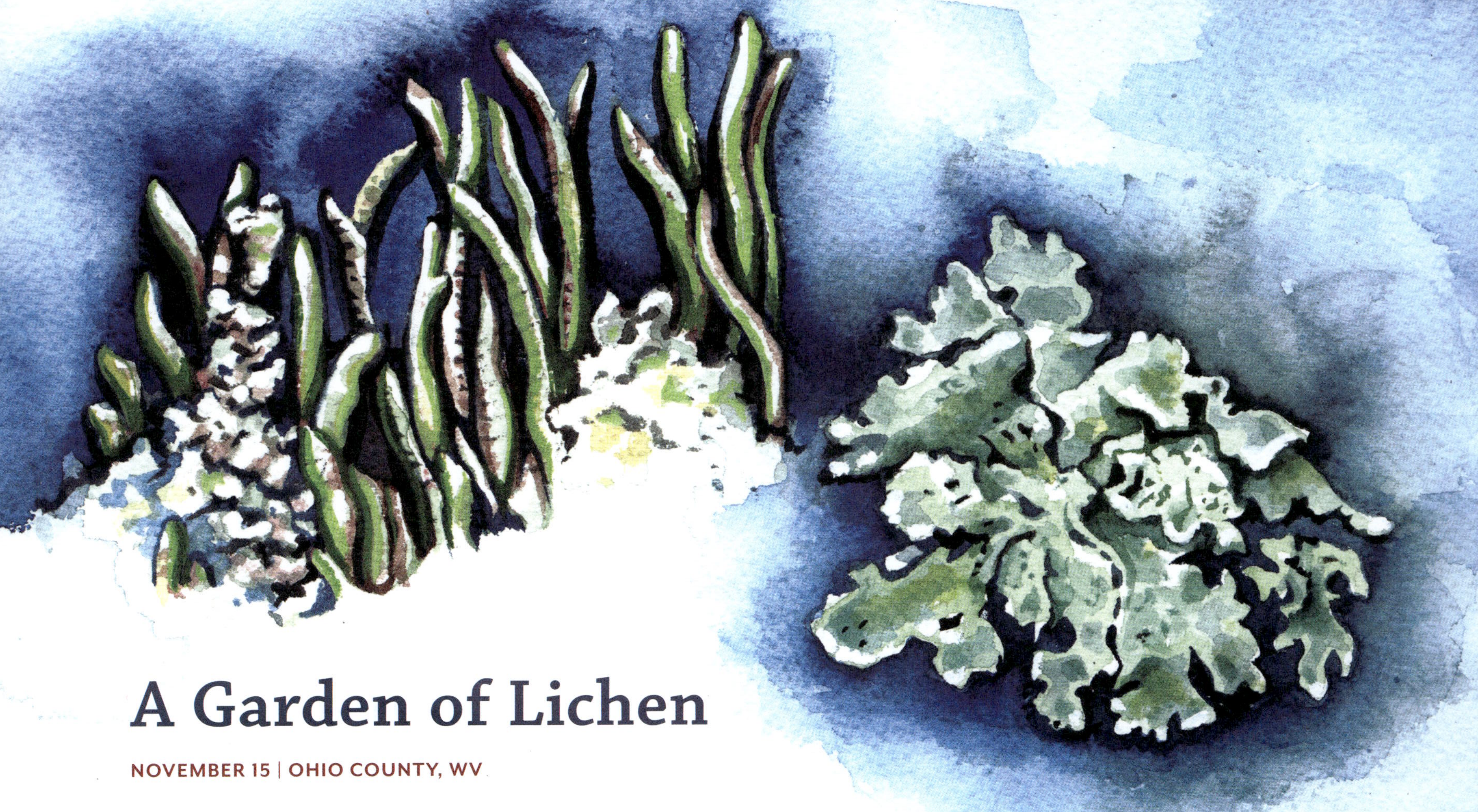

A Garden of Lichen

While walking over a footbridge in the center of town, I spot a delicate lichen garden growing in the most unexpected place—atop a snow-covered wooden post on the bridge. I gently blow away the powdery blanket to get a better look. Among the several varieties growing here is British soldier lichen, with its knobby green branches topped with bright red fruiting bodies that resemble miniscule hats. I squeal with delight—I love this kind of lichen, one I normally see only in forest settings. The funky shapes and textures remind me of a coral reef, and for a moment, as I imagine miniature tropical fishes flitting about among the colorful structures, I'm transported out of my wintery reality.

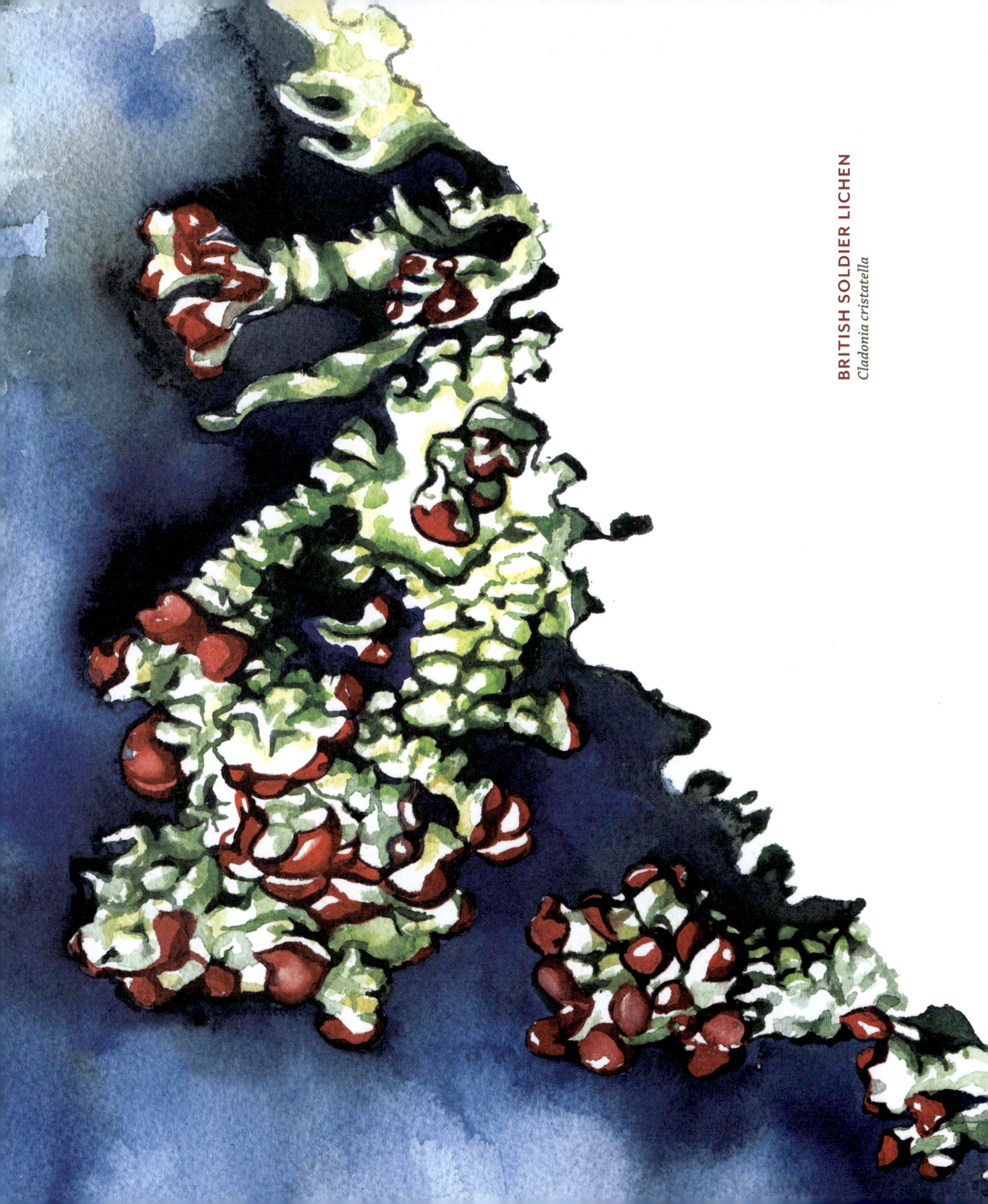

BRITISH SOLDIER LICHEN
Cladonia cristatella

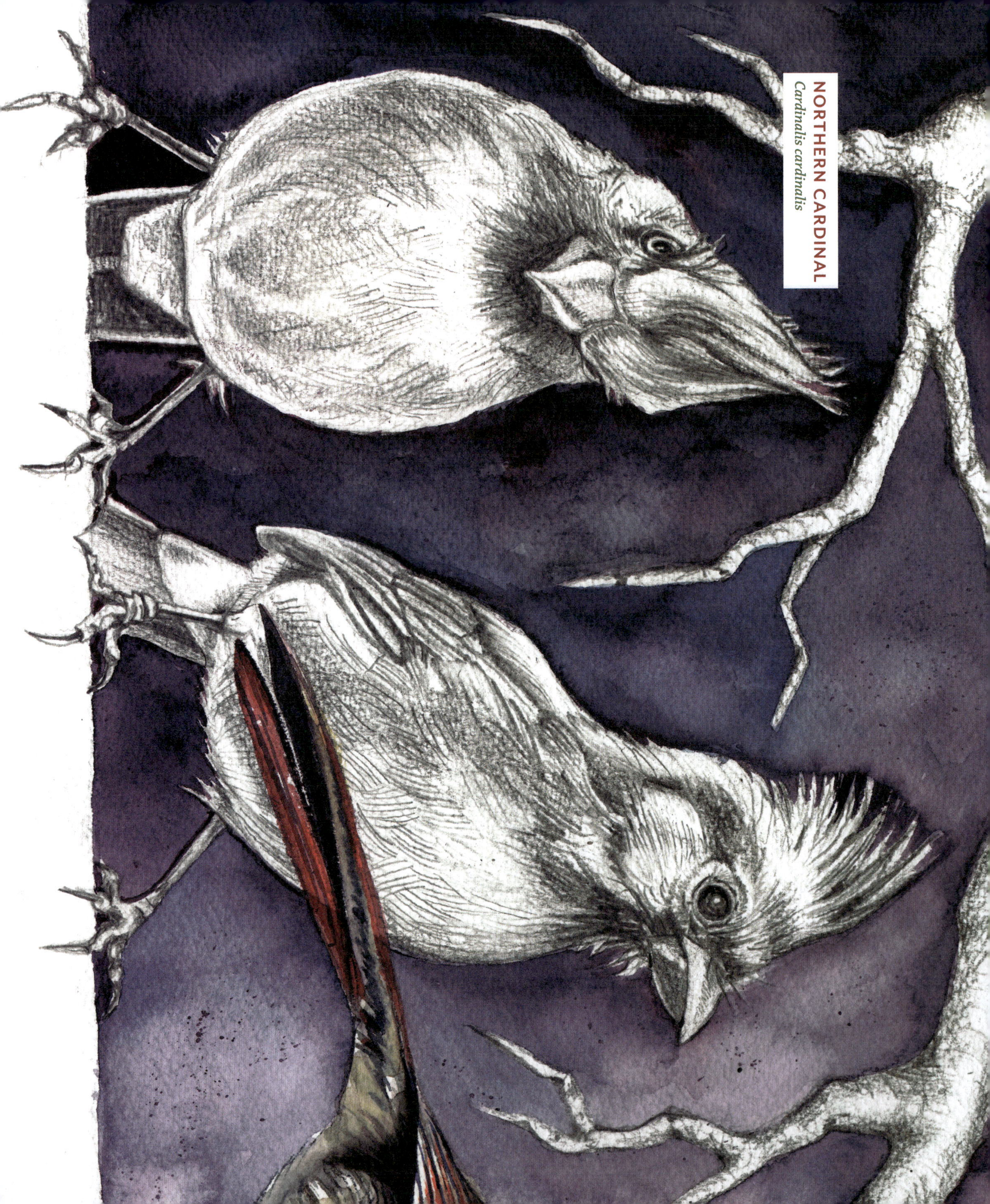
NORTHERN CARDINAL
Cardinalis cardinalis

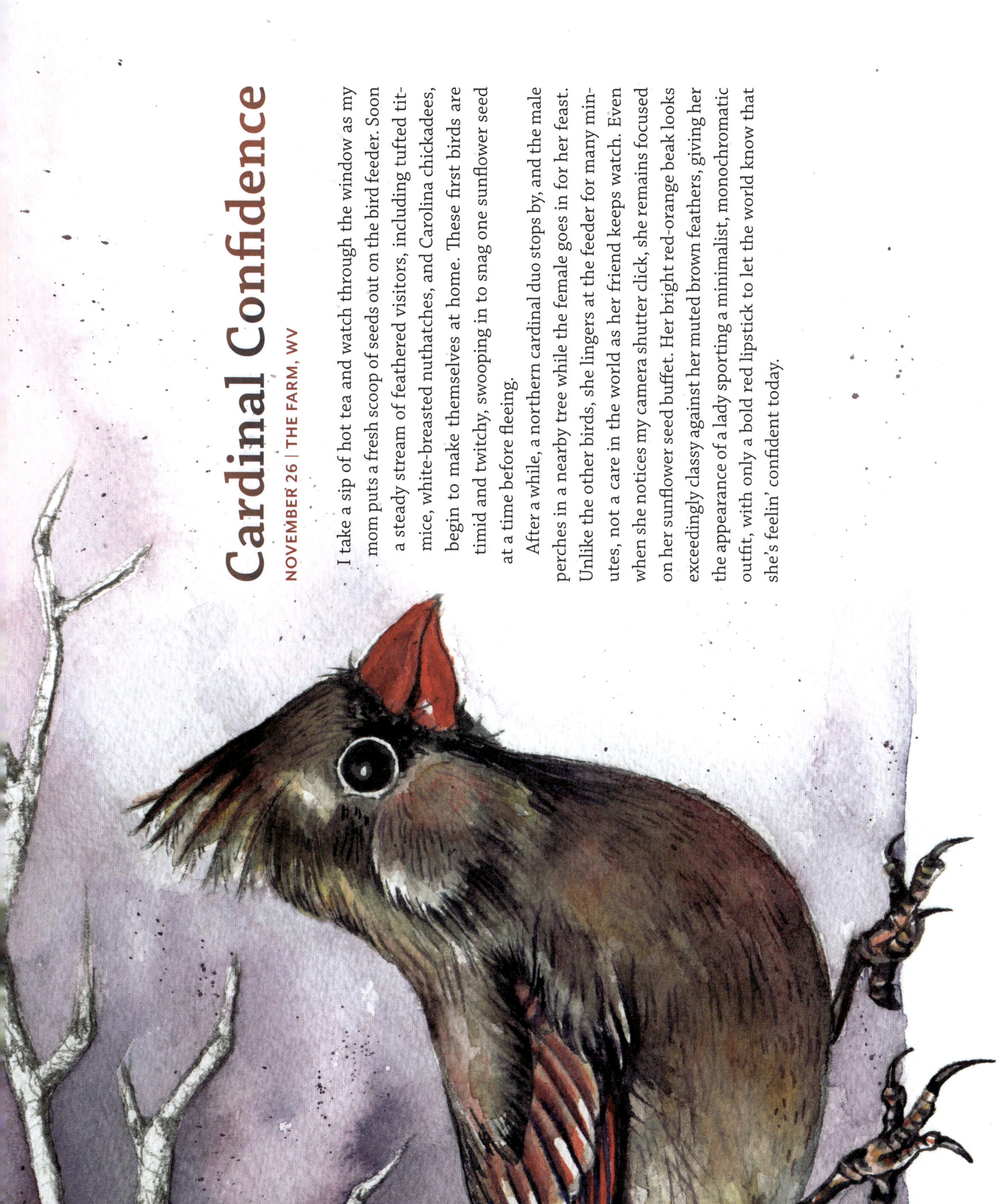

Cardinal Confidence

I take a sip of hot tea and watch through the window as my mom puts a fresh scoop of seeds out on the bird feeder. Soon a steady stream of feathered visitors, including tufted titmice, white-breasted nuthatches, and Carolina chickadees, begin to make themselves at home. These first birds are timid and twitchy, swooping in to snag one sunflower seed at a time before fleeing.

After a while, a northern cardinal duo stops by, and the male perches in a nearby tree while the female goes in for her feast. Unlike the other birds, she lingers at the feeder for many minutes, not a care in the world as her friend keeps watch. Even when she notices my camera shutter click, she remains focused on her sunflower seed buffet. Her bright red-orange beak looks exceedingly classy against her muted brown feathers, giving her the appearance of a lady sporting a minimalist, monochromatic outfit, with only a bold red lipstick to let the world know that she's feelin' confident today.

Spark Owl

I'm out for a slow walk after sitting and working on a painting all day. Not really looking to make nature observations, I wander down to the fields where I played soccer as a kid. The town's population continues to dwindle, so the fields are no longer in use, but a few toppled goal frames remain among the weeds, surfacing hazy memories of childhood matches.

Sycamores line the perimeter, and out of habit I glance up at a cavity in one of the trees, squinting to see if there is anything living inside it. I expect to find the usual vacant squirrel or bird's nest, but today I notice what at first appears to be a very symmetrical, stubby stick poking out from the hollow. Then it begins to look *too* symmetrical. Is that a beak . . . and a pair of eyes? For a few seconds, I can't believe there's something

looking back at me from within the tree. I tiptoe closer and see that it is a small eastern screech owl, no taller than seven inches. His eyes are shut and his orange feathers are puffed out in a cozy yet regal fashion, how I imagine a British king looks while he sleeps. I snap a distant photo on my phone, but I can't get a good look without binoculars, so I hurry home after a few moments to retrieve them. By the time I return, the owl has vanished. I stand there for a while, willing him to reappear. But he is nowhere to be found.

The next day, I beeline it back to the tree. The owl is fast asleep in the exact same spot, perched at the entrance to the hollow as though it's his throne. Over the following weeks and months, I return often to check on the owl. I find him there about 90 percent of the time, sleeping in the same position. Eastern screech owls are usually monogamous, and the males often stake their claim to a cavity nesting site during winter, with the female joining him in early spring to lay her eggs. Will

I stop by the tree one day to find that this owl's mate has moved in? Will I find chicks living inside in a couple of months? The unknowns capture my imagination; I could easily pull up a folding chair and binge-watch this owl like my favorite television show.

If you hang around avid bird-watchers for a while, you'll hear someone refer to their "spark bird," the specific bird that first sparked their interest in bird-watching. I had always enjoyed birds but never felt a uniquely deep connection to any one species. But now things are different, and I find myself obsessively looking in every tree cavity I pass for other screech owls. My owl fever soon expands to include other birds of prey that I observe this winter, from the dainty kestrel perched high on a telephone wire to the majestic bald eagle that is currently building a hefty nest high in the treetops on my parents' farm. I head out on even the coldest days to find birds, and whenever I get tired of walking the short trails around my hometown, I think about the owl—a reminder that this familiar place will always hold new discoveries.

Unable to keep the owl to myself, I offer to take friends and family to see my muse. Soon I am routinely making my pilgrimages there with new people in tow, and everyone seems to walk away just as amazed as I am, even if they aren't usually interested in birds.

One day, I take my nephew to see the owl. Milo is six and, like me, he loves to hunt for interesting things outside. His collection of shiny rocks and shells is getting so expansive that he recently used his birthday money to invest in a tackle box to keep everything organized. We walk through the muddy soccer field and raise our binoculars to the owl's tree. Once he catches sight of the bird, he lowers his binoculars to reveal eyes as wide as saucers. His missing-tooth smile turns my heart to liquid. "That's the most prettiest bird I've ever seen!" he exclaims.

It's easy to feel helpless over the ongoing destruction of critical plant and animal habitat in the name of development and progress. But I am starting to realize that there are actions within my power: I can go outside and delight in what I see, and I can tell others about the wonders that I find. I can take my nephew by the hand and show him my spark owl, and maybe he will grow up with a similar affinity for the voiceless.

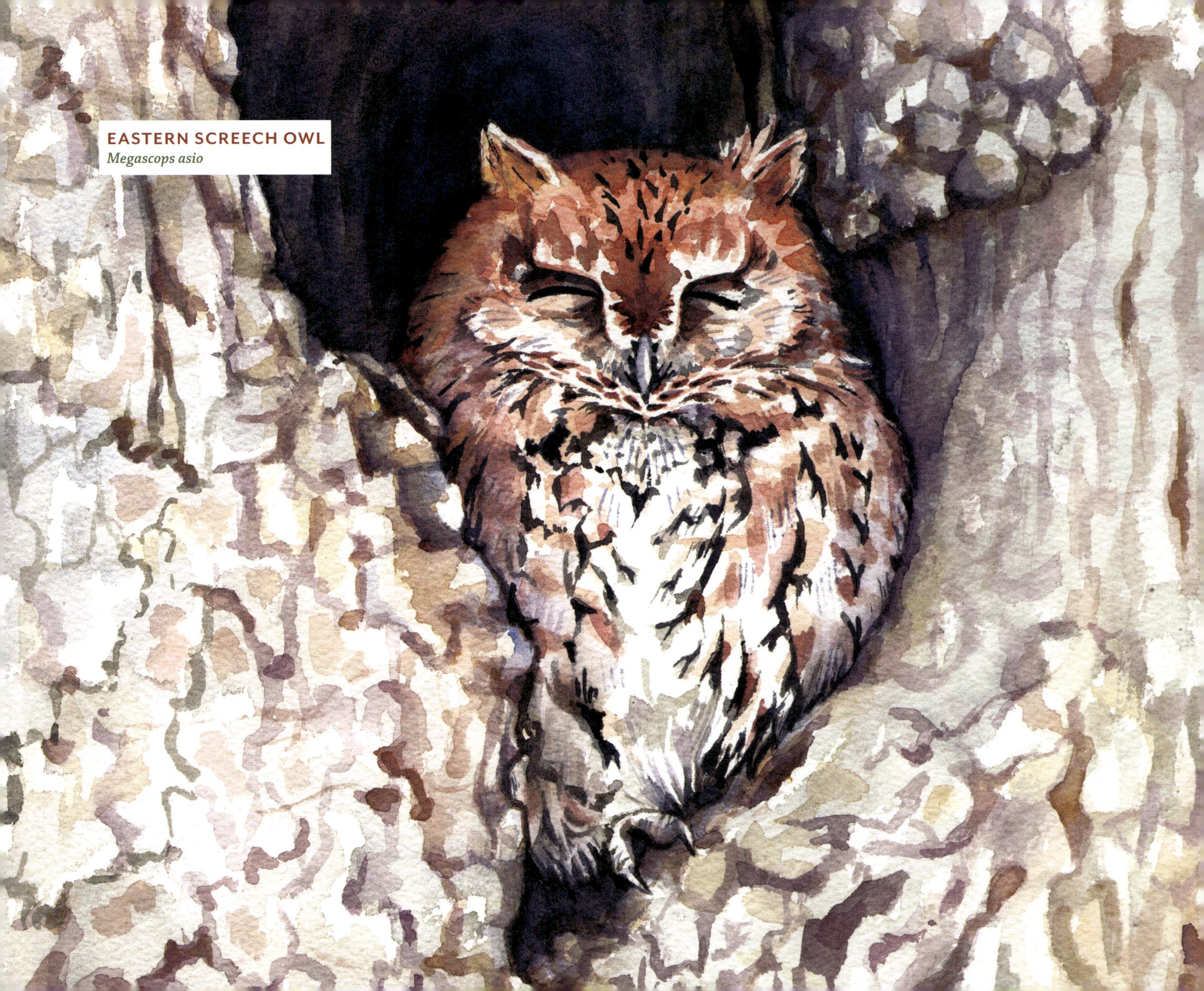

EASTERN SCREECH OWL
Megascops asio

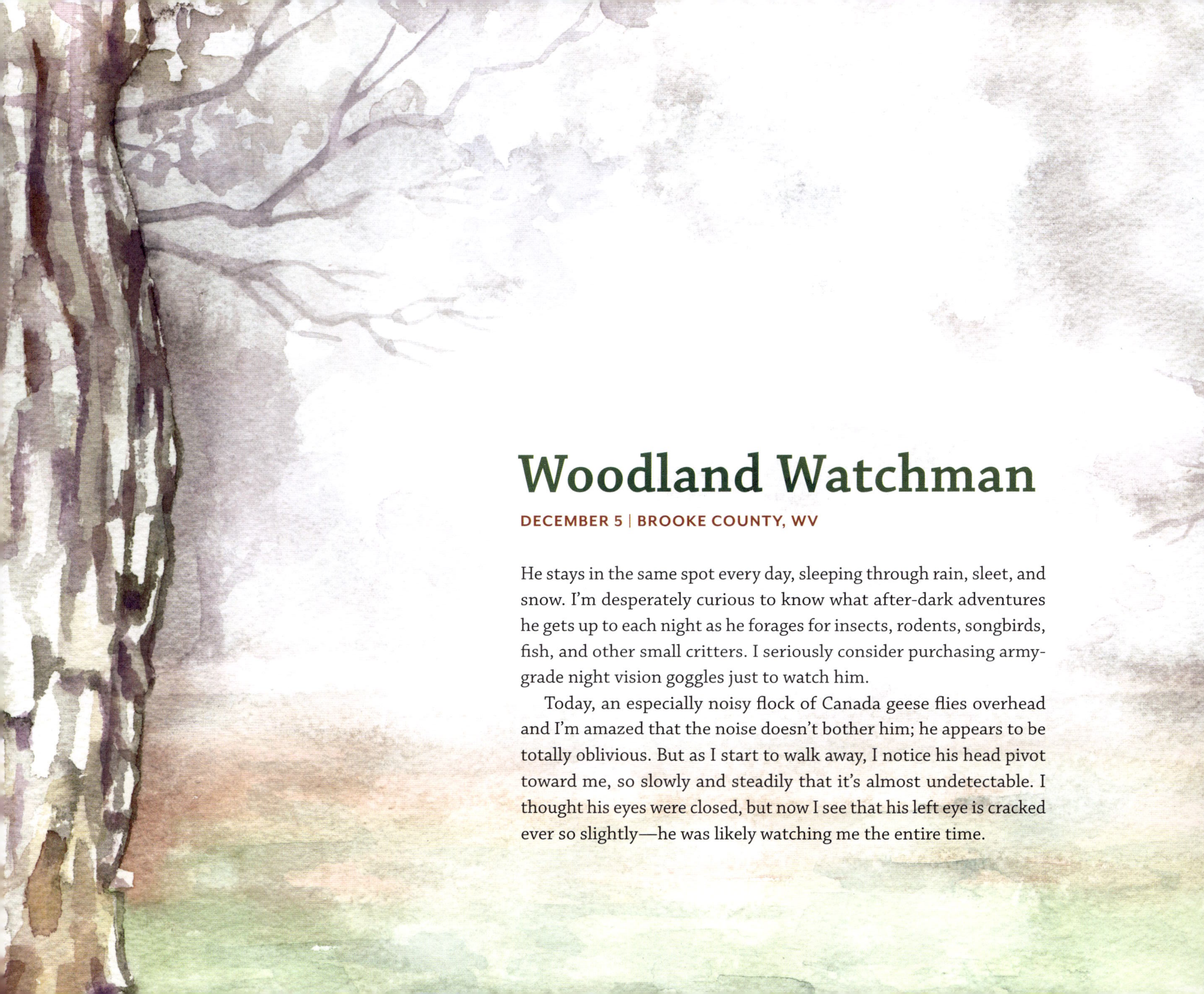

Woodland Watchman

DECEMBER 5 | BROOKE COUNTY, WV

He stays in the same spot every day, sleeping through rain, sleet, and snow. I'm desperately curious to know what after-dark adventures he gets up to each night as he forages for insects, rodents, songbirds, fish, and other small critters. I seriously consider purchasing army-grade night vision goggles just to watch him.

Today, an especially noisy flock of Canada geese flies overhead and I'm amazed that the noise doesn't bother him; he appears to be totally oblivious. But as I start to walk away, I notice his head pivot toward me, so slowly and steadily that it's almost undetectable. I thought his eyes were closed, but now I see that his left eye is cracked ever so slightly—he was likely watching me the entire time.

It's so good to be home! And just in time for my favorite month, too. I'm sitting on Dad's little rock bench by the creek after finishing up a paddleboard sit. The water's so still right now that I was able to float downstream and paddle back upstream super easily.

I was able to see lots of stuff — crayfish, sculpins, bullfrog tadpoles + minnows. Outside of the water, I saw several Kingfishers, heard a red-bellied woodpecker that yipped like a dog, and got to watch some sort of comma butterfly for a while.

Morning on the Creek

This morning, I've been thinking about how even after spending the last 6 months learning about my surroundings, there's so much that I still don't know. But I'm learning to accept the way my brain works and lean into the wonder of not having all the answers. I think we need both — Knowledge helps us dig in a little deeper and wonder keeps us coming back for more!

- October 11, 2022
- 8 – 10ish am
- The farm, along the creek
- 68°F + sunny

NATURE JOURNAL IDEAS

ORDINARY AWE

There are dozens of animals that I used to think of as too common and ordinary to take special interest in, like cardinals and chipmunks. But how much do we really know about the critters that are our closest neighbors? Go for a walk in a nearby park and be on the lookout for an animal that you'd normally walk right past. Think of all the things you already know about that animal, then consult a field guide to fill in the gaps in your knowledge. Does this animal migrate seasonally, or is it a year-round resident? How long does it live? What plants or other animals compose its diet? Where does it sleep, and at what times of day or night is it most active?

HOLLOW HUNT

Hollows are safe havens for all sorts of wildlife—I've found screech owls, sleeping raccoons, squirrels feasting on nuts, and cozy songbird nests. Grab your binoculars and go for a walk in a wooded area with the intention of looking in all the hollows and cavities of the trees you pass. Even if you don't find an animal, do you notice anything else while paying attention to this very specific part of the trees that you see?

SHARE THE SPARK

Talking about our experiences in nature can help us become more awake to the nonhuman world and can even promote collective stewardship of our planet. Come up with a list of ways that you can share the gift of paying attention. Maybe you can invite your neighbor to come look at the robin's nest that's in your yard, or perhaps you can share some interesting photos of your observations online. Consider hosting a nature journaling meetup in a nearby park where you simply gather for an hour or two to pay close attention and record what you see on paper—no art skills required!

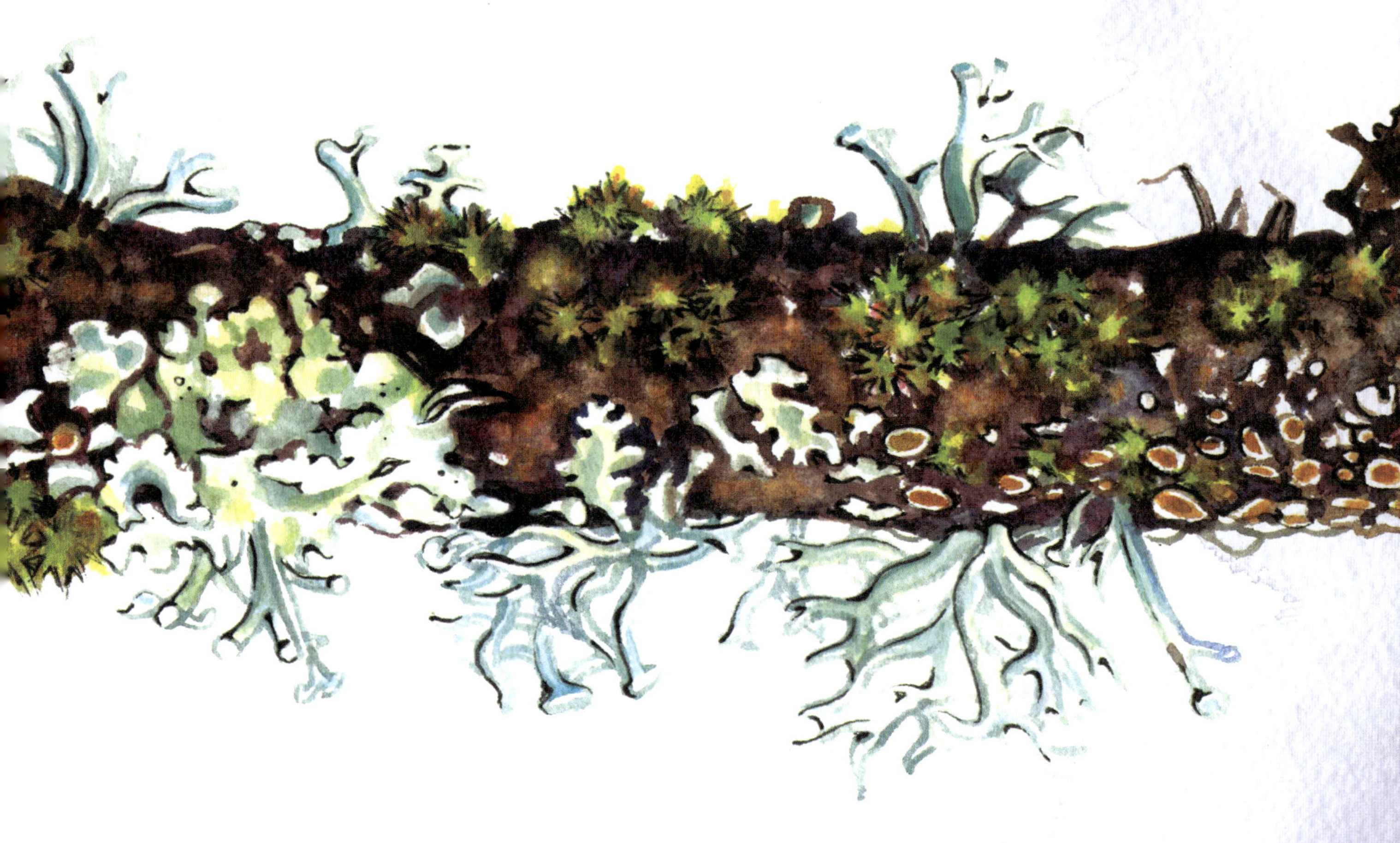

May your days be filled with a million moments of delight and adventure, no matter how tiny.

A LIVING COLLABORATION

As I bring this book to a close, part of me feels that my project is incomplete. My motivation as I set out on my Appalachian journey was to gain a clearer picture of my home region. I wanted to see everything, and I wanted to paint it all. While I found tiny worlds on each outing, and created hundreds of illustrations, I observed many more enchanting plants and animals than I was able to record—far more than I could paint in a single lifetime. And then there are the thousands of species that I didn't even get a chance to see . . .

It's only now that I realize that this whole endeavor is really a collaboration with you, my readers. This book encompasses a sampling of what I found as I moved through the Appalachian Mountains. But now it's your turn to see this project through to its fullest potential by stepping outside and recording the amazing wonders that *you* discover, in whatever corner of the world that you call home. This isn't a collaboration in the traditional sense of the word, since our work won't be displayed side by side. Instead, it's a collective effort to become more awake to our surroundings.

In June, I attended a community nature journaling meetup at a wetland in Pennsylvania. There were about fifteen of us there, all somewhere between the ages of five and seventy-five. We spread out on a short boardwalk that led through the bog, quietly sketching and taking notes on whatever captured our interest. When we reconvened a couple of hours later, we laid our journals out in a circle. Some had referenced field guides to record interesting facts about what they saw; others had used watercolors to capture the wetland landscape as a whole. One person wrote a short poem to describe the way the bog felt, and still others sketched specific plants that caught their attention. One person didn't record anything at all; he just mentally took it all in. I was struck by how unique our journal pages were—how each of us responded to the same setting in different, equally beautiful ways.

If you'd like to share photos of your tiny discoveries or your nature journal pages with me and other readers, use the hashtag #tinyworldsjournal across social media.

Fostering a deeper connection to nature isn't about the accumulation of knowledge. It's a way of being. It's about waking up to the beauty that surrounds us and treating each day like a scavenger hunt. It's about acknowledging the sacred thread that binds our existence, our experiences, to all those who cross our path, whether human or nonhuman. It's about appreciating the richness of this life.

—Rosalie

ACKNOWLEDGMENTS

I have to begin by thanking my husband, Ben, who not only helped plan and accompanied me on our great Appalachian road trip, but also patiently listened to me ramble on about this book every single day for two years. There were many nights when I tapped him on the shoulder to wake him up because my anxious book-related thoughts were running away with me, and his gentle reassurance helped me feel confident that everything would work out. You add boundless adventure, fun, and fullness to my life, dear.

Thank you to my family for your sincere interest in all my projects and for giving your feedback on everything from the title to which species I should feature. Mom, thanks for keeping my online shop up and running while we traveled—you're my lifelong MVP! Dad, I'll never forget the conversation we had when I was in college when I expressed doubts about what my future would hold if I pursued an art career. "Why don't you just see how far you can go?" you asked, and that statement prompted me to dream bigger dreams. And thanks for liking all my social media posts in the middle of the night when you have insomnia. My gratitude also goes out to my sister Clara, who read through my work many times and provided valuable input on my writing. I'm sometimes resistant to taking advice from my baby sister, but I know that everything you said made my work infinitely better.

Several wonderful organizations came together to make this project a reality. Thank you to the Eckelberry Fellowship from Drexel University's Academy of Natural Sciences and the West Virginia Department of Arts, Culture and History for providing grant funding for my travels. Blue Ridge Discovery Center and Shaver's Creek Environmental Center generously hosted me during the trip and let me tag along on nature outings. Tin Mountain Conservation Center, Squam Lakes Natural Science Center, and North Branch Nature Center also offered me warm welcomes when I came to visit.

Thank you to the following scientists and naturalists who gave of their time by either meeting with me on my travels or providing expert feedback as I worked: Kristen Wickert, Brent and Angela Martin, Emmie Cornell, Scott Schuette, Chelsea Clarke Sawyer, Amber Wiewel, Alexa Sarussi, Jon Kauffman, Aaron Floyd, Frank Gebhard, Jerry Hassinger, Betty Gatewood, Scott Weidensaul, Ed Schwartzman, John Burkhart, and Michael Burzynski.

I also am deeply grateful to the caring crew of teachers and professors who went out of their way to encourage my creative instincts. My earliest art teachers, Irene Simmons and Russell Shaffer, worked with me as an elementary-school-aged kid and modeled what it is to be a working artist who loves what they do. Guy Gellner and Michael Doig created true communities in their classrooms and demonstrated that art has the power to bring people together from all backgrounds. In college, my professors Brian Fencl, Lambros Tsuhlares, Robert Villamagna, Sarah Davis, and Moonjung Kang nurtured many facets of my creativity, and I think about the specific instruction each of you provided on a regular basis.

To my editing team at Mountaineers Books—thank you for choosing to work with me out of the many submissions that come your way! To my acquisitions editor, Emily, you immediately understood my vision for this book and have been lovely to work with. Jen, Janet, and Laura—I know that your gentle shepherding and beautiful attention to detail will help my art and words reach the reader with greater clarity and impact. Thank you for investing your energy and creativity into this project.

Lastly, I want to thank my readers and all the people who have supported my work over the years. Every time I get the chance to do in-person events and meet the folks who enjoy my art, I am blown away by just how kind and cool you all are. In a world filled with media coming at you from every direction, I am indebted to each of you for paying attention to my work and rooting for me as I create new things.

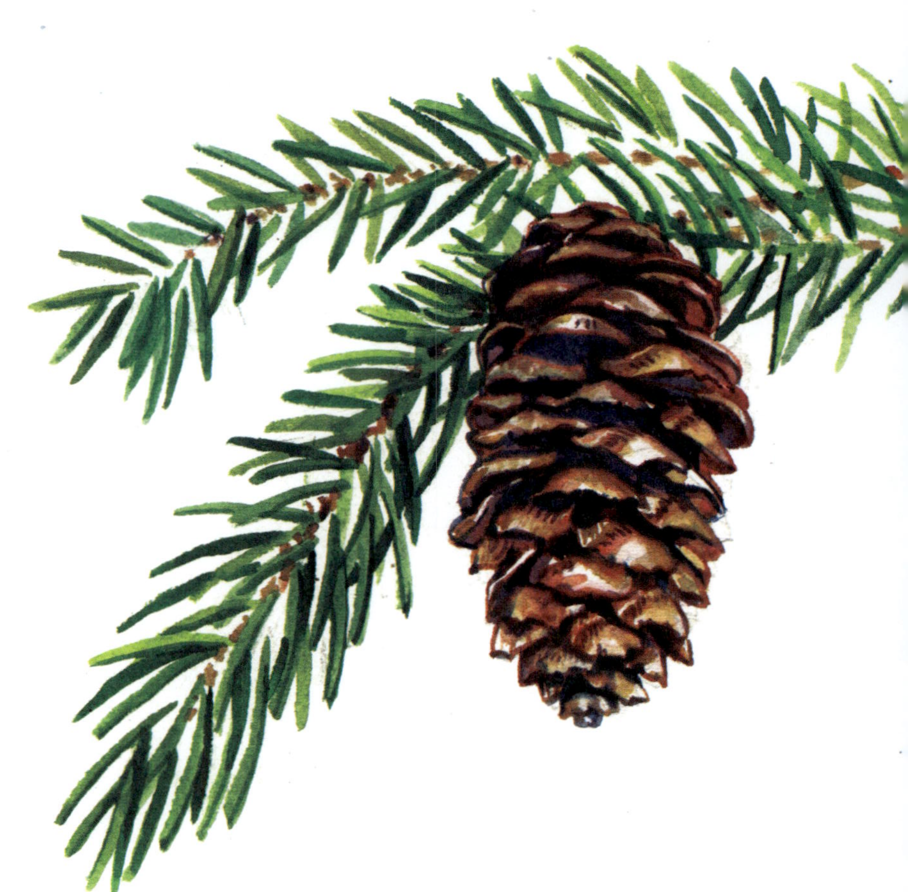

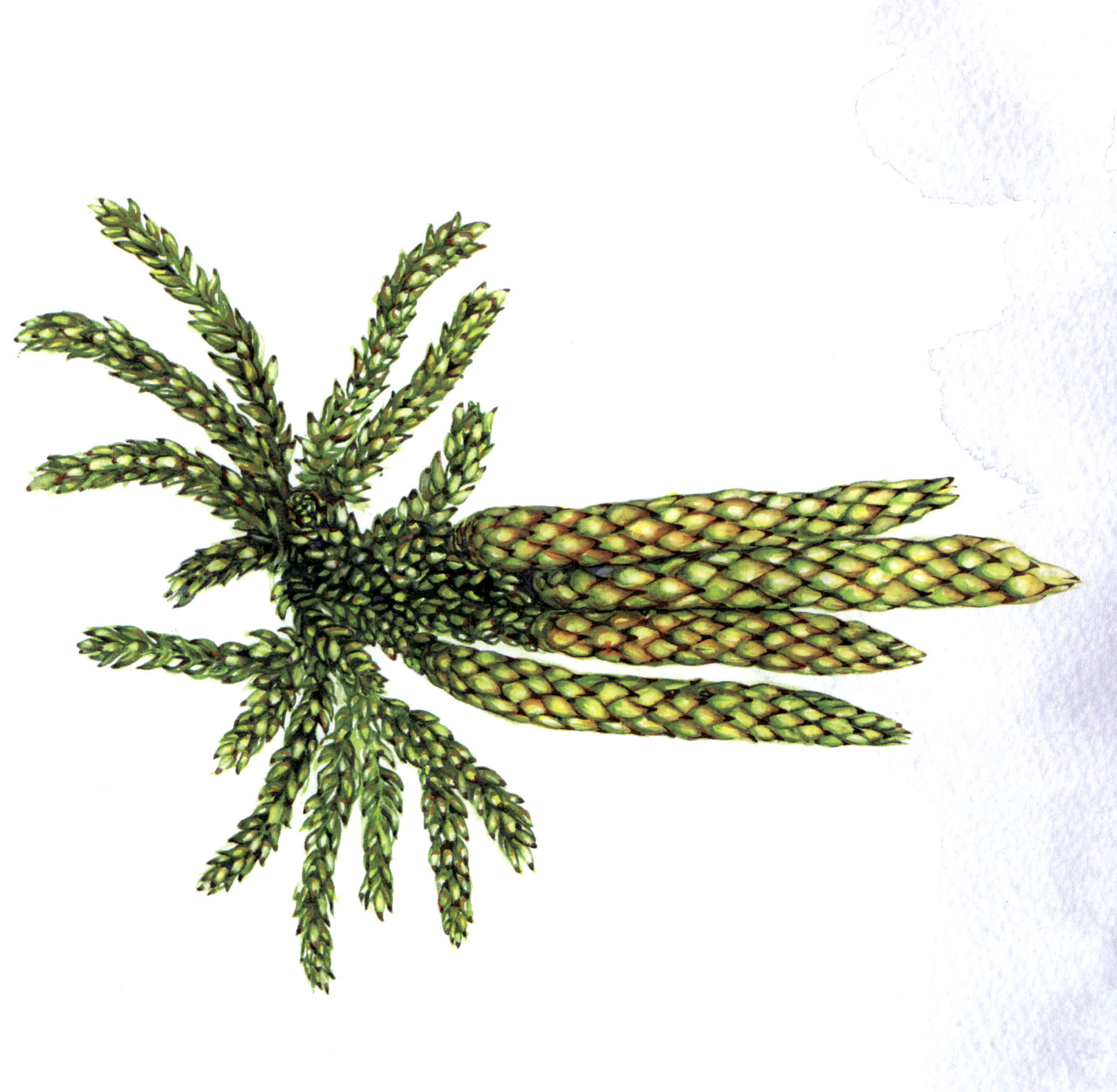

FURTHER READING

ON THE APPALACHIAN MOUNTAINS

Appalachian Mushrooms: A Field Guide, by Walter E. Sturgeon

Appalachian Wildflowers, by Thomas E. Hemmerly

Mountains of the Heart: A Natural History of the Appalachians, by Scott Weidensaul

The Plants of the Appalachian Trail, by Kristen Wickert

Sibley Birds East: Field Guide to Birds of Eastern North America, 2nd edition, by David Allen Sibley

ON NATURE JOURNALING AND ART

The Artist's Way, 30th anniversary edition, by Julia Cameron

Keeping a Nature Journal: Deepen Your Connection with the Natural World All Around You, 3rd edition, by Claire Walker Leslie

The Laws Guide to Nature Drawing and Journaling, by John Muir Laws

Watercolor in Nature: Paint Woodland Wildlife and Botanicals with 20 Beginner-Friendly Projects, by Rosalie Haizlett

ON BECOMING A NATURALIST

The Naturalist at Home: Projects for Discovering the Hidden World Around Us, by Kelly Brenner

A PEEK INSIDE MY ARTISTIC PROCESS

My nature illustration process always begins on the trail. I scrawl notes in my notebook or type them into my phone; I take photos and videos with either my wildlife camera or phone; and I render quick sketches with ink, pencil, and sometimes a quick wash of watercolor.

The next phase involves learning about my subject. I'll get out a field guide and also look up information online to help me understand what I've observed. With that greater context, I head to my studio to consider how I might pair some of that newly acquired information with my own observations to create a painting that captures the totality of the experience.

When I'm traveling, my painting studio takes the form of a dining room table pushed up against a window for natural light, a kitchen counter, or even a picnic table under the trees. I've learned that a few simple touches can turn any environment into an inspiring workspace—a few fresh wildflowers in a jar, the glow of a candle, a cozy mug of

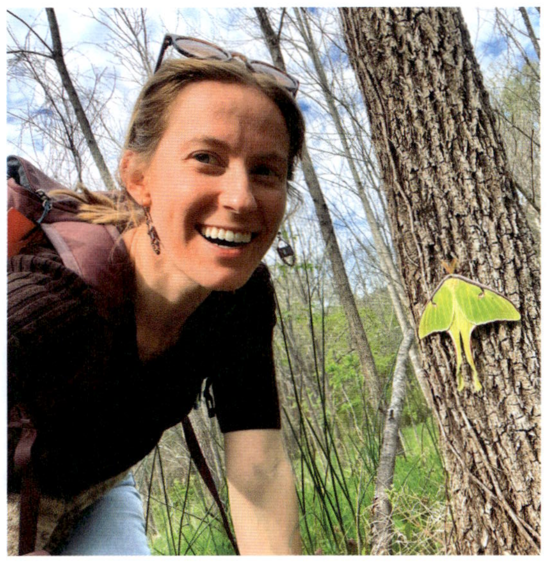

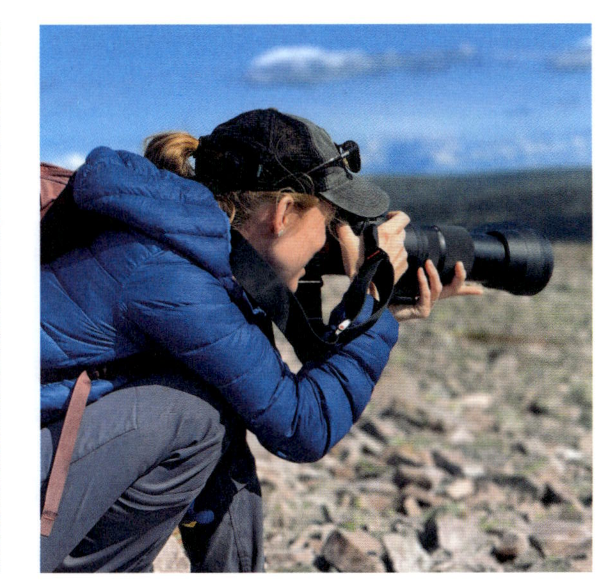

coffee or tea, or music that makes me feel calm and happy. It's freeing to realize that having the perfect studio isn't essential to making great work—you can be creative in any situation with the right mindset.

Next, I do a series of quick pencil sketches in a small notebook to come up with a general composition for the piece. I usually create three to four sketches, erasing and reworking to land on a pleasing layout. Then, I lightly sketch out my composition on thick, 300-pound Arches watercolor paper. I typically represent my subjects significantly larger than life-size so that I can capture all their intricate details. While I work, I either reference printed, enlarged photos or pull up the images on my tablet. I also draw from my imagination to weave together elements in an aesthetically pleasing way, or perhaps adjust certain colors to make the scene feel more whimsical.

Once I have my pencil sketch on the watercolor paper, I'll add either Pigma Micron permanent ink outlines or thin watercolor outlines. This really depends on the subject matter—if it's an extremely detailed subject or an illustrated map, I'll use the ink pen since it's easier to control. If not, I'll jump straight in with watercolor. I use all sorts of watercolor brands—for the illustrations in this book, I primarily worked with Daniel Smith watercolors. You may notice that starting in the Virginia section, however, some of my illustrations take on a more intensely vibrant color scheme; that's because I decided to experiment with a set of Dr. Ph. Martin's Hydrus liquid watercolors.

Once a painting is complete, I scan it into my computer and, if necessary, adjust the color and values in Adobe Photoshop so that the scan accurately reflects the original. If I was a bit messy

and there are paint splotches that need to be cleaned up, I can do that there as well. Then . . . *ta-da*! My painting is digitized and ready to be included in this book.

For a much deeper dive into my watercolor process, check out my other book, *Watercolor in Nature: Paint Woodland Wildlife and Botanicals with 20 Beginner-Friendly Projects*.

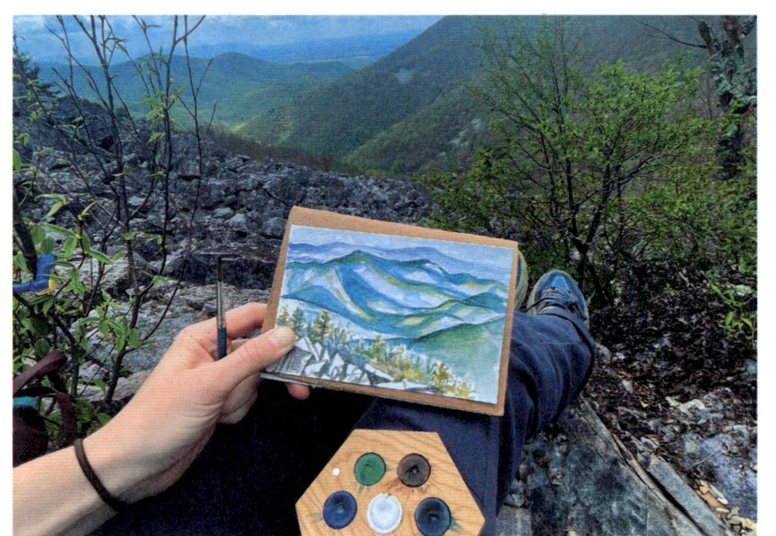

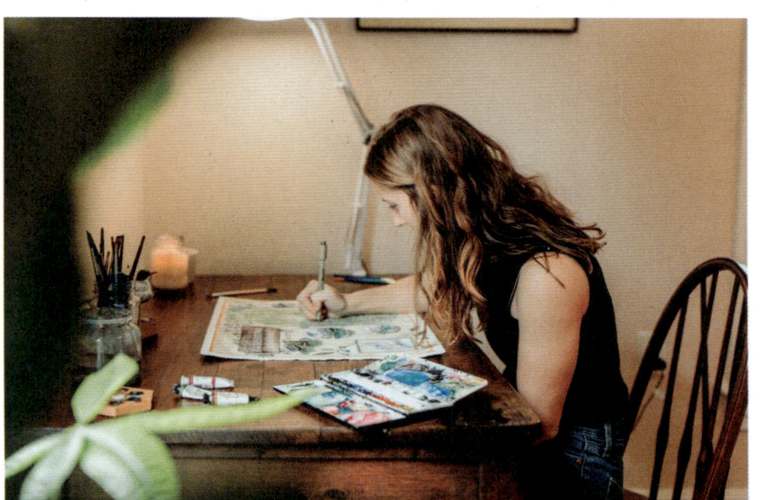

About the Author

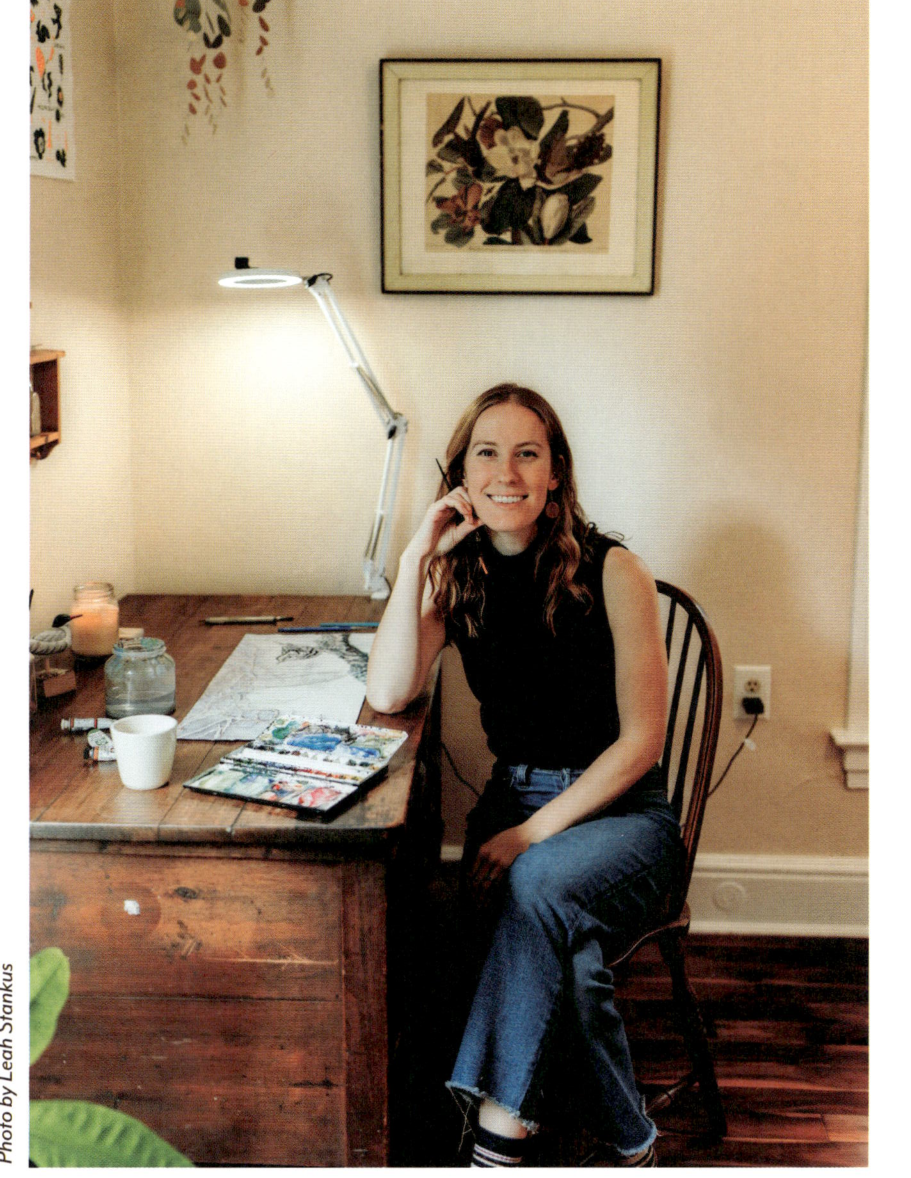

Photo by Leah Stankus

ROSALIE HAIZLETT is an illustrator from West Virginia. Her mission is to notice and creatively celebrate the often-overlooked wonders of the natural world through hand-illustrated maps and vibrant watercolor paintings featuring the flora, fauna, and fungi that she finds on her hikes.

Rosalie is the author of *Watercolor in Nature: Paint Woodland Wildlife and Botanicals in 20 Beginner-Friendly Projects* (Page Street Publishing, 2021) and is the former artist-in-residence at Great Smoky Mountains National Park, the National Audubon Society's Hog Island, and the Roger Tory Peterson Institute of Natural History. She was awarded the 2022 Eckelberry Fellowship for distinguished wildlife illustrators and has collaborated with organizations like the US Fish and Wildlife Service, the Smithsonian Center for Folklife and Cultural Heritage, and the US Forest Service. As a Top Teacher on Skillshare, she has taught over 60,000 students techniques for drawing creative inspiration from the natural world.

Find her artwork, books, and online classes at www.rosaliehaizlett.com, and keep up with her adventures on social media: @rosaliehaizlett.

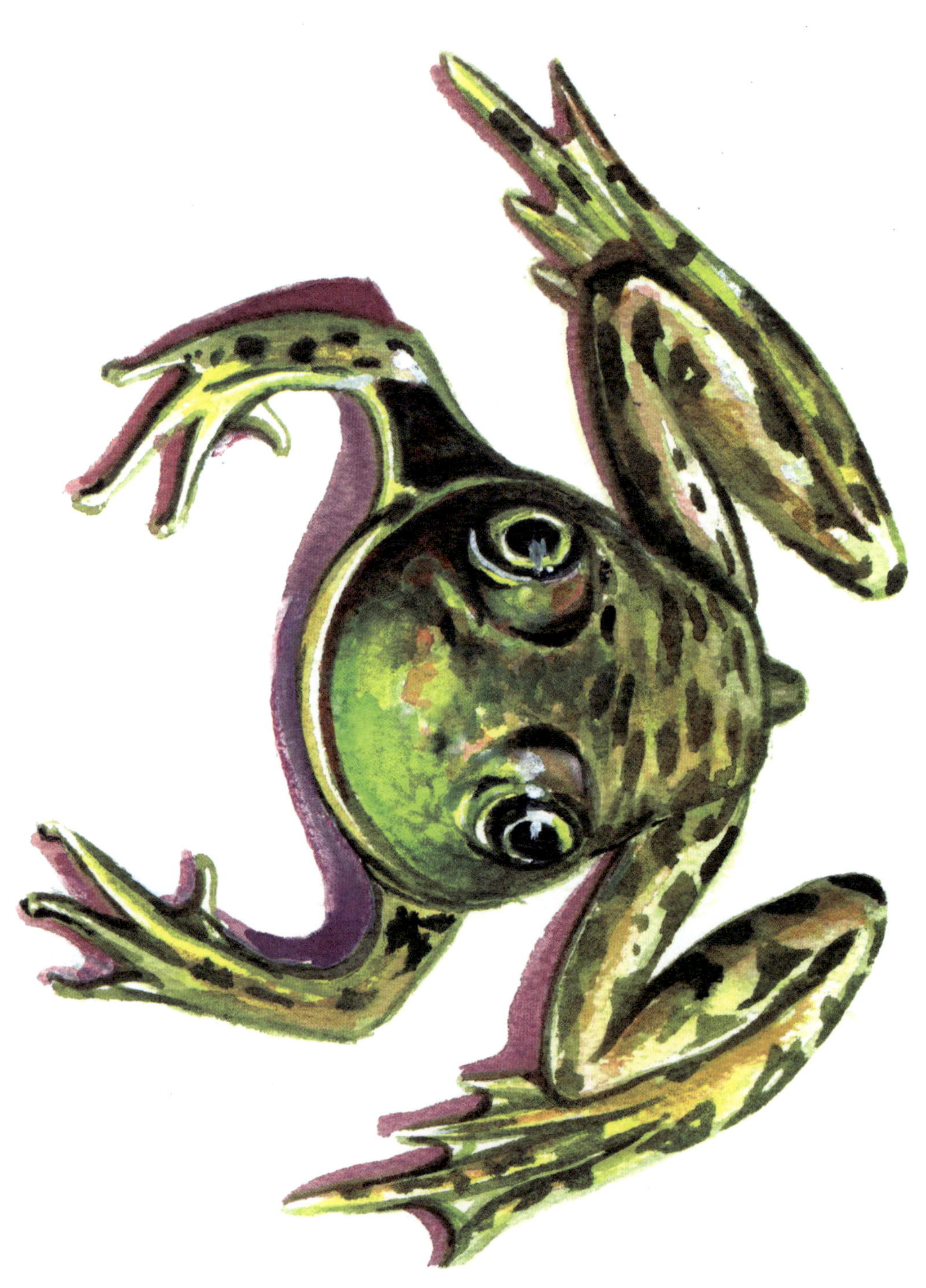